BIOMAT 2005

Proceedings of the International Symposium on
Mathematical and Computational Biology

BIOMAT 2005

Proceedings of the International Symposium on
Mathematical and Computational Biology

Rio de Janeiro, Brazil, 3-8 December 2005

edited by

Rubem P Mondaini
(Universidade Federal do Rio de Janeiro, Brazil)

Rui Dilão
(Instituto Superior Ténico, Portugal)

World Scientific

NEW JERSEY • LONDON • SINGAPORE • BEIJING • SHANGHAI • HONG KONG • TAIPEI • CHENNAI

Published by

World Scientific Publishing Co. Pte. Ltd.

5 Toh Tuck Link, Singapore 596224

USA office: 27 Warren Street, Suite 401-402, Hackensack, NJ 07601

UK office: 57 Shelton Street, Covent Garden, London WC2H 9HE

British Library Cataloguing-in-Publication Data
A catalogue record for this book is available from the British Library.

BIOMAT 2005
Proceedings of the International Symposium on Mathematical and Computational Biology

ISBN 981-256-797-6

Printed in Singapore by B & JO Enterprise

Preface

The BIOMAT 2005 International Symposium on Mathematical and Computational Biology, together with the Fifth Brazilian Symposium on Mathematical and Computational Biology, was held in the city of Petrópolis, state of Rio de Janeiro, Brazil, from the 3rd to the 8th December 2005. The atmosphere of the symposium was informal and the approach interdisciplinary, with the contribution of the expertise of fifteen keynote speakers from different fields and backgrounds.

In the proceedings of BIOMAT 2005, there are state of the art research papers in the mathematical modelling of cancer development, malaria and aneurysm development, among others. Models for the immune system and for epidemiological issues are also analyzed and reviewed. Protein structure prediction by optimization and combinatorial techniques (Steiner trees) are explored. Bioinformatics questions, regulation of gene expression, evolution, development, DNA and array modelling, small world networks are other examples of topics covered in the BIOMAT 2005 symposium.

The diversity of topics and the combination of original with review approaches make BIOMAT Symposia important events for graduate students and researchers.

This Symposium would never have taken place without the generous contribution of all the sponsoring agencies. Our first thanks go to the Brazilian agencies CAPES and FINEP and their Board of Trustees. We deeply thank the support of CENPES-PETROBRAS, the Research Centre of the Brazilian Oil Company and the world leader of research in deep sea waters, and the support to the Fogarty International Centre, Harvard Medical School, USA, through the grant number #1 D43 TW7015-01.

We particularly thank the directors and representatives of these institutions: Dr. Geova Parente from CAPES; Dr. Henrique A. C. Santos, Dr. Gina Vasquez and Miss Raquel Prata from CENPES-PETROBRAS; Dr. Lucila Ohno-Machado, Dr. Eduardo P. Marques, Prof. Eduardo Massad and Dr. Heimar Marin from the Harvard Medical School.

We would also like to thank Prof. M. A. Raupp, Director of the National Laboratory of Scientific Computing (LNCC), at Petrópolis, for his invitation to host the BIOMAT Symposium at the LNCC. We are indebted to the members of the local Organizing Committee, Dr. Maurício V. Kritz, Dr. Luiz Bevilacqua and Dr. Marcelo T. Santos for their collaboration and effort in the local organization of the conference and the support of its social program. We also thank the partial support of FCT (Fundação para a Ciência e a Tecnologia, Portugal) for the edition of these proceedings.

Finally, on behalf of the Scientific Program Committee and the Editorial Board of the BIOMAT Consortium, we thank all the participants and authors of BIOMAT 2005 for keeping the tradition of the BIOMAT Symposia.

Rubem P. Mondaini and Rui Dilão

Rio de Janeiro, December 2005

Contents

Protein Structure

Bioinformatics

MODELLING ASPECTS OF VASCULAR CANCER DEVELOPMENT

PHILIP K. MAINI

Centre for Mathematical Biology, Mathematical Institute
University of Oxford, 24-29 St Giles', Oxford OX1 3LB, United Kingdom
E-mail: maini@maths.ox.ac.uk

TOMÁS ALARCÓN

Bioinformatics Unit, Department of Computer Science
University College London, Gower Street, London WC1E 6BT
United Kingdom

HELEN M. BYRNE

Centre for Mathematical Medicine, Division of Applied Mathematics
University of Nottingham, Nottingham NG7 2RD
United Kingdom

The modelling of cancer provides an enormous mathematical challenge because of its inherent multi-scale nature. For example, in vascular tumours, nutrient is transported by the vascular system, which operates on a tissue level. However, it also affects processes occurring on the molecular level. Molecular and intra-cellular events in turn affect the vascular network and therefore the nutrient dynamics. Our approach is to model, using partial differential equations, processes on the tissue level, and couple these to the intra-cellular events (modelled by ordinary differential equations) via cells modelled as automaton units. Thus far, within this framework, we have investigated the effects on tumour cell dynamics of structural adaptation at the vessel level, have explored certain drug protocol treatments, and have modelled the cell cycle in order to account for the possible effects of p27 in hypoxia-induced quiescence in cancer cells. We briefly review these findings here.

1. Introduction

Cancer is one of the biggest killers in the Western World. There has been a huge amount of experimental and medical research into this disease and for certain cancers cure rates have improved. Unfortunately, however, we still do not have an understanding of how this disease progresses and how the myriad processes involved conspire to initiate cancer and the growth of tumours. In comparison to experimental research in this area, there has been relatively little theoretical work on cancer growth. It is now slowly being recognised that mathematical modelling may help us to extract the full potential of the vast amounts of data being generated in the laboratory and provide a framework in which to interpret these results[1]. Modelling cannot find a cure for cancer, but it may allow experimental work to be directed in more efficient ways.

The ultimate challenge in the modelling of biological systems in general is to integrate the huge amount of experimental information being generated at the many different scales that make up a biological system. The traditional "top-down" approach does not capitalise on lower level data, while the "bottom-up" approach runs the risk of being too unwieldy and simply replacing a biological system we do not understand by a computational system we do not understand. Moreover, we must take into account the reality that many parameters are unknown and information is only partial. At the moment, it is an open question as to whether mathematics can meet this challenge. Equally, the best way to implement such an approach remains to be established.

In this paper we briefly review our recent attempt to build an integrated model of tumour growth. In Section 2 we present a very brief overview of tumour growth and then in Section 3 we outline our modelling approach, which uses a hybrid cellular automaton framework. Our philosophy is to start with a model which is comprised largely of "black boxes" or modules, which are represented at the outset by simple imposed rules. This is very much a macroscale level approach. We then aim to "zoom in" on particular modules as more experimental data becomes available and develop more realistic models. We illustrate this in Section 4 with a model for the G1/S transition in the cell cycle and in Section 5 with a simple model for pH.

2. Brief biological background

Under normal conditions, cell division and growth are tightly regulated by proliferation (division) and apoptotic (self-induced cell death) signals. However, in cancer, it is thought that a series of mutations (see, for example, Michor *et al.*[2]) within a cell leads to it escaping from these controls and this, in turn, can lead to an uncontrolled growth of tissue. Initial growth of a tumour has been studied in the laboratory using multi-cellular spheroids. The growth of this tissue is diffusion-limited as its main nutrient is oxygen and it has no active transport mechanisms. It develops a pattern typically composed of an inner necrotic (dead) core, surrounded by a quiescient region (live cells which are not dividing), and an outer rim of proliferative cells. The growth rate greatly diminishes when the spheroid reaches about 1 mm in diameter and at this stage, if the tumour is to continue to grow significantly it needs a vascular system to provide it with nutrient. There is now quite a substantial amount of literature on the mathematical modelling of avascular tumour growth, ranging from very simple models which consider the dynamics of cell populations, to more sophisticated models ranging from those which delve into the microscopic levels of biochemical control of nutrient uptake, to those which consider the tumour mass as a multi-phase material modelled via the techniques of continuum mechanics. Other approaches include individual-based-models which consider cells as independent units and define equations or rules on how each unit grows, divides, moves, etc. References are too numerous to mention here so we simply refer the interested reader to the review by Roose *et al.*[3] and references therein.

To gain access to more nutrient, the tumour cells secrete what are known as Tumour Angiogenesis Factors (TAFs) which diffuse into the surrounding normal tissue and, on reaching normal blood vasculature, initiate a series of events, the net result of which is that cells lining the vessel walls break away and begin to migrate chemotactically towards the tumour. On approaching the tumour they join up via the process of anastomosis establishing a blood supply for the tumour. This was first shown by the classical experiment of Folkman[4]. As with avascular tumours, there is now a quite substantial amount of modelling literature on the interaction of TAFs with the vessel lining, the formation of the angiogenic network and its chemotactic response. We refer the reader to the review by Mantzaris *et al.*[5].

As the tumour mass now begins to grow out further it produces proteases that can degrade the extracellular material surrounding it, giving the tumour space to move. Cells can also break off from the main (or primary) tumour mass and enter the blood supply, leading to the process of metastasis and the formation of often fatal secondary tumours.

There are several reviews describing the process of nutrient consumption and diffusion inside tumours and we refer the reader to the papers[6,7].

3. Cellular automaton model

As mentioned in the previous section, there is a growing literature on the mathematical modelling of various aspects of tumour growth. However, there is little theoretical work to date on how blood is delivered to tissue, how tissue demands are met by the structural adaptation of the blood network, and how spatial heterogeneity affects tumour dynamics. If we wish eventually to develop a model which allows us to explore different drug delivery protocols for therapy, then it is important that we understand these aspects. This was the motivation for developing the modelling framework below (we refer the reader to the original paper[8] for full details).

We consider for simplicity a vascular structure which is composed of a regular hexagonal network embedded in a two-dimensional NxN lattice composed of normal cells, cancer cells, and space into which cells can divide. We impose a pressure drop across the vasculature, assuming that blood flows into the idealised "tissue" through a single inlet vessel and drains through a single outlet vessel. To compute the flow of blood through each vessel we use the Poiseuille approximation, and, given the initial network configuration (that is, radii and lengths) we compute the flow rates through, and pressure drops across, each vessel using Kirchoff's laws. To calculate the radii, we begin by assuming that all vessels have the same radius, but assume that vessels undergo structural adaptation. We follow the work of Pries *et al.*[9] by assuming that the radius $R(t)$ of a vessel, is modified as follows:

$$R(t + \Delta t) = R(t) + R\Delta t \left(\log\left(\frac{\tau_w}{\tau(P)}\right) + k_m \log\left(\frac{\dot{Q}_{ref}}{\dot{Q}H} + 1\right) - k_s \right) \quad (1)$$

where Δt is the time scale, $\dot{Q}$ is the flow rate, $\dot{Q}_{ref}$, k_m and k_s are constants, H is the haematocrit (red blood cell volume), $\tau_w = R\Delta P/L$ is the wall shear stress acting on a vessel of length L. P is the transmural pressure, and $\tau(P)$ the magnitude of the corresponding "set point" value of the wall shear stress obtained from an empirical fit to experimental data. The second term on the right-hand side represents the response to mechanical or haemodynamic stimuli. The third term on the right-hand side is the metabolic stimulus and increases with decreasing red blood cell flux. The constant k_s represents the so-called shrinking tendency, that is, without the mechanical and metabolic stimuli, the vessel would atrophy.

Blood viscosity is a complex function of H and R and this is taken from empirical studies, while the distribution of haematocrit at branch points is assumed to be proportional to the flow velocity along each adjoining vessel[10]. Pries *et al.* found that for efficient structural adaptation a third stimulus (the so-called conducted stimulus) was required. We omit this from our model because it is well-known that tumour vasculature does not adapt as well as normal vasculature.

With the above equation we can now iterate our scheme until we reach a steady state and a vascular network with a distribution of different radii. We now use this to conduct nutrient into the tissue. Assuming, for simplicity, that the only nutrient is oxygen, we calculate the nutrient distribution by solving the diffusion equation with the cells as sinks for uptake (we take the adiabatic approximation) with internal boundary conditions representing diffusion of oxygen out of the blood vessels. We impose zero flux boundary conditions at the edge of the tissue.

To model the cell dynamics we assume that if the oxygen level is sufficiently high then cells will divide if there is space (or die otherwise) while if the oxygen level is too low then cells die. However, we assume that for intermediate values of oxygen, cancer cells can undergo quiescence and survive for a certain amount of time, whereas normal cells cannot (see Section 4). We further assume that the threshold levels of oxygen below which cells die is dependent on cell type and on the type of neighbouring cells. For example, if a normal cell is surrounded by cancer cells, then we raise the threshold level (that is, the cell is more likely to die). This is a very crude attempt to model the effects of pH (see Section 5).

A typical solution for the resultant oxygen profile is shown in Figure 1. One sees regions of very high oxygen levels interspersed with regions of hypoxia (low oxygen). Clearly, the system has not adapted well and this is reminiscent qualitatively of oxygen distributions within tumours.

Figure 2 shows the spatio-temporal and temporal evolution of cancerous cells for the case above, compared with the case where we do not assume any structural adaptation but instead impose the condition that the oxygen is distributed uniformly throughout the tissue. We see that spatial inhomogeneity has a significant effect on tumour dynamics by actually lowering the total cancer cell population. This is because there is not an efficient use of nutrient.

Furthermore, we see that the shape of tumour predicted has "finger-like" protrusions similar to those observed in some spreading cancers. This structure has arisen in this model simply because of the spatial heterogeneity in the nutrient distribution. Indeed, closer inspection reveals that one or two parts of the tumour have almost "broken away". This cannot actually happen in this model because we have not included cell motion but we can imagine that if we did include motion towards areas of high nutrient concentration, then this may be a mechanism for metastasis (Alarcón *et al.*

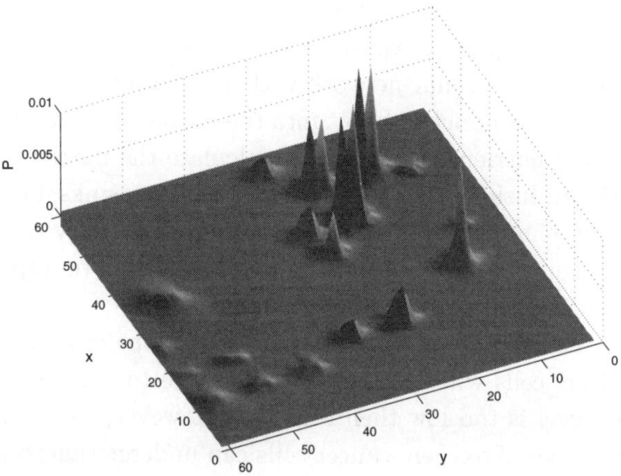

Figure 1. First 3 normalised frequencies versus release location for clamped simply supported beam with internal slide release.

in prep.).

4. Effects of hypoxia on cell cycle dynamics

In the above model we assumed that in hypoxic conditions, cancer cells can undergo quiescience whereas normal cells cannot (in fact, they undergo hypoxia-induced arrest leading to apoptosis). Whereas in the above we simply included this as a rule, here we aim to understand what is the mechanistic underpinning of this phenomenon.

The cell cycle is composed of 4 stages, G1, S, G2, M, with occasionally a GO phase (see, for example, Alberts *et al.*[11]). There have been a number of models proposed to account for the G1/S and for the G2/M transitions. The G1/S transition is particularly important because once a cell has passed through this checkpoint it is almost certain to divide. We chose to focus on this transition because some experimentalists felt that cells under hypoxic conditions may be inhibited from making this transition[12].

The cell cycle is controlled by a complex series of coordinated molecular events, with the central components of this interacting network being the two families of proteins, the cyclin-dependent kinases (CDKs) and the cyclins. During G1, the cyc-CDK complexes have low activity, which becomes high after transition. Coupled to this is the activity of the anaphase protein complex (APC) and the protein Cdh1 which both begin at high levels in

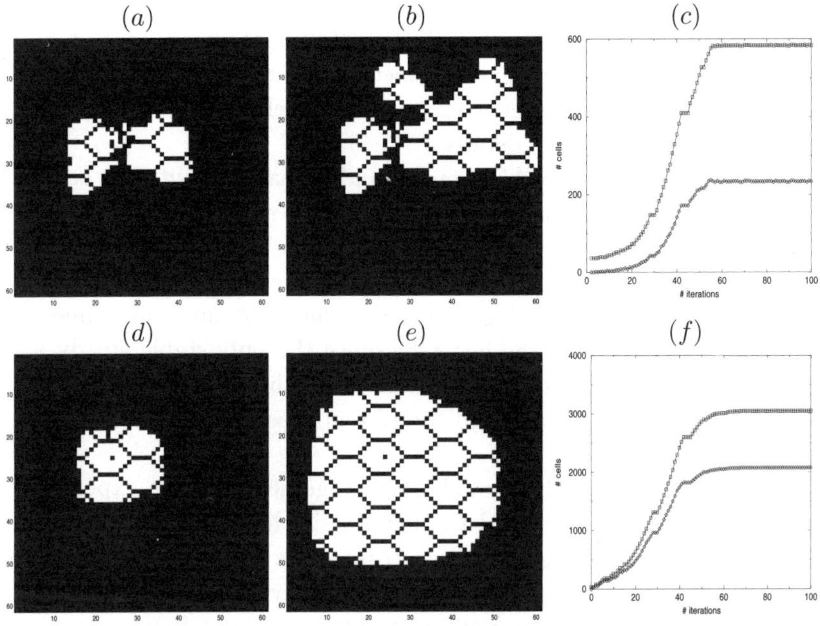

Figure 2. Series of images showing the evolution of the spatial distribution of cells for growth in inhomogeneous (panels a and b), and homogeneous environments (panels d and e). In panels (a), (b), (d), and (f) white spaces are occupied by cancer cells, whereas black spaces are either empty or occupied by vessels. Panels (c) and (f) show the time evolution of the number of (cancer) cells for the heterogeneous and homogeneous cases, respectively. Squares represent the total number of cancer cells (proliferating + quiescent). Diamonds correspond to the quiescent population.

G1 but fall to low levels of activity after the G1/S transition. There are a number of models of this process spanning a large range of detail (from 2 equations to over 60) but for our purposes we consider the model of Tyson and Novak[13], which captures the essence of the problem. The model takes the form

$$\frac{dx}{dt} = \frac{(k_3' + k_3''A)(1-x)}{J_3 + 1 - x} - \frac{k_4 m y x}{J_4 + x}, \tag{2}$$

$$\frac{dy}{dt} = k_1 - (k_2' + k_2'' x)y, \tag{3}$$

$$\frac{dm}{dt} = \mu m \left(1 - \frac{m}{m_*}\right), \tag{4}$$

where $x \equiv [\text{Cdh1}]$ is the concentration of active Cdh1/APC complexes,

$y \equiv [\text{Cyc}]$, is the concentration of cyclin-CDK complexes[a], and m is the mass of the cell. The parameters k_i $(i = 1,2,3,4)$ and J_i $(i = 3,4)$ are positive constants. A represents a generic activator. In Eq. (4), μ is the cell growth rate and m_* is the mass of an adult cell. We refer the reader to Tyson and Novak[13] for full details.

The above model can exhibit mono- and bi-stability, with the cell mass m as a bifurcation parameter. For low values of m there is a single stable steady state with a high value of x and a low value of y - this would correspond to G1. As m increases, we enter the bistable regime, with a new stable steady state arising at a high value of y and a low value of x. For a critical value of m the latter becomes the only stable steady state and the system switches to this state, corresponding to the S phase. After the cell divides, m decreases, and the system is set back to the "G1 phase steady state".

We take this as our base model and, together with the experimental results in Gardner *et al.*[12] and the hypothesis that under hypoxic conditions the expression of the regulator p27 increases (in fact due to decreased degradation), which in turn inhibits Cdh1 activity, we derive the (non-dimensionalised) model (see Alarcon *et al.*[14] for full details):

$$\frac{dx}{d\tau} = \frac{(1 + b_3 u)(1 - x)}{J_3 + 1 - x} - \frac{b_4 m x y}{J_4 + x}, \tag{5}$$

$$\frac{dy}{d\tau} = a_4 - (a_1 + a_2 x + a_3 z) y, \tag{6}$$

$$\frac{dm}{d\tau} = \eta m \left(1 - \frac{m}{m_*}\right), \tag{7}$$

$$\frac{dz}{d\tau} = \chi(m) - c_2 \frac{P}{B + P} z, \tag{8}$$

$$\frac{du}{d\tau} = d_1 - (d_2 + d_1 y) u, \tag{9}$$

where P is the oxygen tension, z is the p27 concentration and u is the concentration of phosphorylated retinoblastoma (RB).

We make the following assumptions: for normal cells, p27 activity is regulated by cell size, that is, $\chi(m) = c_1(1 - \frac{m}{m_*})$, but for cancerous cells, this size-regulation is lost, that is $\chi(m) = c_1$. We make the further assumption

[a] In Tyson and Novak[13], [Cyc] corresponds to the concentration of the specific complex cyclinB-CDK. Here we simply consider a generic cyclin-CDK complex in order to keep our model as simple as possible.

that c_1 (maximum rate of synthesis of p27) is larger in normal cells than in cancer cells – this we do to account for the observation of low p27 levels in cancer cells compared to normal cells (see, for example, Philipp-Staheli *et al.*[15]). Using other parameter values from Tyson and Novak[13] we find that assuming these two phenomena characterise the differences between the regulation of p27 in cancer and in normal cells is sufficient to account for hypoxia-induced quiescience in the former, and hypoxia-induced arrest in the latter. Our numerical simulation results are supported by an analytic study of the bifurcation structure of the model (see Alarcon *et al.*[14] for details).

5. The role of acidity

In the cellular automaton model of Section 2 we imposed a rule in which the fate of cells depended on their neighbours. This was motivated by the work of Gatenby and Gawlinski[16,17]. They proposed a reaction-diffusion model for interaction between tumour cells and normal cells and hypothesised that when tumour cells undergo anaerobic metabolism (which they do even under normoxic conditions) the by-product of lactic acid lowers the pH into a regime where the tumour cells can "over-power" the neighbouring normal cells and invade the tissue simply because of their ability to tolerate more acidic conditions. Their model predicted that there should be a gap between the advancing tumour front and the regressing normal tissue and, indeed, they later observed this phenomenon experimentally.

A drawback in their model was that it predicted either a travelling wave of tumour invasion, or total clearance of tumour cells. It could not predict the formation of a benign tumour. This problem can be overcome if one considers a very simple model in which tumour cells produce acid and the tumour grows but also loses cells via necrosis if the acid level is too high. The resultant coupled system of ordinary differential equations yields three different types of behaviour: saturated (benign) growth of avascular tumours; benign growth of vascular tumours which can become invasive (malignant) as a key dimensionless parameter passes through a critical value (see Smallbone *et al.*[18] for full details).

6. Discussion

We have presented results from our recent research into the growth of vascular tumours. Our approach to incorporating processes occurring on very different length scales is to use a hybrid cellular automaton framework[19,20].

Our very preliminary work in this area has already revealed some experimentally testable predictions. Our model shows that nutrient heterogeneity can have a significant effect on the spatio-temporal dynamics of tumour growth. In particular, it shows that it may be the cancerous cells' exploitation of high nutrient sources that causes an initial homogeneously growing tumour to begin to break up. We are in the process of incorporating cell movement into our model to see if this can lead to metastasis. We have recently shown that in some cases anti-angiogenesis treatment could actually enhance tumour growth due to the modified vasculature being more efficient at delivering nutrient[21].

Our modelling framework allows for detailed sub-models to be included for processes occurring on a specific scale. Thus, for example, our simple rule for the signal for cell division can be expanded to incorporate a model of this process. In doing so, we have generated a hypothesis as to how cancerous cells can undergo hypoxia-induced quiescience while normal cells undergo hypoxia-induced arrest. We propose that p27 plays a key role in this but we must be aware that this is still controversial[22]. An important point here is that if we were to include a full model for the cell cycle into the cellular automaton model, the resultant model would require a huge amount of computational power to solve and would be so complicated that it would be difficult to gain insight into the phenomena observed from the model. Therefore we must reduce the model and indeed one can do this by taking a caricature model of only a few equations which aims to capture the essence of the full cell cycle model. In this case, however, the question of whether our results are artifacts because of the simplifications we made arises and this is a crucial problem facing all theoreticians working in the Life Sciences, namely, how robust are the models that we generate?

The simple model presented in Section 3 proves inadequate if we want to use it to explore the effects of drug treatment where a drug acts on cells in a certain part of the cell cycle. In this case, we need to incorporate cell cycle models of the form proposed in Section 4, or we can use a probabilistic approach based on empirical data to determine the probability that a certain cell is in a certain phase of its cell cycle at a particular time. The latter approach was used to examine the effects of Doxorubicin treatment on non-Hodgkin's lymphoma to determine the optimal dosage protocol[23].

In Section 5 we explored in more detail the effects of acidity. This simple model isolates a single nondimensional bifurcation parameter which determines whether or not a tumour will grow in an uncontrolled fashion. This raises a number of possible control mechanisms, including the counterintuitive prediction that increasing the acidity may eliminate the tumour. This model prediction remains to be tested.

Future work in this area must address the underlying biochemistry of many of the processes we mentioned above and incorporate the mechanical aspects involved in tumour growth. We have recently incorporated rules for production of the growth factor VEGF in response to hypoxic conditions, computed its spatio-temporal distribution by solving a reaction-diffusion model, and modified the vessel structural adaptation equation accordingly[24]. While this allows us to capture the initial vessel dilation in response to VEGF, it only in a very crude way accounts for the angiogenic response. We are presently incorporating growth of new vasculature into the model.

A crucial aspect of all this work will be model reduction so that the resultant model is computationally tractable and understandable. Only then can mathematical modelling gain useful insights to help direct medical research.

Acknowledgments

TA thanks the EPSRC for financial support under grant GR/509067. HMB thanks the EPSRC for funding as an Advanced Research Fellow. This work has been supported in part by NIH grant CA 113004. The authors wish to acknowledge the support provided by the funders of the Integrative Biology project: The EPSRC (ref no: GR/S72023/01) and IBM.

References

1. Gatenby, R.A., Maini, P.K. (2003), "Mathematical oncology: Cancer summed up," *Nature 421*, 321.
2. Michor, F., Iwasa, Y., Nowak, M.A. (2004), "Dynamics of cancer progression," *Nature Reviews, Cancer 4*, 197-205.
3. Roose, T., Chapman, S.J., Maini, P.K. (2005), " Mathematical models of avascular tumour growth," (submitted)
4. Folkman, J. (2003), "Fundamental concepts of the angiogenic process, " *Curr. Mol. Med. 3*, 643-651.
5. Mantzaris, N., Webb, S., Othmer, H.G. (2004), "Mathematical modeling of tumor-induced angiogenesis," *J.Math.Biol. 95*, 111-187.
6. Adam, J.A. (1996), "Mathematical models of perivsacular spheriod development and catastrophe-theoretic description of rapid metastatic growth/tumor remission," *Invasion and Matastasis 16*, 247-267.
7. Araujo, R.P., McElwain, D.L.S. (2004), "A history of the study of solid tumor growth: the contribution of mathematical modelling," *Bull. Math. Biol. 66*, 1039-1091.

8. Alarcón, T., Byrne, H.M., Maini, P.K. (2003), "A cellular automaton model for tumour growth in inhomogeneous environment," *J.theor.Biol. 225*, 257-274.
9. Pries, A.R., Secomb, T.W., Gaehtgens, P. (1998), "Structural adaptation and stability of microvascular networks: theory and simulations," *Am. J. Physiol. 275*, H349-H360.
10. Fung, Y.C. (1993), "Biomechanics," *Springer, New York.*
11. Alberts, B., Bray, D., Lewis, J., Raff, M., Roberts, K., Watson, J.D. (1994), "Molecular Biology of the Cell," 3rd edition *Garland Publishing, New York, USA.*
12. Gardner, L.B., Li, Q., Parks, M.S., Flanagan, W.M., Semenza, G.L., Dang, C.V. (2001), "Hypoxia inhibits G_1/S transition through regulation of p27 expression," *J. Biol. Chem. 276*, 7919-7926.
13. Tyson, J.J., Novak, B. (2001), "Regulation of the eukariotic cell-cycle: molecular anatagonism, hysteresis, and irreversible transitions," *J. Theor. Biol. 210*, 249-263.
14. Alarcón, T., Byrne, H.M., Maini, P.K. (2003), "A mathematical model of the effects of hypoxia on the cell-cycle of normal and cancer cells," *J. Theor. Biol. 229*, 395-411.
15. Philipp-Staheli, J., Payne, S.R., Kemp, C.J. (2001), "p27(Kip1): regulation and function of haploinsufficient tumour suppressor and its misregulation in cancer," *Exp. Cell. Res. 264*, 148-168.
16. Gatenby, R.A., Gawlinski, E.T. (1996), "A reaction-diffusion model of cancer invasion," *Cancer Res. 56*, 5745-5753.
17. Gatenby, R.A., Gawlinski, E.T. (1996), "The glycolytic phenotype in carcinogenesis and tumor invasion: insights through mathematical models," *Cancer Res. 56*, 5745-5753.
18. Smallbone, K., Gavaghan, D.J., Gatenby, R.A., Maini, P.K. (2005), "The role of acidity in solid tumour growth and invasion," J. Theor. Biol. 235, 476-484.
19. Patel, A.A., Gawlinsky, E.T., Lemieux, S.K., Gatenby, R.A. (2001), "Cellular automaton model of early tumour growth and invasion: the effects of native tissue vascularity and increased anaerobic tumour metabolism," *J. Theor. Biol. 213*, 315-331.
20. Moreira, J., Deutsch, A. (2002), "Cellular automaton models of tumor development: A critical review," *Adv. in Complex Systems 5*, 247-267.
21. Alarcón, T., Byrne, H.M., Maini, P.K. (2004), "Towards whole-organ modelling of tumour growth," *Prog. Biophys. Mol. Biol. 85*, 451-472.
22. Green, S.L., Freiberg, R.A., Giaccia, A.(2001), "$p21^{Cip1}$ and $p27^{Kip1}$ regulate cell cycle reentry after hypoxic stress but are not necessary for hypoxia-induced arrest," *Mol. & Cell. Biol. 21*, 1196-1206.
23. Ribba, B., Marron, K., Agur, Z., Alarcón T., Maini, P.K. (2005), "A mathematical model of Doxorubicin treatment efficacy for non-Hodgksin's lymphoma: investigation of the current protocol through theoretical modelling results," *Bull. Math. Biol. 67*, 79-99.
24. Alarcón, T., Byrne, H.M., Maini, P.K. (2005), "A muliple scale model for tumour growth," *SIAM J. Multiscale Mod. & Sim. 3*, 440-475.

MODELLING COOPERATIVE PHENOMENA IN INTERACTING CELL SYSTEMS WITH CELLULAR AUTOMATA

ANDREAS DEUTSCH

Center for Information Services and High Performance Computing
Dresden University of Technology
Zellescher Weg 12, D-01062 Dresden, Germany
E-mail: andreas.deutsch@tu-dresden.de

Cellular automata can be viewed as simple models of spatially extended decentralized systems made up of a number of individual components (e.g. biological cells). The communication between constituent cells is limited to local interaction. Each individual cell is in a specific state which changes over time depending on the states of its local neighbors. In particular, cellular automata have been proposed as models for cooperative phenomena arising in ecological, epidemiological, ethological, evolutionary, immunobiological and morphogenetic systems. Here, we present an overview of cellular automaton models of cooperative phenomena in interacting cell systems with a focus on spatio-temporal pattern formation. Finally, we introduce a specific example – avascular tumour growth – and introduce a cellular automaton model for this phenomenon which is able to lead to testable biological hypotheses.

1. Introduction: roots of cellular automata

The notion of a cellular automaton originated in the works of John von Neumann (1903-1957) and Stanislaw Ulam (1909-1984). Cellular automata as discrete, local dynamical systems can be equally well interpreted as a mathematical idealization of natural systems, a discrete caricature of microscopic dynamics, a parallel algorithm or a discretization of partial differential equations. According to these interpretations distinct roots of cellular automata may be traced back in biological modeling, computer science and numerical mathematics which are well documented in numerous and excellent sources[5,10,45,53a].

The basic idea and trigger for the development of cellular automata as biological models was a need for non-continuum concepts. There are

[a]The journal Complex Systems is primarily devoted to cellular automata.

central biological problems in which continuous (e.g. differential equation) models do not capture the essential dynamics. A striking example is provided by self-reproduction of discrete units, the cells. In the forties John von Neumann tried to solve the following problem: which kind of logical organization makes it possible that an automaton (viewed as an "artificial device") reproduces itself? John von Neumann's lectures at the end of the forties clearly indicate that his work was motivated by the self-reproduction ability of biological organisms. Additionally, there was also an impact of achievements in automaton theory (Turing machines) and Gödel's work on the foundations of mathematics, in particular the incompleteness theorem ("There are arithmetical truths which can, in principle, never be proven."). A central role in the proof of the incompleteness theorem is played by self-referential statements. Sentences as "This sentence is false" refer to themselves and may trigger a closed loop of contradictions. Note that biological self-reproduction is a particularly clever manifestation of self-reference[45]. A genetic instruction as "Make a copy of myself" would merely reproduce itself (self-reference) implying an endless doubling of the blueprint, but not a construction of the organism. How can one get out of this dilemma between self-reference and self-reproduction?

The first model of self-reproduction proposed by von Neumann in a thought experiment (1948) is not bound to a fixed lattice, instead the system components are fully floating. The clue of the model is the two-fold use of the (genetic) information as uninterpreted and interpreted data, respectively, corresponding to a syntactic and semantic data interpretation. The automaton actually consists of two parts: a flexible construction and an instruction unit refering to the duality between computer and program or, alternatively, the cell and the genome[45]. Thereby, von Neumann anticipated the uncoding of the genetic code following Watson's and Crick's discovery of the DNA double helix structure (1953) - since interpreted and uninterpreted data interpretation directly correspond to molecular translation and transcription processes in the cell. Arthur Burks, one of von Neumann's students, called von Neumann's first model the kinematic model since it focuses on a kinetic system description. It was Stanislaw Ulam who suggested a "cellular perspective" and contributed with the idea of restricting the components to discrete spatial cells (distributed on a regular lattice). In a manuscript of 1952/53, von Neumann proposed a model of self-reproduction with 29 states. The processes related to physical motion in the kinematic model are substituted by information exchange of neighboring cells in this pioneer cellular automaton model. Chris Langton, one of the pioneers of artificial life research, reduced this self-reproducing automaton model drastically[35]. Meanwhile, it has been shown that the cellular automaton idea is a useful modeling concept in many further biological situations.

2. Cellular automaton definition

Cellular automata are defined as a class of spatially and temporally discrete dynamical systems based on local interactions. A cellular automaton can be defined as a 4-tuple (L, S, N, F), where

- L is an infinite regular lattice of cells/nodes (discrete space),
- S is a finite set of states (discrete states); each cell $i \in L$ is assigned a state $s \in S$,
- N is a finite set of neighbors, indicating the position of one cell relative to another cell on the lattice L; Moore and von Neumann neighborhoods are typical neighborhoods on the square lattice,
- F is a map

$$F : S|N| \rightarrow S \tag{1}$$

$$\{s_i\}_{i \in N} \mapsto s, \tag{2}$$

  which assigns a new state to a cell depending on the state of all its neighbors indicated by N (local rule).

The evolution of a cellular automaton is defined by applying the function F synchronously to all cells of the lattice L (homogeneity in space and time).

The definition can be varied, giving rise to several variants of the basic cellular automaton definition. In particular:

- *Probabilistic cellular automaton*: F is not deterministic, but probabilistic, i.e.

$$F : S|N| \rightarrow S \tag{3}$$

$$\{s_i\}_{i \in N} \mapsto s_j \text{ with probablity } p_j, \tag{4}$$

  where $p_j \leq 0$ and $\sum_j p_j = 1$.
- *Non-homogeneous cellular automaton*: transition rules and/or neighborhoods are allowed to vary for different cells.
- *Asynchronous cellular automaton*: the updating is not synchronous.
- *Coupled map lattice*: the state set S is infinite, e.g. $S = [0, 1]$.

3. Cellular automaton models of cell interaction

Cellular automaton models have been proposed for a large number of biological applications including ecological, epidemiological, ethological (game theoretical), evolutionary, immunobiological and morphogenetic aspects. Here, we give an overview of cellular automaton models of pattern formation in interacting cell systems. While von Neumann did not consider the spatial aspect of cellular automaton patterns per se – he focused on the pattern as a unit of self-reproduction – we are particularly concerned with the spatio-temporal dynamics of pattern formation.

Various automaton rules mimicking general pattern forming principles have been suggested and may lead to models of (intracellular) cytoskeleton and membrane dynamics, tissue formation, tumor growth, life cycles of microorganisms or animal coat markings. Automaton models of cellular pattern formation can be roughly classified according to the prevalent type of interaction. Cell-medium interactions dominate (nutrient-dependent) growth models while one can further distinguish direct cell-cell and indirect cell-medium-cell interactions. In the latter communication is established by means of an extracellular field. Such (mechanical or chemical) fields may be generated by tensions or chemoattractant produced and perceived by the cells themselves.

3.1. *Cell-medium or growth models*

Growth models typically assume the following scenario: A center of nucleation is growing by consumption of a diffusible or non-diffusible substrate. Growth patterns typically mirror the availability of the substrate since the primary interaction is restricted to the cell-substrate level. Bacterial colonies may serve as a prototype expressing various growth morphologies in particular dendritic patterns. Various extensions of a simple diffusion-limited aggregation (DLA) rule can explain dendritic or fractal patterns[52]. In addition, quorum-sensing mechanisms based on communication through volatile signals have recently been suggested to explain the morphology of certain yeast colonies[50].

A cellular automaton model for the development of fungal mycelium branching patterns based on geometrical considerations is suggested in Deutsch (1993)[13]. Recently, various cellular automata have been proposed as models of tumor growth[22,39]. Note that cellular automata can also be used as tumor recognition tools, in particular for the detection of genetic disorders of tumor cells[38].

3.2. *Cell-medium-cell interaction models*

Excitable media and chemotaxis Spiral waves can be observed in a variety of physical, chemical and biological systems. Typically, spirals indicate the excitability of the system. Excitable media are characterized by resting, excitable and excited states. After excitation the system undergoes a recovery (refractory) period during which it is not excitable. Prototypes of excitable media are the Belousov-Zhabotinskii reaction and aggregation of the slime mould *Dictyostelium discoideum*[12]. A number of cellular automaton models of excitable media have been proposed which differ in state space design, actual implementation of diffusion and in the consideration of random effects[36]. A stochastic cellular automaton was constructed as a model of chemotactic aggregation of myxobacteria[47]. Here, a nondiffusive chemical, the slime, and a diffusive chemoattractant are assumed in order to arrive at realistic aggregation patterns.

Turing systems Spatially stationary Turing patterns are brought about by a diffusive instability, the Turing instability[49].sec:chemotaxis The first (two-dimensional) cellular automaton of Turing pattern formation based on a simple activator-inhibitor interaction was suggested by Young[54]. Simulations produce spots and stripes (claimed to mimic animal coat markings) depending on the range and strength of the inhibition. Turing patterns can also be simulated with appropriately defined reactive lattice-gas cellular automata[20].

Activator-inhibitor automaton models might help to explain the development of ocular dominance stripes[48]. Ermentrout et al. introduced a model of molluscan pattern formation based on activator-inhibitor ideas[25]. Further cellular automaton models of shell patterns have been proposed (e.g. Kusch and Markus[34]). An activator-inhibitor automaton proved also useful as a model of fungal differentiation patterns[13].

3.3. *Cell-cell interaction models*

Differential adhesion In practice, it is rather difficult to identify the precise pattern forming mechanism, since different mechanisms (rules) may imply phenomenologically indistinguishable patterns. It is, particularly, difficult to decide between effects of direct cell-cell interactions and indirect interactions via the medium. For example, one-dimensional rules based on direct cell-cell interactions have been suggested as an alternative model of animal coat markings[29]. Such patterns have been traditionally explained with the help of reaction-diffusion systems based on indirect cell interaction. A remarkable three-dimensional automaton model based on cell-cell interaction by differential adhesion and chemotactic communication via a diffusive signal molecule is able to model aggregation, sorting out, fruiting body formation and motion of the slug in the slime mould *Dictyostelium discoideum*[44].

Alignment, swarming While differential adhesion may be interpreted as a density-dependent interaction one can further distinguish orientation-dependent cell-cell interactions. An automaton model based on alignment of oriented cells has been introduced in order to describe the formation of fibroblast filament bundles[24]. An alternative model of orientation-induced pattern formation based on the lattice-gas automaton idea has been suggested[14]. Within this model the initiation of swarming can be associated with a phase transition[8]. A possible application is street formation of social bacteria (e.g. Myxobacteria). We have previously also introduced a cellular automaton model for Myxobacterial rippling pattern formation based on cellular collisions[6].

3.4. *Cytoskeleton organization, differentiation*

Beside the spatial pattern aspect a number of further problems of developmental dynamics has been tackled with the help of cellular automaton models. The organization of DNA can be formalized on the basis of a one-dimensional cellular automaton[7]. Microtubule array formation along the cell membrane is in the focus of models suggested by Smith et al.[46]. Understanding microtubule pattern formation is an essential precondition for investigations of interactions between intra- and extracellular morphogenetic dynamics.

In Nijhout et al.[40] a rather complicated cellular automaton model is proposed for differentiation and mitosis based on rules incorporating morphogens and mutations. Another automaton model addresses blood cell differentiation as a result of spatial organization[37]. It is assumed in this model that spatial structure of the bone marrow plays a key role in the control process of hematopoiesis. The problem of differentiation is also the primary concern in a stochastic cellular automaton model of the intestinal crypt[41].

It is typical of many of the automaton approaches sketched in this short overview that they lack detailed analysis, the argument is often based on the sole beauty of simulations - for a long time people were just satisfied with the simulation pictures. This simulation phase in the history of cellular automata characterized by an overwhelming output of a variety of cellular automaton rules was important since it triggered a lot of challenging questions, particularly related to the quantitative analysis of automaton models. We have shown that in some cases the basic characteristics of the pattern formation dynamics can be grasped by a mean-field theory[6].

4. An example: a cellular automaton model of a vascular tumor growth

4.1. *Avascular tumor growth*

[b]Tumor growth always starts from a small number of malignantly proliferating cells, the tumor cells. The initial avascular growth phase can be studied *in vitro* by means of multicellular spheroids. In a typical experiment, tumor cells are grown in culture and are repeatedly exposed to fresh nutrient solution. Interestingly, after an initial exponential growth phase which implies tumor expansion, growth saturation is observed even in the presence of a periodically applied nutrient supply[26]. A section of the tumor spheroid shows a layered structure: A core zone composed mainly of necrotic material is surrounded by a thin layer of quiescent tumor cells and an outer ring of proliferating tumor cells (Fig. 1). A better understanding

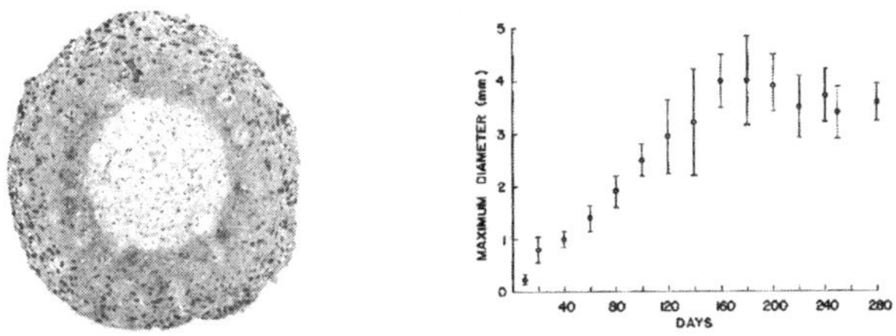

Figure 1. Folkman and Hochberg (1973) studied the growth of isolated spheroids from V-79 Chinese hamster lung cells, repeatedly transfered to new medium. Left: a cross section of a V-79 spheroid is shown, 1.0 mm in diameter and 20 days old. Viable cells are labeled with [^{3}H]thymidine; right: mean diameter and standard deviation of 70 isolated spheroids of V-79 cells.

of the processes which are responsible for the growth of a layered and saturating tumor is crucial. It has been realized that mathematical modeling

[b]Parts of this section have been published in Dormann and Deutsch[19].

can contribute to a better understanding of tumor growth[28]. In particular, various models have been suggested for the avascular growth phase[23]. We show here with a hybrid cellular automaton model that the layered pattern can be explained solely by the self-organized growth of an initially small number of tumor cells. A better knowledge of the spatio-temporal tumor dynamics should allow to design treatments which transfer a growing tumor into a saturated (non-growing and undangerous) regime by means of experimentally tractable parameter shifts.

A realistic model of avascular solid tumor growth should encompass mitosis, apoptosis and necrosis, processes which are particularly depending on growth factors and nutrient concentrations (cp. Fig. 2). Growth inhibitors

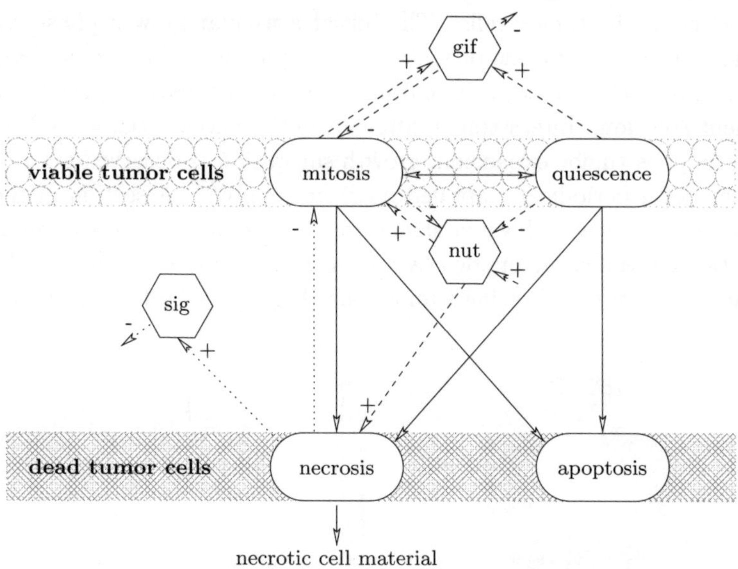

Figure 2. Cell dynamics for solid *in vitro* tumor growth; nut: nutrient dependency, gif: growth inhibitor factor dependency, sig: necrotic signal dependency.

play an important regulative role during tumor growth. Several models suggest that diffusible inhibitors are produced internally (e.g. metabolic waste products) and that mitosis is completely inhibited if the concentrations are too large. With increasing size and cell number, the spheroid requires more energy (nutrient). Since the nutrient concentration is lowest in the center of the avascular nodule, cells will starve here at first and may eventually die (necrosis). Under necrosis cells swell and burst, forming a necrotic site. There is experimental evidence that toxic factors are released or activated in this region and alter the microenvironment of the viable cells[27]. Contrarily, cells, which exceed their natural lifespan (apoptosis), shrink and are rapidly digested by their neighbors or by other specialized cells (macrophages)[4].

Traditional mathematical models of avascular solid tumor growth are formulated as deterministic (integro-)differential equations incorporating mitosis, apoptosis and necrosis inside the tumor (e.g. [1,11,31,42]). These models are based on the assumptions (i) that the tumor is spherically symmetric at all times and (ii) that the tumor sphere comprises a multi-layered structure, particularly a central necrotic core surrounded by an outer ring of proliferating tumor cells. Tumor growth is modeled by following the translocation of the outer radii of these layers. A cell-based Monte-Carlo approach has been introduced as a model of the initial exponential growth phase[21].

Here, we ask how the saturation of growth can be explained and the layered tumor structure can form. We present a two-dimensional hybrid cellular automaton model for the avascular growth phase defined in terms of lattice-gas terminology[20]. As we show in this book cellular automata allow for the systematic analysis of cooperative effects in interacting cell systems. In contrast to differential equation models, it is possible to follow the fate of individual cells. All cells are subject to identical interaction rules. Every cell can proliferate, be quiescent or die due to apoptosis and necrosis depending on its microenvironment.

Experimental work indicates that there are not only cells moving towards the periphery but that a significant number of proliferative and quiescent tumor cells is moving from the periphery towards the core area[18,17]. This inward cell flow is a necessary condition for the growth saturation characterizing multicellular spheroids. If there would be no cell flow towards the center but only resting cells and cells moving in the direction of the periphery, constant nutrient delivery would imply unbounded tumor growth. Accordingly, two oppositely-moving cell populations have to be considered. In the model, it is assumed that migration of cells depends on a chemical signal emitted by cells when they become necrotic. The chemotactic motion induces an antagonistic process to tumor expansion since some cells will migrate into the opposite direction, namely into the direction of the necrotic center. Based purely on local cell dynamics formation of a two-dimensional multi-layered tumor can be observed. We will also present results of statistical analysis of simulation runs. A different type of hybrid model has been previously introduced as a model of angiogenetic pattern formation which can follow the avascular growth phase[3]. Note that in Alarcon et al.[2] a hybrid cellular automaton has been suggested for the vascular growth phase.

4.2. *A hybrid lattice-gas cellular automaton model for tumor growth*

In the tumor lattice-gas cellular automaton model, cells represent tumor cells (C) and necrotic cells (N) which reside on the same two-dimensional square lattice ($b = 4$). With each lattice node ($r = (r_1, r_2)$) four velocity channels $c_1 = (1,0), c_2 = (0,1), c_3 = (-1,0), c_4 = (0,-1)$ and one resting channel $c_5 = (0,0)$ are associated, i.e. $\tilde{b} = 5$. Each channel can be occupied by at most one tumor (C) or necrotic (N) cell (Fig. 3). The von Neumann

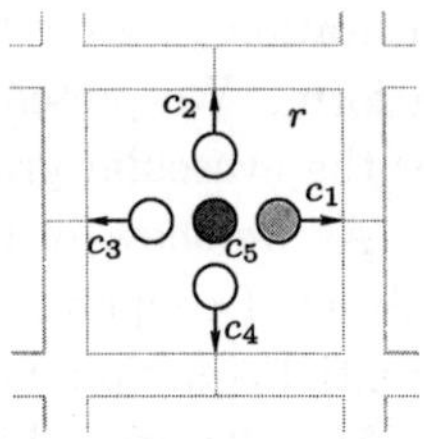

Figure 3. Example of a cell configuration at a lattice node r. The dark-gray dot and the light-gray dots denote the presence of a tumor and a necrotic cell, respectively.

interaction neighborhood is considered.

Furthermore, **diffusion of chemicals** (nutrient and necrotic signal) is modeled explicitly. Nutrient is consumed by proliferating and quiescent tumor cells. When tumor cells become necrotic they burst, leaking cell contents and necrotic signal into the surrounding tissue.

Mitosis, apoptosis and necrosis The model is based on conventional cell kinetics. Mitosis ($\bar{p}_m(r)$), apoptosis ($\bar{p}_d(r)$) and necrosis ($\bar{p}_n(r)$) rates depend on the local nutrient concentration ($c_{nut}(r)$) and local cell density (node configuration). They are defined as[c]

[c]Mostly, $\bar{p}_m + \bar{p}_d + \bar{p}_n \leq 1$; if this is not the case, the parameters are normalized.

$$\bar{p}_m(r) := \begin{cases} \frac{p_m}{\mathrm{n}_C(r)} \left(\frac{c_{nut}(r) - t_{nut}}{1 - t_{nut}} \right) & \text{if} \quad \mathrm{n}_N(r) = 0 \ \wedge \ c_{nut}(r) > \mathrm{n}_C(r)\, t_{nut} \\ 0 & \text{else,} \end{cases}$$

$$\bar{p}_d(r) := \begin{cases} p_d \, \mathrm{n}_C(r) & \text{if} \quad c_{nut}(r) > \mathrm{n}_C(r)\, t_{nut} \\ 0 & \text{else,} \end{cases}$$

$$\bar{p}_n(r) := \begin{cases} 0 & \text{if} \quad \mathrm{n}_N(r) = 0 \ \wedge \ c_{nut}(r) > \mathrm{n}_C(r)\, t_{nut} \\ p_n & \text{if} \quad \mathrm{n}_N(r) > 0 \ \wedge \ c_{nut}(r) > \mathrm{n}_C(r)\, t_{nut} \\ 1 & \text{else,} \end{cases}$$

where $p_m, p_d, p_n < 1$, $\mathrm{n}(r) = \mathrm{n}_C(r) + \mathrm{n}_N(r)$ denotes the number of all cells at r and $t_{nut} \leq 1$ the critical nutrient concentration for necrosis.

According to these rates each tumor cell at a node either proliferates (i.e. divides, if unoccupied channels exist), remains quiescent, dies or becomes necrotic. Nutrient is consumed by proliferating and quiescent tumor cells at a constant rate ($\bar{c}_{nut}$). Note, that the presence of necrotic material at a node leads to a complete inhibition of mitosis and might even act toxic for all tumor cells present at that node. This assumption is based on evidence that cell quiescence is due to factors other than nutrients such as for example cell contact effects[9].

All cells propagate simultaneously according to their orientation – only cells residing in "rest channels" do not move. Redistribution of cells at each lattice node, is defined by rules specifying

(1) adhesion,
(2) contact inhibition: cells are moving towards neighborhoods with low cell density and
(3) chemotactic motility: tumor cells move into the direction of the maximal signal gradient.

The following two node configurations can be distinguished:

1. Presence of tumor cells but no necrotic cells: One tumor cell always occupies the rest channel, if the outer interaction neighborhood contains at least one tumor cell; the remaining tumor cells are placed at channels which point to low density neighboring nodes mimicking the influence of contact inhibition. Thus, cells follow a track of least resistance (*passive motion*). Note, that in this model, the *density* of a node is assumed to be the number of tumor cells (n_C) plus a third of the number of necrotic cells (n_C). This models the smaller volume of necrotic cells viewed as burst tumor cells. The spatial scale of the lattice (i.e. the area of a lattice node) is chosen such that contact inhibition movement is induced up whenever more than one cell is present at a node.

In addition, we assume that the chemotactic response to the chemical signal contributes to the motility of tumor cells (*active motion*). This assumption is inspired by the experimental observation that there is a significant number of cells which drift from the viable rim of spheroids to the necrotic core[18,17].

In order to specify the impact of active and passive motion the following rules for the successive occupation of velocity channels are defined: First, the four neighboring nodes are ordered according to the chemical signal concentration. The density of cells together with the order number of a neighboring node $\bar{r}$ define an interval from which a *preference weight* (pw) is randomly chosen (cp. Fig. 4).

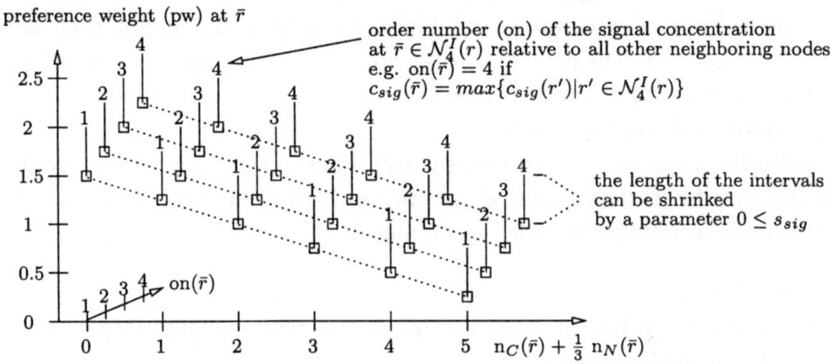

$$\text{pw}(\bar{r}) = 0.25\,(5 - n_C(\bar{r}) - \tfrac{1}{3}\,n_N(\bar{r}) + \text{on}(\bar{r})) + s_{sig}\,rnd, \quad rnd \in [0,1]$$

Figure 4. Interval of preference weights (pw) for each neighborhood configuration of a node r, concerning number of cells and signal concentration (c_{sig}). **Example**: If $s_{sig} = 0$ then a neighboring node $\bar{r}$ with no cells and maximal signal concentration (on($\bar{r}$) = 4) always receives the highest preference weight (pw($\bar{r}$) = 2.25), while a neighboring node r' with no cells and an order number of 1 obtains a smaller weight (pw(r') = 1.5) than a node r'' with either one tumor cell or three necrotic cells and order number 3 (pw(r'') = 1.75). Compare also Fig. 5.

Finally, the velocity channels of node r are ordered according to the magnitude of preference weights (the highest value is first) of the neighboring nodes to which they point, and the remaining tumor cells are sequentially placed on the channels. Figure 5 shows an example of this process. A

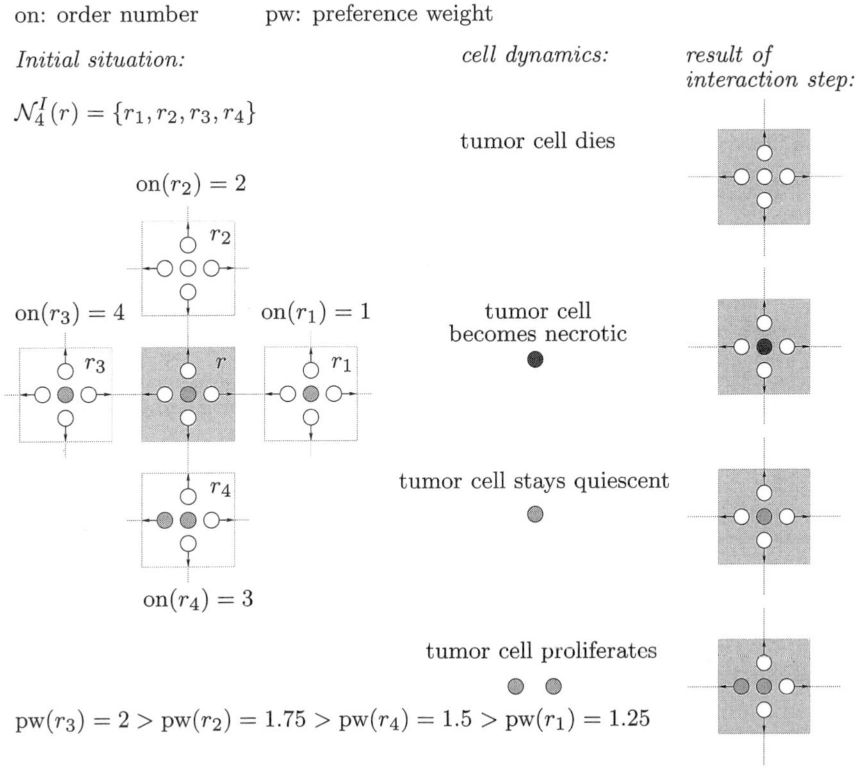

Figure 5. Example of the redistribution of cells at a lattice node. Since the focal (gray) node possesses more than one tumor cell in its outer interaction neighborhood the rest channel always gains a tumor cell. With $s_{sig} = 0$ the preference weights are uniquely determined (cp. Fig. 4). Channel c_3 is associated with the maximal weight.

special situation occurs if no signal and no tumor cells are present in the nearest neighborhood of r. Then, all cells are redistributed randomly on the channels. Accordingly, the cells perform a random walk.

2. No tumor cells but presence of necrotic cells: Necrotic cells are always distributed at first among the channels. If necrotic cells reside at a lattice node, then the rest channel receives a necrotic cell and the remaining necrotic cells are distributed at the velocity channels according to a line of least resistance with respect to the densities of the corresponding neighbor nodes.

If tumor cells are simultaneously present at the node, they are placed at the remaining channels according to their preference weights. This rule mimics that necrotic cell material which is in contact with tumor cells decreases the adhesivity of the cells.

The model dynamics is summarized in Fig. 6.

The automaton is *scaled* as follows:

- tumor cell size: tumor cells have a volume of about $3.36 \times 10^{-5}\ mm^3$ (V-79 cells[26]), necrotic cells are assumed to occupy one third of this volume. It is supposed that cells are "packed" in the volume of a cubic lattice node which is chosen as $2 \times$ volume of one tumor cell ($6.7 \times 10^{-5}\ mm^3$). Accordingly, the length of a square lattice area is $\Delta l := 0.04\, mm$,
- time steps: for cell dynamics $\Delta k = 1\ h$, for chemical diffusion $\Delta k_d = 1\ min$,
- diffusion coefficients of nutrient and necrotic signal: $D = 10^{-6}\ \frac{cm^2}{s} = 3.64\ \frac{\Delta l^2}{min}$.

4.3. *Simulations*

We have performed simulations starting from a small number of active tumor cells[d] and applying realistic parameter sets (Fig. 7). Parameters taken from the literature and incorporated in the automaton rules are glucose uptake rate, critical glucose concentration and doubling times for V-79 cells[27,32,51]:

- **glucose uptake rate**: Investigations with V-79 cell cultures[32] indicated that, if the external glucose concentration is approx. $1.15\,10^{-5}\ \frac{mg}{mm^3}$ ($\rightarrow 7.7 \cdot 10^{-8}\ \frac{mg}{\Delta l^3}$), then the consumption rate of glucose can be determined as $7.2 \cdot 10^{-8}\ \frac{mg}{cell\ h}$. Hence, during one hour all available nutrient is consumed, i.e. $\bar{c}_{nut} = 1\ \frac{1}{cell\ h}$,
- **critical glucose concentration**: The critical glucose concentration is about $1.4 \cdot 10^{-4}\ \frac{mg}{mm^3}$ ($\rightarrow 9.38 \cdot 10^{-9}\ \frac{mg}{\Delta l^3}$), hence $t_{nut} = 0.12$,
- **doubling times**:
Doubling times for V-79 cells are measured to be $10 - 19$ hours[27,51]. Assuming an initial doubling time of 16 hours, the growth rate of the initial exponential growth period is $p_m - p_d = \frac{ln(2)}{16}\ h^{-1}$, hence

[d]The initial number of tumor cells is always 44.

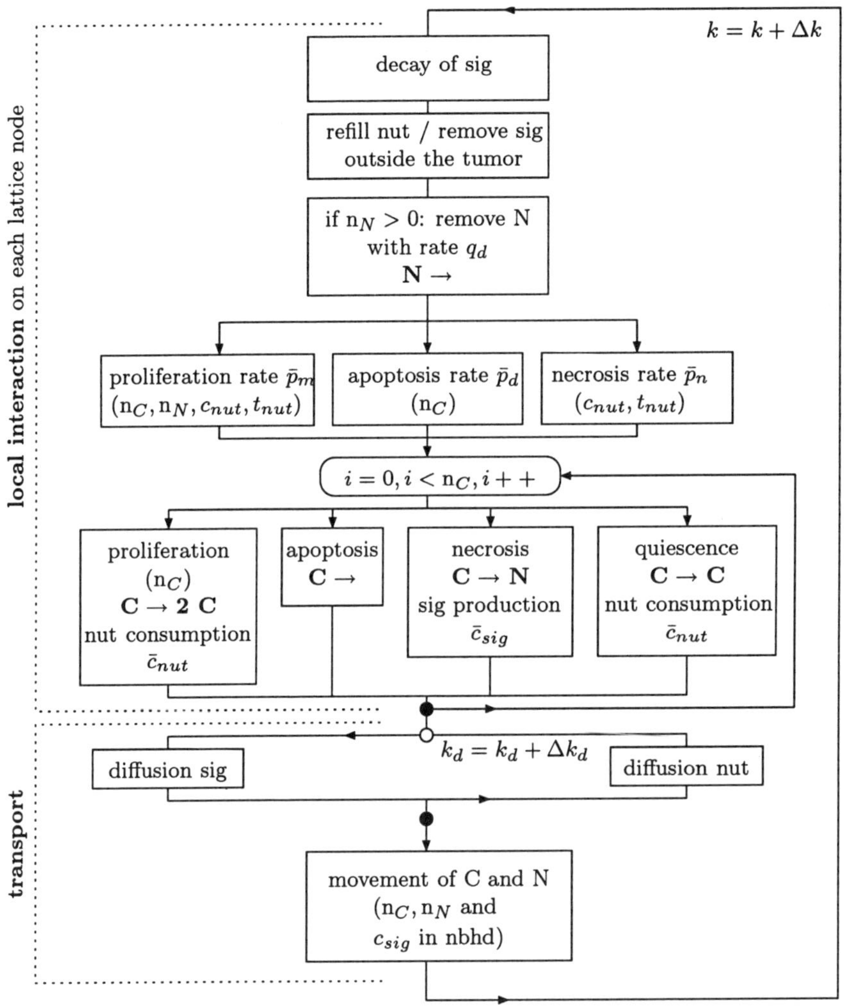

Figure 6. Schematic representation of the model dynamics; parameters: n_C: number of tumor cells and n_N: number of necrotic cells at a node; c_{nut}: nutrient and c_{sig}: signal concentration at a node, t_{nut}: critical nutrient concentration, $\bar{c}_{nut}$: nutrient consumption of a tumor cell, $\bar{c}_{sig}$: signal production during necrosis, nbhd: neighborhood.

$$p_m - p_d = 0.04\, h^{-1} .$$

In the simulations nutrient is regularly applied and the chemical signal regularly removed (every hour) outside of the tumor, i.e. at nodes which have no tumor or necrotic cell material in the Moore neighborhood with range 3 (48 empty neighboring nodes). Furthermore, the size of the lattice (200×200 nodes $\approx 8 \times 8$ mm) is chosen sufficiently large such that the boundaries do not influence the tumor growth within the considered time interval. The formation of a layered pattern comprised of a central necrotic core, a rim of quiescent tumor cells and an outer thin ring of proliferating cells can be observed. After an initial exponential growth phase, growth significantly slows down (Fig. 8). This is accompanied by the increase of the necrotic cell population and simultaneous decrease of the tumor cell number.

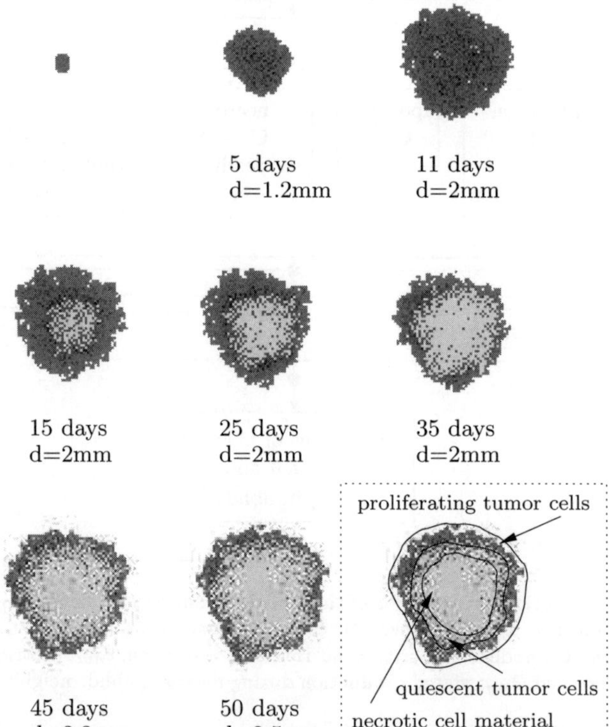

Figure 7. Simulation of tumor growth with a cellular automaton. A layered tumor forms, comprised of necrotic cell material, quiescent and proliferating tumor cells; parameters: mitosis rate $p_m = 0.05$, apoptosis rate $p_d = 0.01$, necrosis rate $p_n = 0.008$, rate for N dissolution $q_d = 0.0005$, production rate for chemical signal $\bar{c}_{sig} = 1$, decay rate for chemical signal 0.8, strength of chemical signal $s_{sig} = 0.4$, lattice size $|\mathcal{L}| = 200 \times 200$.

The nutrient concentration in the tumor decreases until the onset of necrosis and increases afterwards since the necrotic core does not consume nutrient (Fig. 9). For comparison, we performed simulations without considering a necrotic signal (i.e. no chemotactic influence on the tumor cells). The result is an unlimited growth of the spheroid (Fig. 10).

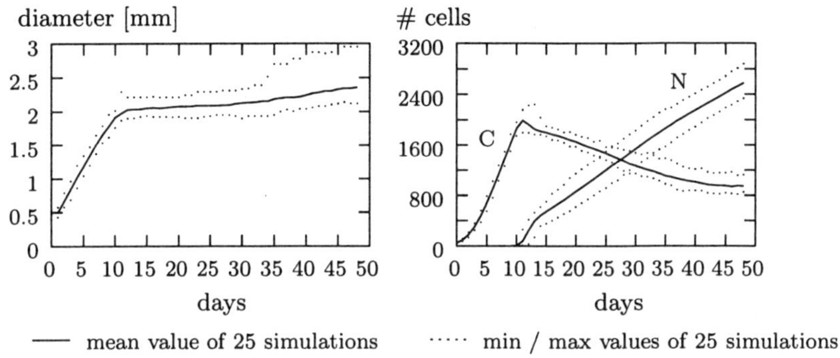

Figure 8. Simulation of diameter and cell number of 25 tumor growth simulations with "necrotic signaling" (C: tumor cells, N: necrotic cells).

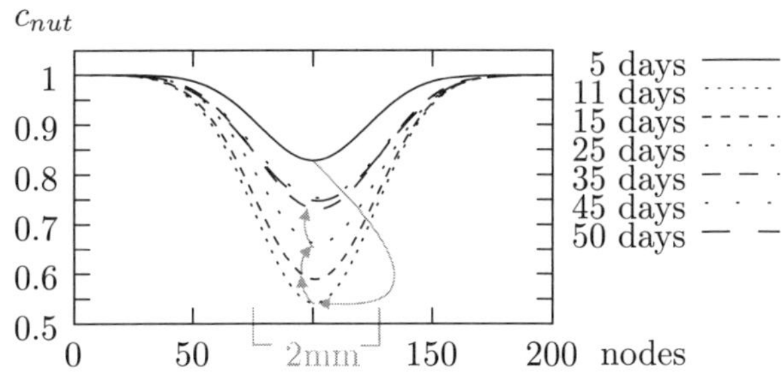

Figure 9. Simulation of nutrient concentration in a system of 200 × 200 lattice nodes (cp. Fig. 7). The figure shows a section of the lattice at row 100.

The cellular automaton introduced here reproduces experimental results, particularly the formation of a layered structure and growth saturation observed in multicellular spheroids[26]. Purely local rules (cell-cell interactions) allow for the transition from an initially small number of tumor cells to the final structured tumor. There are other cellular automaton models of avascular tumor growth but these are based on non-local rules [33,43]. [33] uses a Delaunay triangulation instead of a regular lattice. The

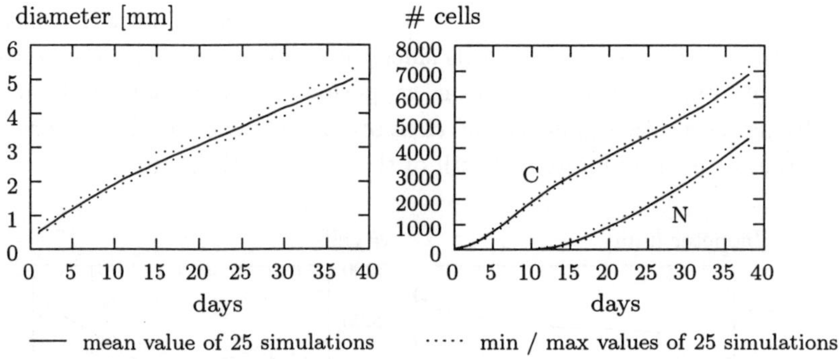

Figure 10. Simulation of diameter and cell number of 25 tumor growth simulations without a necrotic signal (C: tumor cells, N: necrotic cells).

hybrid cellular automaton approach presented here incorporates both the dynamics of discrete cells and the dynamics of chemical concentrations.

A sufficient condition for growth saturation during avascular growth even in the case of periodic nutrient supply is to guarantee a tendency of tumor cell motion into the direction of the necrotic core. Otherwise the tumor would continue to expand until finally the tumor compound would break up as a result of necrotic material dissolution. The "antagonistic growth direction" is established in the simulations by the chemotactic migration of tumor cells into the direction of the maximum necrotic signal gradient. Accordingly, in the model it is assumed that a diffusible signal emitted by bursting tumor cells is attracting living tumor cells. This mechanism produces a cell flow towards the center. Initially, the inward flow is small since the necrotic core is not existing or small. Accordingly, the outmoving cell population dominates, i.e. the tumor expands. But, later in development if the necrotic core has reached a critical size the inward flow takes over which limits further growth.

Our cellular model in principle allows to manipulate single cells or the microenvironment and to simulate the consequences of various treatments. For example, tumor growth can be followed after parts of the tumor have been "surgically" removed (Fig. 11a). Tumor spread is observed if the cell-cell adhesion is lowered by some substance (Fig. 11b). Finally, even if the cells have been manipulated such that they become necrotic (i.e. burst) survival of some tumor cells might occur (cp. Fig. 11c). Particularly, Fig. 11b demonstrates that a lowering of cellular adhesivity might have important consequences for the onset of tumor invasion. It is also straightforward to incorporate interactions with the immune system in the model.

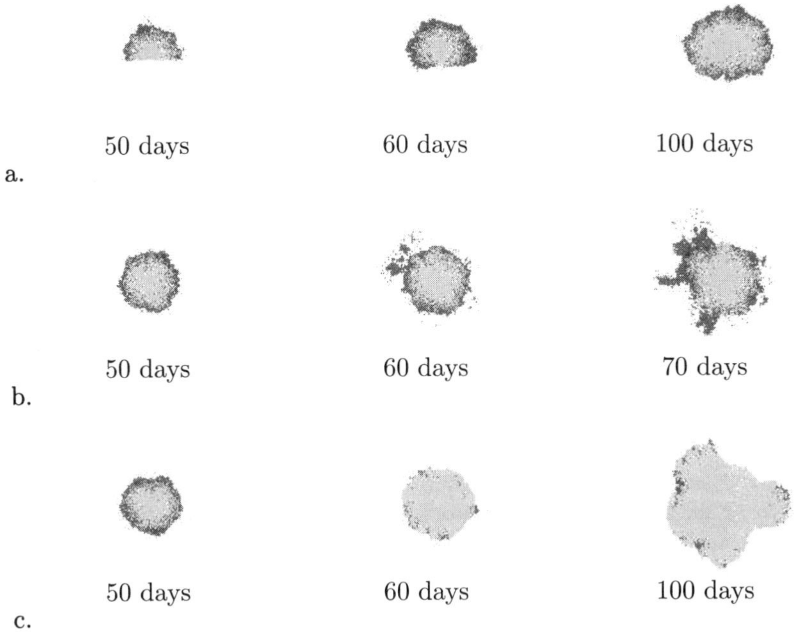

Figure 11. Simulation of tumor growth as in Fig. 7. Various "treatments" are simulated: a. After 50 days one half of the tumor is removed. The tumor recovers from this surgery. b. After 50 days the cell-cell adhesion is lowered. c. After 50 days the necrosis rate is magnified by a factor 10 ($p_n = 0.08$). However, tumor cells still survive.

5. Discussion

After giving a short overview of cellular automaton models for various cell interactions, we focused on the presentation of a specific example. We introduced a lattice-gas cellular automaton model for the avascular tumor growth phase which can be analyzed in multicellular spheroids. The lattice-gas cellular automaton model is hybrid – it represents biological cells as discrete entities and molecular concentrations as continuous states. The example illustrates the potential of cellular automaton modeling of interacting cell systems. As a cell-based model, a cellular automaton allows manipulation of individual cells, which enables particularly the simulation of the introduction of mutated (cancer) cells. Parallelization of the algorithms is straightforward for synchronous cellular automata, simulations are fast and allow the follow-up of large cell numbers. In simplified cell interaction models, stability analysis can be performed [8,6,15]. Cell size and the fastest biological process to be modeled determine the resolution of the cellular automaton model. The potential and versatility of cellular automaton models along with the availability of more and more "cellular data" (at the genetic and proteomic level) offer a promising approach to analyse self-organization in interacting cell systems.

Acknowledgments

The author greatfully acknowledges Andreas Dress (Bielefeld, Leipzig) for his introduction to the world of cellular automata and Sabine Dormann (Cologne) who designed, simulated and analyzed the lattice-gas cellular automaton introduced in this article.

References

1. J. A. Adam and N. Bellomo, editors. *A survey of models for tumor immune system dynamics.* Birkhauser, Boston, 1996.
2. T. Alarcon, H. Byrne, and P. K. Maini. *J. Theor. Biol.*, 225(2): 257-274.
3. M. Andrecut. A simple three-states cellular automaton for modelling excitable media. *Intern. J. Mod. Phys. B*, 12 (5): 601–607, 1998.
4. M. J. Arends and A. H. Wyllie. Apoptosis: mechanisms and roles in pathology. *Inter. Rev. Experim. Pathol.*, 32: 223–254, 1991.
5. F. Bagnoli. In F. Bagnoli, P. Lio, and S. Ruffo, editors, *Dynamical modelling in biotechnologies.* World Scientific, Singapore, 1998.
6. U. Börner, A. Deutsch, H. Reichenbach and M. Bär. *Phys. Rev. Lett.*, 89: 078101, 2002.
7. C. Burks and D. Farmer. In D. Farmer, T. Toffoli, and S. Wolfram, editors, *Cellular automata: Proceedings of an interdisciplinary workshop, New York, 1983*, pp. 157–167. North-Holland Physics Publ., Amsterdam, 1984.
8. H. Bussemaker, A. Deutsch, and E. Geigant. *Phys. Rev. Lett.*, 78: 5018–5021, 1997.
9. J. J. Casciari, S. V. Sotirchos, and R. M. Sutherland. *J. Cell. Physiol.*, 151: 386–394, 1992.
10. J. L. Casti. *Alternate realities.* John Wiley, New York, 1989.
11. M. A. J. Chaplain. *Math. Comput. Modell.*, 23: 47–87, 1996.
12. J. C. Dallon, H. G. Othmer, C. v. Oss, A. Panfilov, P. Hogeweg, T. Höfer, and P. K. Maini. In W. Alt, A. Deutsch, and G. Dunn, editors, *Dynamics of cell and tissue motion*, pp. 193–202. Birkhäuser, Basel, 1997.
13. A. Deutsch. In L. Rensing, editor, *Oscillations and morphogenesis*, chapter 28, pp. 463–480. Marcel Dekker, New York, 1993.
14. A. Deutsch. *J. Biol. Syst.*, 3: 947–955, 1995.
15. A. Deutsch and S. Dormann. *Cellular automaton modeling of biological pattern formation.* Birkhauser, Boston, 2004.

16. A. Deutsch, M. Falcke, J. Howard, and W. Zimmermann, editors. *Function and regulation of cellular systems: experiments and models*, Basel, 2003. Birkhauser.
17. M. Dorie, R. Kallman, and M. Coyne. *Exp. Cell Res.*, 166: 370–378, 1986.
18. M. Dorie, R. Kallman, D. Rapacchietta, D. van Antwerp, and Y. Huang. *Exp. Cell Res.*, 141: 201–209, 1982.
19. S. Dormann and A. Deutsch. *In Silico Biol.*, 2: 0035, 2002.
20. S. Dormann, A. Deutsch, and A. T. Lawniczak. *Fut. Comp. Gener. Syst.*, 17: 901–909, 2001.
21. D. Drasdo. In G. Beysens and G. Forgacs, editors, *Networks in biology and medicine*, pp. 171–185. Springer, New York, 2000.
22. D. Drasdo. In W. Alt, M. Chaplain, M. Griebel, and J. Lenz, editors, *Models of polymer and cell dynamics.* Birkhäuser, Basel, 2003.
23. D. Drasdo, S. Höhme, S. Dormann, and A. Deutsch. In A. Deutsch, M. Falcke, J. Howard, and W. Zimmermann, editors, *Function and regulation of cellular systems: experiments and models.* Birkhauser, Basel, 2003.
24. L. Edelstein-Keshet and B. Ermentrout. *J. Math. Biol.*, 29: 33–58, 1990.
25. B. Ermentrout, J. Campbell, and G. Oster. *The Veliger*, 28(4): 369–388, 1986.
26. J. Folkman and M. Hochberg. *J. Experim. Med.*, 138: 745–753, 1973.
27. J. P. Freyer. *Cancer Res.*, 48: 2432–2439, 1988.
28. R. Gatenby and P. K. Maini. *Nature*, 421:321, 2003.
29. G. Gocho, R. Pérez-Pascual, and J. Rius. *J. Theor. Biol.*, 125(4): 419–435, 1987.
30. N. S. Goel and R. L. Thompson. *Computer simulation of self-organization in biological systems.* Croom, Melbourne, 1988.
31. H. P. Greenspan. *Stud. Appl. Math.*, 51: 317–340, 1972.
32. L. Hlatky, R. K. Sachs, and E. L. Alpen. *J. Cell. Physiol.*, 134: 167–178, 1988.
33. A. R. Kansal, S. Torquato, G. R. Harsh, E. A. Chiocca, and T. S. Deisboeck. *J. Theor. Biol.*, 203: 367–382, 2000.
34. I. Kusch and M. Markus. *J. Theor. Biol.*, 178: 333–340, 1996.
35. C. G. Langton. *Physica10 D*, 10: 135–144, 1984.
36. M. Markus and B. Hess. *Nature*, 347(6288): 56–58, 1990.
37. R. Mehr and Z. Agur. *Bio Syst.*, 26: 231–237, 1992.
38. J. H. Moore and L. W. Hahn. In *Proc. of the Genetic and Evolutionary Computation Conference (GECCO-2001)*, p. 1452, 2001.
39. J. Moreira and A. Deutsch. *Adv. Compl Syst. (ACS)*, 5(2): 1–21, 2002.
40. H. F. Nijhout, G. A. Wray, C. Krema, and C. Teragawa. *Syst. Zool.*, 35: 445–457, 1986.
41. C. S. Potten and M. Löffler. *J. Theor. Biol.*, 127: 381–391, 1987.
42. L. Preziosi, editor. *Cancer modelling and simulation.* Chapman Hall/CRC Press, Boca Raton, Florida, USA, 2003.
43. Ah-Shen Qi, Xiang Zheng, Chan-Ying Du, and Boa-Sheng An. *J. theor. Biol.*, 161: 1–12, 1993.
44. N. J. Savill and P. Hogeweg. *J. Theor. Biol.*, 184: 229–235, 1997.

45. K. Sigmund. *Games of life – explorations in ecology, evolution, and behaviour.* Oxford University Press, Oxford, 1993.
46. S. A. Smith, R. C. Watt, and R. Hameroff. *Physica 10 D*, 10: 168–174, 1984.
47. A. Stevens. In W. Alt and G. Hoffmann, editors, *Biological Motion*, Lecture Notes in Biomathematics, pp. 548–555. Springer, Berlin, Heidelberg, 1990.
48. N. V. Swindale. *Proc. Roy. Soc. London Ser. B*, 208: 243–264, 1980.
49. A. M. Turing. *Phil. Trans. R. Soc. London*, 237: 37–72, 1952.
50. T. Walther, A. Grosse, K. Ostermann, A. Deutsch, and T. Bley. *J. Theor. Biol.*, 229(3): 327-338, 2004.
51. J. P. Ward and J. R. King. *IMA J. Math. Appl. Medic. & Biol.*, 14: 39–69, 1997.
52. T. A. Witten and L. M. Sander. *Phys. Rev. Lett.*, 47(19): 1400–1403, 1981.
53. S. Wolfram, editor. *Theory and applications of cellular automata*, Singapore, 1986. World Publishing Co.
54. David A. Young. *Math. Biosci.*, 72: 51–58, 1984.

A MATHEMATICAL ANALYSIS OF CYLINDRICAL SHAPED ANEURYSMS*

TOR A. KWEMBE AND SHATONDRIA N. JONES

Department of Mathematics, College of Science, Engineering and Technology, Jackson State University, P. O. Box 17610, Jackson, MS 39217,USA
E-mail: tor.a.kwembe@jsums.edu; Mstjoness@aol.com

Various investigations using the analysis of linearized models or experiments with rubber models suggest that aneurysmal walls are dynamically unstable. They resonate in response to pulsating blood flow. Recent mathematical models describing pressure and stress on the complement of spherical shaped intracranial saccular aneurysm suggests stability of aneurysmal wall. However, there is a need for more critical analysis of other regimes of observation and shapes of aneurysms. To this end, we have derived here a new non-linear equation of motion for pulsating cylindrical shaped aneurysms whose material fluid content behavior is described by a Fung-type pseudo-strain energy function that fits data on human lesions. We have used mathematical methods of non-linear dynamics to examine in theory the stability and instability of such lesions. This approach differs from other investigations, both from the geometry of the aneurysm and the regime of investigation. Our investigation deals with the interior walls of the aneurysm as opposed to the exterior or the surroundings of the aneurysmal wall. The results found here may help improve the prediction of the rupture potential of saccular lesions.

1. Introduction

Although there have been various extensive research applications involved with aneurysm development, specific conditions surrounding their genesis, enlargement, rupture, and management are highly controversial. An aneurysm is an abnormal localized sac or irreversible dilation of an artery, caused by decreased elastin of the arterial wall. These lesions can be located in the aorta, brain, leg, intestine and sometimes the splenic artery[7, 33]. The enlargement of these arteries are classified as fusiform, saccular, or dissecting. Saccular aneurysms bulge from one side of the artery while

*This work is supported by jackson state university computational center for molecular structure interactions (ccmsi)

fusiform bulges in all directions. Dissecting aneurysms are the result of a traumatic tear of an artery and are the most likely to cause mortality.

In general, it is believed that an aneurysm occurs when the pressure of blood passing through part of a weakened artery forces the vessel to bulge outward, forming a blister. Not all aneurysms are life-threatening, but it has been hypothesized that if the bulging stretches the artery too far the vessel may rupture and cause fatality[11]. An aneurysm that bleeds into the brain can lead to stoke or mortality. The genesis of an aneurysm is contingent on any condition that causes the walls of the arteries to weakened. Sekhar and Heros discuss in [25] competing hypotheses on the pathogenesis and Humphrey in [13] reviewed biomechanical factors that are implicated in lesion development. There are many hypotheses regarding aneurysm enlargement and rapture[8, 11, 17, 18, 22, 23, 25, 34], however, those important to our study involving saccular aneurysms are: (i) limit point instabilities [1, 2], (ii) the equilibrium wall stress exceeding the strength of the wall [3, 21, 23], or (iii) the dynamic behavior of the wall is unstable in response to pulsating blood flow [17, 25, 31].

In [19] Kyriacou and Humphrey showed that saccular aneurysms more than likely do not experience limit point instabilities. In [27], Shah and Humphrey examined the third hypothesis by proposing an idealized nonlinear equation of motion for a pulsating spherical aneurysm that is surrounded by celebral spinal fluid and whose behavior is described by a Fung-type pseudostrain-energy function that fits data on human lesions. They employed the methods of nonlinear dynamics to examine the stability of such lesions against perturbations to both in vivo and in vitro conditions. The numerical results suggest that the sub-class of lesions examined is dynamically stable. They suggested, however, that further study of the mechanics of saccular aneurysms be done and focused on quasi-static stress analysis investigating the roles of lesion geometry and material properties which include growth and remodeling. David and Humphrey gave further evidence for the dynamic stability of intracranial saccular aneurysms in [5].

In response to the call, Shah et al[26] studied further the roles of geometry and properties in the mechanics of saccular aneurysms. In this paper, though,we have derived a mathematical model of a sub-class of aneurysms having an un-deformed cylindrical geometry of radius R and wall thickness H relating pressure and the stretched ratio of the aneurysmal wall. An idealized assumption is made that allows lesions to experience time varying radial loads with deformed configuration remaining cylindrical. We use the model to theoretically postulate the following: (i) When the intramural pressure and the aneurysmal wall pressure are in balance, the stretch ratio λ is equal to one and the lesion is dynamically stable, (ii) if the intramural and aneurysmal wall pressures are not in balance and the stretch ratio is in the interval $0 < \lambda < 1$, the lesion is dynamically unstable, (iii) when the stretch ratio λ is greater than one and the transmural pressure is dominated by the stresses and strains, the lesion is stable and when the transmural pressure dominates the stresses and strains, the lesion is unstable.

The paper is arranged as follows: In section 2, we derived the mathematical model, in section 3, we have given the mathematical analysis by way of lemmas and theorems and in section 4 we discussed the model and its implications.

2. The Mathematical Model

We shall derive here a mathematical model of a sub-class of aneurysms having an un-deformed cylindrical geometry of radius R and wall thickness H relating pressure and the stretched ratio of the aneurysmal wall. We shall use this equation to analyze the stability and the instability of a pulsating aneurysm with a nonlinear isotropic material behavior. In line with Shah and Humphrey [27], we begin by considering relations describing the effect of stress, strain, and stretch ratio on aneurysmal walls. The differences between our model and that of Shah and Humphrey is in the geometric shape of the aneurysm and the regime of observation. They considered a sub-class of saccular aneurysms with spherical geometry and their regime of observation is outside of the aneurysm from the deformed radius to infinity (rupture). Here we have considered a sub-class of aneurysms with cylindrical geometry and our regime of observation is within the aneurysm up to the deformed radius.

Thus, we consider that the aneurysmal wall is of a nonlinear isotropic material defined by a Fung-type strain-energy function of the form [4, 5, 12, 13, 14, 15, 19, 24, 25, 27]

$$\omega = c[e^Q - 1]; Q = c_1(E_1^2 + E_2^2) + 2c_2E_1E_2 \tag{1}$$

where c, c_1, c_2 are material parameters and E_α; $\alpha = 1, 2$ are principal Green strains. We shall also consider the physical components of the Cauchy stress resultant tensor (the force per deformed length) T given in [14] by

$$T_{\alpha\beta} = \frac{1}{J} F_{\alpha p} F_{\beta q} \frac{\partial w}{\partial E_{pq}};\ \alpha,\ \beta,\ \mathrm{p},\ \mathrm{q} = 1,\ 2 \tag{2}$$

where $\mathbf{F}$ is the 2-D deformation gradient, J = det $\mathbf{F}$. Here too, we shall allow lesions to experience only time-varying radial loads with deformed configuration remaining cylindrical. Hence $\mathbf{F} = \begin{bmatrix} \lambda(t) & 0 \\ 0 & \lambda(t) \end{bmatrix}$ and $\lambda = \frac{\mathrm{a(t)}}{\mathrm{R}}$, where a is the deformed radius at time t. Thus, $\mathrm{J} = \lambda^2$ and $\mathrm{E}_\alpha = \frac{1}{2}$ (J-1) $= \frac{1}{2}$ (λ^2 - 1). Since $\mathrm{F}_{12} = \mathrm{F}_{21} = 0$ and $\mathrm{F}_{11} = \mathrm{F}_{22} = \lambda$ a direct computation from (1) and (2) gives that $\mathrm{T}_{11} = \mathrm{T}_{22} = \mathrm{c}e^{Q}(\mathrm{c}_1 + \mathrm{c}_2)(\lambda^2 - 1)$; $\mathrm{Q} = \frac{1}{2}(\mathrm{c}_1 + \mathrm{c}_2)(\lambda^2 - 1)^2$. Hence, the spatially uniform principal stress resultant T is

$$T(\lambda) = (c_1 + c_2) c e^{Q} (\lambda^2 - 1) \tag{3}$$

and ω in terms of λ is $\omega(\lambda) = \mathrm{c}[\exp\{\frac{1}{2}(c_1 + c_2)(\lambda^2 - 1)^2\} - 1]$. From here on references to equation (1) means $\omega(\lambda)$.

Linear momentum balance yields the associated equation of motion of the familiar form,

$$\rho_m h R \frac{d^2\lambda}{dt^2} = P(t) - 2\frac{T(\lambda)}{\lambda R} \tag{4}$$

where ρ_m is the constant mass density of the membrane and h its deformed thickness, assuming incompressibility, $\mathrm{h} = \frac{\mathrm{H}}{\lambda^2}$. P(t) is the time-varying transmural pressure, i.e., inner minus outer fluid pressure. Equation (4) was first used by Knowles in [19] to study finite amplitude oscillations of nonlinear solids for cylindrical tubes and then by Wang in [32] to study oscillations for spherical thin shells. Thus, we use it here to study cylindrical shaped aneurysms.

We shall assume that the behavior of the material content in the aneurysm is incompressible and Newtonian. Hence, the spatial balance of mass and Navier- Stokes equations govern its response. In the absence of body forces we have the divergence free equation

$$\rho \frac{D\mathbf{v}}{Dt} = -\nabla P + \mu \nabla^2 \mathbf{v} \tag{5}$$

$$\nabla \bullet \mathbf{v} = 0 \tag{6}$$

where **v**, ρ, μ, and P are the velocity, constant mass density of the aneurysmal fluid, viscosity, and pressure. $\frac{\mathrm{D}\mathbf{v}}{\mathrm{Dt}}$ is the acceleration. Assuming flow is in the radial direction r, we then have the nonaxial version of equations (5) and (6) for cylindrical coordinates in the form

$$\rho(\frac{\partial v^{(r)}}{\partial t} + v^{(r)}\frac{\partial v^{(r)}}{\partial r}) = -\frac{\partial P}{\partial r} + \mu[\frac{1}{r}\frac{\partial}{\partial r}(r\frac{\partial v^{(r)}}{\partial r}) - \frac{1}{r^2}v^{(r)}] \tag{7}$$

$$\frac{1}{r}\frac{\partial}{\partial r}(rv^{(r)}) = 0 \tag{8}$$

where $\mathrm{v}^{(\mathrm{r})}$ is the radial component of velocity. A direct integration of (8) gives

$$v^{(r)}(t) = \frac{g(t)}{r} \tag{9}$$

The function g(t) is found by matching the fluid and membrane velocities at r=a, the deformed radius. The membrane velocity is given by

$$\frac{d}{dt}(a(t) - R) = \frac{da(t)}{dt} = R\frac{d\lambda}{dt} \tag{10}$$

At balance $\frac{\mathrm{g(t)}}{\mathrm{a}} = \mathrm{R}\ \frac{\mathrm{d}\lambda}{\mathrm{dt}}$, which gives

$$g(t) = aR\frac{d\lambda}{dt} = R^2\lambda\frac{d\lambda}{dt} \tag{11}$$

Thus,

$$v^{(r)}(t) = \frac{R^2\lambda}{r}\frac{d\lambda}{dt} \tag{12}$$

upon substituting (12) into (7), we have

$$\frac{\rho R^2}{r}[\lambda\frac{d^2\lambda}{dt^2} + (\frac{d\lambda}{dt})^2] - \frac{\rho R^4\lambda^2}{r^3}(\frac{d\lambda}{dt})^2 = -\frac{\partial P}{\partial r} \tag{13}$$

We now integrate (13) over r from (r = R) to (r = a) and obtain an expression for the pressure exerted on the membrane by the material content of the aneurysm as:

$$P_R = P_a + \rho R^2\{\lambda Ln(\lambda)\frac{d^2\lambda}{dt^2} + [Ln(\lambda) - \frac{\lambda^2}{2} + \frac{1}{2}](\frac{d\lambda}{dt})^2\}, \quad (14)$$

where P_a, the pressure at the aneurysmal wall can be constant or time varying. On taking the radial stress as a dynamic boundary condition, we have

$$t_{rr}\mid_R = -P\mid_R + 2\mu D_{rr}\mid_R = -P_R + 2\mu\frac{\partial v^{(r)}}{\partial r}\mid_R = -P_R - 2\mu\lambda\frac{d\lambda}{dt} \quad (15)$$

where D is the stretching tensor. Equations (4), (14), and (15) gives

$$\begin{aligned} P_i(t) = P_a + [\tfrac{\rho_m HR}{\lambda^2} + \rho R^2\lambda Ln(\lambda)]\tfrac{d^2\lambda}{dt^2} + \rho R^2[Ln(\lambda) - \tfrac{\lambda^2}{2} + \tfrac{1}{2}](\tfrac{d\lambda}{dt})^2 \\ + 2\mu\lambda\tfrac{d\lambda}{dt} + \tfrac{2T(\lambda)}{\lambda R} \end{aligned} \quad (16)$$

Equation (16) is the nonlinear ordinary differential equation for a pulsating, nonlinear elastic, cylindrical membrane holding an incompressible Newtonian fluid. $T(\lambda)$ is given in equation (3) and $P_i(t)$ is the pressure inside the aneurysm. To solve equation (16) completely, the aneurysmal wall will be subjected to the initial conditions on λ and the stretch rate $\frac{d\lambda}{dt}$. However, as it is the custom with most physical problems, we shall first non-dimensionalize equations (16). To that end, we let $\tau = t\left(\frac{c}{\rho_m R^2 H}\right)^{\frac{1}{2}}$ be the non dimensional time, then the non-dimensional version of equation (16) is

$$(\frac{1}{x^2} + bxLn(x))\ddot{x} + b(Ln(x) - \frac{x^2}{2} + \frac{1}{2})\dot{x}^2 + 2mx\dot{x} + \frac{2f(x)}{x} = F(\tau) \quad (17)$$

where $b = \frac{\rho R}{\rho_m H}$, $m = \frac{\mu}{(\rho c H)^{\frac{1}{2}}}$, $F = \frac{R}{c}\,(P_i(t) - P_a)$, $f = \frac{T}{c}$, and $x = \lambda$, where in R, $\sqrt{\frac{\rho R^2 H}{c}}$, $\rho_m R^2 H$ are respectively the length, time, and mass scales. We now decompose equation (17) into the following system of first order nonlinear ordinary differential equations to enable us provide a dynamical analysis of the aneurysmal wall and the impact the pressure flow has on the wall:

$$\frac{dy_\circ}{d\tau} = y_1 \quad (18)$$

$$\frac{dy_1}{d\tau} = [y_\circ^{-2}+by_\circ Ln(y_\circ)]^{-1}[F(\tau)\text{-}b\{Ln(y_\circ)\text{-}\frac{y_\circ^2}{2}+\frac{1}{2}\}y_1^2\text{-}2my_\circ y_1\text{-}\frac{2f(y_\circ)}{y_\circ}] \tag{19}$$

where $y_\circ \equiv x$; $y_1 \equiv \frac{dx}{d\tau} = \dot{x}$

3. The Dynamical Properties of the Model

Shah and Humphrey [27] gave some numerical solutions of their nonlinear ordinary differential equation. Haslach, Jr. [12] gave a qualitative analysis of the Shah and Humphrey model. In this section, we examine the qualitative properties of the model and establish necessary conditions for achieving asymptotic stability and the instability of the fixed points and hence of the aneurysmal wall. Numerical results and simulations fitting empirical data depicting lesion responses are provided in a paper in preparation.

The model as we have derived here is physically realistic for stretch ratios $\lambda > 0$, the mathematical properties examined here are broken into three categories of λ, namely, when (i) $0 < \lambda < 1$, (ii) $\lambda = 1$, and (iii) $\lambda > 1$. The nonlinear dynamical model is then analyzed in three stages of increasing complexity: the intramural pressure p_i and the pressure $p_a$on the aneurysmal wall are in balance; when the pressure difference is dominated by the resultant stress, strain-energy function; and when the pressure difference dominates the resultant stress, strain-energy function. In all cases, we considered a viscous Newtonian fluid content of the aneurysm (that is $\mu > 0$). In the discussion section, we discussed the case of inviscid flow and compare our results with those of Haslach, Jr. for the Shah-Humphrey model[12].

3.1. *The case $P_i(\tau) = P_a$*

The differential equivalent of the system of equations (18) and (19) is:

$$\left[b\{y_0^2\ln(y_0)+\frac{y_0^3}{2}-\frac{y_0^2}{2}\}y_1^2+2my_0^3y_1+\frac{2}{c}\frac{d\omega}{dy_0}\right]dy_0+y_1[1+by_0^3\ln(y_0)]dy_1=0 \tag{20}$$

A careful examination reveals that equation (20) is not exact and therefore the orbits are not easily determined as level curves on the surface of the integral energy function. Hence, it is best to study the dynamic behavior of the system via the classification of fixed points.

Lemma 3.1. *Let c, k be the material parameters of the Fung type strain-energy function ω, ρ the constant mass density of the aneurysm fluid content, H the wall thickness of the un-deformed aneurysm and $\mu > 0$, the viscosity of the fluid content of the aneurysm. If $P_i(\tau) = P_a$, then (1,0) is a fixed point of the nonlinear dynamical system. It is an asymptotically stable node if $\mu^2 > 4kc\rho H$. It is an asymptotically stable spiral point if $\mu^2 < 4kc\rho H$. It is an asymptotically stable proper node if $\mu^2 = 4kc\rho H$.*

Proof. Since $P_i(\tau) = P_a$, we let $F(\tau) = 0$ in the systems of equations (18) and (19) and then solve for the fixed points the equations $\dot{y_0} = \dot{y_1} = 0$ and obtain (-1,0) and (1,0) as fixed points. Since the stretch ratios are positive, we have that (1,0) is the only fixed point. The Jacobian at (1,0) is

$$J = \begin{pmatrix} 0 & 1 \\ -4k & -\frac{2\mu}{(\rho c H)^{\frac{1}{2}}} \end{pmatrix}$$

The eigenvalues $\Lambda = \frac{-\mu \pm \sqrt{\mu^2 - 4Kc\rho H}}{2}$. Since μ, K, c, ρ, and H are all positive, we see immediately that the fixed point is an asymptotically stable node and an asymptotically stable spiral point if $\mu^2 < 4kc\rho H$. It is an asymptotically stable proper node if $\mu^2 = 4kc\rho H$. Hence the lemma follows. ∎

Lemma 3.1 shows that the stretch ratio is 1 when the intramural pressure P_i is in balance with the pressure P_a of the aneurysmal wall. It also tells us that under this situation the aneurysmal wall is stable. In the next section we show that when $P_i \neq P_a$, the nonlinear dynamic system has fixed points $(c_0, 0)$ given that $c_0 \neq 1$. We then analyzed the fixed points when $0 < c_0 < 1$ and when $c_0 > 1$.

3.2. *The case $P_i \neq P_a$*

Lemma 3.2. *Let $F(\tau) \neq 0$. Then, the nonlinear dynamical systems (18) and (19) has a fixed point $(c_0, 0)$; $c_0 \neq 1$ and satisfy the equation $F(\tau) = \frac{2K(c_0^2-1)}{c_0}\exp\{\frac{1}{2}K(c_0^2-1)^2\}$. Furthermore, c_0 is such that $[c_0^{-2} + \frac{\rho R c_0}{\rho_m H}\ln(c_0)] \neq 0$. That is, c_0 is not a singular point of the system.*

Proof. We observe from the statement of the lemma that if $c_0 = 1$, then $F(\tau) = 0$ for all τ. Therefore $F(\tau)$ cannot be zero if $c_0 \neq 1$. Now on solving the equations $\dot{y_0} = \dot{y_1} = 0$ for fixed points, we have that $y_1 = 0$ and $F(\tau) - \frac{2T(y_0)}{y_0} = 0$ provided $[y_0^{-2} + \frac{\rho R y_0}{\rho_m H}\ln(y_0)] \neq 0$. With $K = c_1 + c_2$, we utilized equation (3) to get the desired result. Hence the lemma follows. □

Lemma 3.3. *Suppose that $P_i(\tau) - P_a < \frac{2[K((c_0^2-1)^2+1]\omega'(c_0)}{R(c_0^2-1)}$, where ω is the Fung type strain-energy function and c_0 is such that $(c_0, 0)$ is a fixed point of the nonlinear dynamical system (18) and (19). If $0 < c_0 < 1$, then the fixed point $(c_0, 0)$ is an unstable saddle point. Similarly, if $P_i(\tau) - P_a > \frac{2[K((c_0^2-1)^2+1]\omega'(c_0)}{R(c_0^2-1)}$ and $0 < c_0 < 1$, the fixed point $(c_0, 0)$ is an unstable improper node.*

Proof. Lemma 3.2 gives that for $F(\tau) \neq 0$ and $[c_0^{-2} + \frac{\rho R c_0}{\rho_m H}\ln(c_0)] \neq 0$, $(c_0, 0)$ is a fixed point such that $c_0 \neq 1$. The Jacobian at $(c_0, 0)$ is

$$J = \begin{pmatrix} 0 & 1 \\ \frac{c_0^{-1}[F(\tau)-2f'(c_0)]}{c_0^{-2}+bc_0\ln(c_0)} & \frac{-2mc_0}{c_0^{-2}+bc_0\ln(c_0)} \end{pmatrix}$$

where $f'(c_0) = 2Kc_0[K(c_0^2-1)^2+1]\exp\{\frac{1}{2}K(c_0^2-1)^2\}$. The eigenvalues $\Lambda = \frac{-\mu\pm\sqrt{\mu^2+(\rho cH)c_0^{-3}[c_0^{-2}+b\ln(c_0)][F(\tau)-2f'(c_0)]}}{(\rho cH)^{\frac{1}{2}}[c_0^{-3}+b\ln(c_0)]}$. Since $F(\tau) = \frac{R}{c}(P_i(\tau) - P_a) < \frac{2[K(c_0^2-1)^2+1]\omega'(c_0)}{c(c_0^2-1)} = 2f'(c_0)$, we see that $F(\tau) - 2f'(c_0) < 0$. For $0 < c_0 < 1$, we also see that $c_0^{-3} + b\ln(c_0) < 0$ and $c_0^{-2} + b\ln(c_0) < 0$. Thus, the eigenvalues are real with mixed signs and hence the fixed point $(c_0, 0)$ is an unstable saddle point. On the other hand if $P_i(\tau) - P_a > \frac{2[K(c_0^2-1)^2+1]\omega'(c_0)}{R(c_0^2-1)}$, then $\sqrt{\mu^2+(\rho cH)c_0^{-3}[c_0^{-2}+b\ln(c_0)][F(\tau)-2f'(c_0)]} < \mu$, and the eigenvalues are all positive for viscous flows. Hence, the fixed point is an unstable improper node. □

We observe from the proof of lemma 3.3 that if $\mu^2 + (\rho cH)c_0^{-3}[c_0^{-2} + b\ln(c_0)][F(\tau) - 2f'(c_0)] < 0$ and $0 < c_0 < 1$, the fixed point $(c_0, 0)$ is an unstable spiral point and the state trajectory will experience periodic behavior. We also observe here that in lemmas 3.2 and 3.3, we assume that $P_i(\tau) - P_a$ is a constant. The complex case where $P_i(\tau) - P_a$ is a function of τ with or without periodic behavior will be considered in a separate article.

Lemma 3.4. *Let's suppose that $P_i(\tau) - P_a$ is a constant. Let's suppose further that $P_i(\tau) - P_a < \frac{2[K(c_0^2-1)^2+1]\omega'(c_0)}{R(c_0^2-1)}$, where ω is the Fung type strain-energy function and c_0 is such that $(c_0, 0)$ is a fixed point of the nonlinear dynamical system (18) and (19). If $c_0 > 1$ such that*

$\mu^2 + (\rho cH)c_0^{-3}[c_0^{-2} + b\ln(c_0)][F(\tau) - 2f'(c_0)] < \mu$, then the fixed point ($c_0$, 0) is an asymptotically stable node. If $P_i(\tau)$ - $P_a > \frac{2[K(c_0^2-1)^2+1]\omega'(c_0)}{R(c_0^2-1)}$ and $c_0 > 1$, the fixed point (c_0, 0) is a saddle point and hence unstable.

Proof. From lemma 3.3, we have the eigenvalues of the Jacobian at (c_0, 0) as $\Lambda = \frac{-\mu \pm \sqrt{\mu^2 + (\rho cH)c_0^{-3}[c_0^{-2} + b\ln(c_0)][F(\tau) - 2f'(c_0)]}}{(\rho cH)^{\frac{1}{2}}[c_0^{-3} + b\ln(c_0)]}$. On noting that the condition $P_i(\tau)$ - $P_a < \frac{2[K(c_0^2-1)^2+1]\omega'(c_0)}{R(c_0^2-1)}$ implies that $F(\tau) - 2f'(c_0) < 0$ and that $c_0^{-2} + b\ln(c_0)$ and $c_0^{-3} + b\ln(c_0)$ are positive whenever $c_0 > 1$. Then clearly $\mu^2 + (\rho cH)c_0^{-3}[c_0^{-2} + b\ln(c_0)][F(\tau) - 2f'(c_0)] < \mu$ and hence the eigenvalues Λ are all real and negative. Thus, the fixed point (c_0, 0) is an asymptotically stable node. Similarly, the condition $P_i(\tau)$ - $P_a > \frac{2[K(c_0^2-1)^2+1]\omega'(c_0)}{R(c_0^2-1)}$ implies that $F(\tau) - 2f'(c_0) > 0$. Therefore, for $c_0 > 1$, we observe that $\mu^2 + (\rho cH)c_0^{-3}[c_0^{-2} + b\ln(c_0)][F(\tau) - 2f'(c_0)] > \mu$ and so the eigenvalues Λ are real with alternating signs. Thus, the fixed point (c_0, 0) is and unstable saddle point. □

We observe here too that if $F(\tau) - 2f'(c_0) < 0$ for $c_0 > 1$ and $\mu^2 + (\rho cH)c_0^{-3}[c_0^{-2} + b\ln(c_0)][F(\tau) - 2f'(c_0)] < 0$, then the state trajectory exhibits an oscillatory behavior around the fixed point. That is, it may spiral towards the fixed point or a limit cycle.

From the above lemmas, we have the following concluding theorems whose proofs are a direct consequence of the lemmas.

Theorem 3.1. *For stretch ratios λ such that $0 < \lambda < 1$, the aneurysmal wall is dynamically unstable.*

Proof. The proof is a consequence of Lemma 3.3. If (λ, 0) is within the neighborhood of the fixed point (c_0, 0) such the $0 < \lambda = c_0 < 1$, then regardless of whether the pressure differential dominates the stress and strains impacted on the aneurysmal wall or not, the aneurysmal wall will remain unstable and pulsate if $\mu^2 + (\rho cH)\lambda^{-3}[\lambda^{-2} + b\ln(\lambda)][F(\tau) - 2f'(\lambda)] < 0$. □

Theorem 3.2. *If the intramural pressure of an aneurysm is in balance with the pressure of the aneurysmal wall, then aneurysmal wall is dynamically stable for stretch ratios $\lambda = 1$.*

Proof. The proof is a consequence of Lemma 3.2. □

Theorem 3.3. *For stretch ratios λ, such that $\lambda > 1$ and $P_i(\tau)$ - $P_a < \frac{2[K(\lambda^2-1)^2+1]\omega'(\lambda)}{R(\lambda^2-1)}$. The aneurysmal wall is dynamically stable if $\mu^2 + (\rho cH)\lambda^{-3}[\lambda^{-2} + b\ln(\lambda)][F(\tau) - 2f'(\lambda)] < \mu$. On the other hand if $P_i(\tau)$ - $P_a > \frac{2[K(\lambda^2-1)^2+1]\omega'(\lambda)}{R(\lambda^2-1)}$ for $\lambda > 1$, the aneurysmal wall is dynamically unstable.*

Proof. The proof follows from Lemma 3.4 □

4. Discussion

The rupture of saccular aneurysms is the most common cause of spontaneous subarachnoid hemorrhage which, despite advances in neurosurgery, continues to result in significant morbidity and mortality. Aneurysms are treated surgically. A patch or artificial piece of blood vessel is sewn where the aneurysm was. Treatment usually depends on the size and location of the aneurysm and ones overall health. Even though many large lesions do not rupture and some small ones do, the decision to treat a diagnosed unruptured aneurysm is based on the maximum dimension of the lesion. The critical size to warrant surgery is controversial. Therefore, there is need for better and improved predictors of the rupture potential of lesions. In this paper, we showed through theoretical analysis that saccular aneurysms may expand or rupture based on the imbalance in local states of stress and strain and the transmural pressure. Specifically, we found the following: (i) When the intramural pressure and the aneurysmal wall pressure are in balance, the stretch ratio λ is equal to one and the lesion is dynamically stable. That is the lesion is stable when the deformed radius reaches the size of the un-deformed radius, (ii) if the intramural and aneurysmal wall pressures are not in balance and the stretch ratio is in the interval $0 < \lambda < 1$, the lesion is dynamically unstable. That is at the initial stages of deformation, the lesion is unstable and may or may not rupture, (iii) when the stretch ration λ is greater than one and the transmural pressure is dominated by the stresses and strains, the lesion is stable and when the transmural pressure dominates the stresses and strains, the lesion is unstable. This reveals the important roles of lesion shape[8], material properties, loading conditions and size[9, 11, 17, 18, 21, 28, 30, 31] governing the distributions of stress and strain within and on the aneurysmal wall.

In our study we considered the fluid content of the aneurysm to be viscous and Newtonian. If however, we allow the fluid content to be inviscid, that is, $\mu = 0$ with constant transmural pressure, then the system of equations (18) and (19) is inviscid and has two fixed points, a saddle and a center surrounded by a homoclinic orbit. This characteristics coincide with the Shah and Humphrey model [12].

Acknowledgments

We wish to express our gratitude to the reviewers and the editorial Board for BIOMATV International Symposium on Mathematical and Computational Biology for their useful comments and advice. The summer research experience of Shatondria N. Jones was supported by the Computational Center for Molecular Structure Interactions (CCMSI) an NSF funded program at Jackson State University.

References

1. Akkas, N., Aneurysms as a biomechanical instability problem, In: Mosora, F. (Ed), Biomechanical Transport Process, Plenum Press, New York, (303 - 311) 1990.
2. Austin, G.M., Scievink, W., Williams, R., Controlled pressure volume factors in the enlargement of intracranial saccular aneurysms, Neurosurgery, (24)(722 - 730) 1989.
3. Canham, P.B., and Ferguson, G.G., A mathematical model for the mechanics of saccular aneurysms, Neurosurgery, (17)(291-295) 1985.
4. Crompton, M. R., Mechanism of growth and rupture in cerebral berry aneurysms; Br. Med. J., (1)(1138-1142) 1966.
5. David, G. and Humphrey, J. D., Further evidence for the dynamic stability of intracranial saccular aneurysms; J. Biomech. (36) (1143-1150) 2003.
6. de la Monte, S., Moore, G. W., Monk, M. A. and Hutchins, G. M., Risk factors for the development and rupture of intracranial berry aneurysms; Am. J. Med. (78) (957-964) 1985.
7. Di Martino, E., Mantero, S., Inzoli, F., Melissano, G., Astore, D., Chiesa, R., and Fumero, R., Biomechanics of abdominal aortic aneurysm in the presence of endoluminal thrombus: Experimental characterization and structural static computational analysis; Eur. J. Vasc. Endovasc. Surg. (15) (290-299) 1998.
8. Elger, D. F., Blackletter, D. M., Budwig, R. S., and Johansen, K. H., The influence of shape on the stress in model aneurysms; ASME J. Biomech. Eng. (118) (326-332) 1996.
9. German, W. J. and Black S. P. W., Intra-aneurysmal hemodynamics-jet action; Circ. Res. (3) (463-468) 1955.
10. Glynn, L. E., Medial defects in the circle of Willis and their relation to aneurysm formation; J. Pathol. Bacteriol. (51) (213-222) 1940.
11. Hashimoto, N. and Handa, H., The size of cerebral aneurysms in relation to repeated rupture; Surg. Neurol., (19)(107-111) 1983.

12. Haslach Jr., H. W., A nonlinear dynamical mechanism for bruit generation by an intracranial saccular aneurysm; J. Math. Biol. (45) (441-460) 2002.
13. Humphrey, J.D., Arterial wall mechanics: review and directions, Critical Reviews in Biomedical Engineering, (23)(1-162) 1995.
14. Humphrey, J.D., Computer methods in membrane biomechanics, Computer Methods in Biomechanics and Biomedical Engineering, (1)(171-210) 1998.
15. Humphrey, J. D., and Canham, P. B., Structure, mechanical properties, and mechanics of intracranial saccular aneurysms; J. Elasticity. (61) (49-81) 2000.
16. Inzoli, F., Boschetti, F., Zappa, M., Longo, t., and Fumero, R., Biomechanical factors in abdominal aortic aneurysm rupture; Eur. J. Vasc. Surg. (7) (667-674) 1993.
17. Jain, K. K., Mechanism of rupture of intracranial saccular aneurysms; Surgery. (54) (347-350) 1963.
18. Kassell, N. F. and Torner, J. C. Size of intracranial aneurysms; Neurosurgery, (12) (291-297) 1983.
19. Knowles, J.K., Large amplitude oscillations of a tube of incompressible elastic material, Quarterly of Applied Mathematics, (18)(71-78) 1960.
20. Kyriacou, S. K., Humphrey J.D., Influence of size, shape and properties on the mechanics of axisymmetric saccular aneurysms; J. Biomechanics., Vol. 29 (8) (1015-1022) 1996.
21. Mower, W. R., Baraff, L. nJ,, and Sneyd, J., Stress distributions in vascular aneurysms: Factor affecting risk of aneurysm rupture; J. Surg. Res. (55) (155-161) 1993.
22. Ostergaard, J. R., Risk factors in intracranial saccular aneurysms; Acta. Neurol. Scand. (80) (81-98) 1989.
23. Roach, M. R., A model study of why some intracranial aneurysms thrombose but others rupture; Stroke. (9) (583-587) 1978.
24. Ryan, J. M., and Humphrey, J. D., Finite element based predictions of preferred material symmetries in saccular aneurysms; Annals. Biomed. Engineer. (27) (641-647) 1999.
25. Sekhar, L. N. and Heros, R. C., Orgin, growth and rupture of saccular aneurysms: a review; Neurosurgery. (8) (248-260) 1981.
26. Shah, A.D., Harris, J.L., Kyriacou, S.K., Humphrey, J.D., Further roles of geometry and properties in the mechanics of saccular aneurysms., Computer Methods in Biomechanics and Biomedical Engineering, (1)(109-121)
27. Shah, A. D., and Humphrey J. D., Finite strain elastodynamics of intracranial saccular aneurysms; J. Biomed. (32) (593-599) 1999.
28. Stehbens, W. E., Flow in glass models of arterial bifurications and berry aneruysms at low Reynolds numbers; Quart. J. Exp. Physiol. (60) (181-192) 1975.
29. Stehbens, W. E., Pathology and pathogenesis of intracranial berry aneurysms; Neurol. Res. (12) (29-34) 1990.
30. Steiger, H. J., Poll, A., Liepsch, D.. and Reulen, H. J., Basic flow structure in saccular aneurysms: a flow visualization study; Heart Vessels. (3) (55-65) 1987.
31. Simkins, T.E. and Stehbens, W.E., Vibration behavior of arterial aneurysms,

Letters in Applied and Engineering Sciences, (1)(85 -100) 1973.

32. Wang, C.C, On the radial oscillations of a spherical thin shell in the finite elasticity, Quartery of Applied Mathematics, (23)(270-274) 1965.
33. Watton, P. N., Hill, N. A., and Heil, M., A mathematical model for the growth of the abdominal aortic aneurysm; Biomechan. Model. Mechanobiol. (3) (98-113) 2004.
34. Wiebers, D. O., Whisnant, J. P., Sundt, T. M. and O'Fallon, W. M., The significance of un-ruptured intracranial aneurysms; J. Neurosurg. (66) (23-29) 1987.

ON THE ORIGIN OF METAZOANS

FREDERICK W. CUMMINGS
Professor emeritus, University of California Riverside
136 Calumet Ave., San Anselmo, California, 94960, U.S.A.
fredcmgs@berkeley.edu; fwcummings@earthlink.net

The interaction of two signaling pathways is suggested as the biological basis of animal patterns. Interaction of the patterning mechanism with geometrical changes in a (thick, closed) epithelial sheet leads to a primitive invagination, or gastrulation. A theoretical model of pattern consisting of two morphogens interspersed by a stem cell region is presented. This two-way signaling pathway interaction together with the concomitant invagination is proposed as a key component in the transition from single-celled to multicellular life. Contact of the animal and vegetal poles of the gastrula is the starting point for examining simple boundary conditions giving bifurcation into 'Urcnidaria' and 'Urbilateria'. A remarkable observation not suggested by natural selection is that all of the more than 3000 species of Centipede have an odd number of leg pairs. In the case of Urbilateria, the model provides for such an odd number of proto-leg positions in the case that each segment bears legs, regardless of segment number.

1. Introduction

Impressive progress in unraveling the genomic basis of development and evolution has recently been made. What is further clear is that elucidation of the actual cell shape changes along with changes in cell number is further necessary in order to obtain a fuller grasp of morphogenesis.

It has now become clear that virtually all developmental regulatory genes control several different processes, acquiring new developmental roles. Clusters of Hox genes, as well as Pax-6, Dll and Tinman proteins, proteins which shape the development of animals as diverse as flies and mice, are just a part of the collection of proteins that make up the genetic 'tool kit' for animal development. Transcription factors are proteins that bind to DNA and directly turn gene transcription on or off, and comprise a large fraction of the regulatory tool kit. The presently accepted view is that although developmental regulatory genes are remarkably conserved, their interactions are not[1,2,3].

One point of view is that the genome is alone required for grasping patterning in developmental systems. However, doubts arise in this regard. Consideration of patterning of leafs on plant stem raises questions. The leaf numbers counted before a repeat (a leaf directly above an initial leaf) are commonly two, three, five, etc., in a spiral arrangement. Four 'leafs' is conspicuously very much less common in a simple spiral. Another fact of particular interest here is that all of the more than 3000 species of centipede have odd numbers of leg pairs, from 15 to 191[4]. In both cases, plant pattern or centipede, an argument from selection is not available. A third observation of this same flavor is that the segment size (width) of any animal is invariably less than the corresponding segment circumference.

Natural selection is the dominant driving force of evolution. Might there be, along with selection, generative 'rules', from the very origin of multicellularity, leading to bias or constraint acting on, or alongside of, natural selection? A number of authors have argued that this is the case[4,5,6]. Might remnants of such bias or constraint remain today, even after the extensive elaboration of more than 550 million years of evolution? Evolution over such a vast expanse of time is expected to give rise to such complexity that such originally simple rules have become obscured. Perhaps eukaryotic cells found a way to form multicellulars before the Cambrian, discovering 'rules' that were 'adaptive' at that time, and although extensively elaborated since then, have left clear enough hints as to their form and origin.

Of interest here is to propose a possible link in the chain in the origin of the metazoan. The rules or model proposed in what follows, while no doubt too simplified to be realistic, may it is hoped provide a new angle for viewing metazoan origins and evolutionary change.

In the sections below, the simplest patterning mechanism is proposed, one having its basis and motivation in known crucial developmental mechanisms. Cells have discovered adhesive connections on their lateral surfaces during this single-cell to multi-cell transition. The development of such cells is only followed in this paper after they first form into a hollow spherical closed epithelial sheet. One virtue of the basic patterning 'ansatz' laid out below beyond its simplicity is that is has a plausible interpretation in terms of the interaction of two (which two being variable) signaling pathways. The simplest consequences are developed in what follows, as the animal grows.

The patterning mechanism partitions the epithelia into ever more complex regions. At the most basic level of organization, animals can be divided into the sister groups of 1) bilateral protostomes or deuterostomes with through-guts, often with evident segmentation, and 2) the more primitive diploblastic animals with radial symmetry, the cnidaria.

The origin of multicellular life from single-celled beginnings is one of the most enigmatic of puzzles, and one least likely to be ever 'solved'. However, the question will continue to exercise the imagination, as it has for hundreds of years. Willmer[7] has ably reviewed the multiplicity of theories concerning the origin of multicellularity. Often a zooflagellate colony similar to Volvox is invoked as the earliest ancestor of multicellulars. In this view, the basic metazoan was a pelagic, radially symmetric aggregation of flagellated cells. Such an aggregation of cells has several desired properties, such as a separation between somatic and gametic cells, a blastula-like geometry, and cells having loose connections at their lateral surfaces. There are many colonial flagellate protists, but all existing ones are no doubt plants. So although existing Volvox-like colonies cannot provide a convincing origin, an extinct non-plant version emanating originally from sponges is a possibility.

The work of Haeckel ($\sim$ 1874) on metazoan origins involving blastula-like and gastrula-like stages has long been influential, and since that time there have been numerous alternate proposals[7,8]. Haeckel's ideas, while having many virtues such as simplicity, elegance and orderliness, are open to many objections. Our starting point will also be the blastula stage. Our key assumption will be that a crucial 'discovery' by evolution at or near the acute turning point leading to multicellulars was of the patterning potential of the interaction of two signaling pathways. Concomitant felicitous and necessary environmental conditions, such as appropriate oxygen concentrations, are not discussed further here.

2. Patern Formation

There have been numerous suggestions for pattern formation since Turing's time[9,10]. The present approach envisions two different 'morphogens', along with a propensity of these two morphogens to avoid each other. Each has a threshold for activity, assumed the same for each morphogen for simplicity, and our focus is on the desired steady state configurations.

Surprisingly, only a handful of signaling pathways are involved in embryogenesis, employed repeatedly. The "two-interacting-signaling-pathways" ('ISP') model involves four variables: two ligands, and two active receptors[11,17]. The key elements of the model can be stated simply. Activation of one signaling pathway following attachment of a ligand to its receptor acts to deactivate production of ligand of the second type, while at the same time stimulating production of ligand of similar type. Differential equations for these four quantities may be written immediately, as shown in Appendix A.

Regardless of details of the nonlinearities, and additions and other 'bells and whistles' that may be added, three key properties distinguish this model from others. First, there is spontaneous morphogen activation into two distinct regions, and activation from zero morphogen level is not dependent on the presence of nonlinearities[11]. Nonlinearities are invariably present in any model, but they are not here the source of the basic instability. Two lengths, determined by the parameters of the model, dictate the size of these two regions. A second characteristic of the present model is that the relative sizes of two diffusion constants entering the model are not constrained, e.g., the two may be equal. Recent results show that in at least some cases, lipid transporters may act to ferry the morphogens around the tissue. [12] The length parameters of the model[11,17] (or Appendix A), dependent as they are on these diffusion coefficients, may have larger values than otherwise expected. Third, and importantly, the model is directly motivated by the known involvement of signaling pathways in earliest development.

There has not been to date an empirical characterization of the stimulation of further ligand production due to activation of a receptor by its target ligand, as presently proposed. This must serve as a prediction of the present model. The usual empirical description of signaling pathways leaves the origin of the ligand activating a given receptor unknown. The present assumption is that the amount of emitted ligand from a given cell increases as like receptor stimulation of that same cell increases. This is imagined to occur by a "non-canonical" pathway, one bypassing the nucleus. Such a pathway can lead to more immediate secretion of extracellular ligand than if the emission were to go by way of gene activation, etc.

One possible process of producing ligand upon activation of the cell surface receptor could involves numerous steps, involving (e.g.) gene transcription, the endoplasmic reticulum (ER), the Golgi complex, and finally perhaps secretion from the cell. This time is expected to be considerable compared to the time for a free ligand in a given spatial region to become attached to its receptor and to activate the pathway. However, it is supposed here instead that $R_{1,2}$ (the two activated receptor densities) act downstream to release already stored ligand, by a route that bypasses the nucleus. Such ligands are supposed stored at, e.g., a constant rate by an unspecified cellular mechanism. The cell maintains a relatively constant store of ligand awaiting a release signal $\sim R$ analogous (in this respect only!) to the situation of neurotransmitters in neurons. The two times (a: emission time interval between receptor activation and like ligand emission, and b: empty receptor uptake of ligand L) can thus be comparable. This is the situation envisioned here, and will have to serve as a prediction of the model at this point: the activated receptor $R_{1,2}$ releases ligand already stored in vesicles, so that this time is appreciably shorter than ligand production and storage via gene activation, ER and Golgi. This provides for an oft invoked 'pre-pattern'. Importantly for the model, Wnt has at least two modes of action, one that bypasses the nucleus, and a second 'canonical' pathway leading to gene activation via stabilization of nuclear β-catenin. The former 'non-canonical' path bypassing the nucleus acts (at least in part) to release stored Wnt ligand relatively rapidly, or so the model predicts.

A second requirement of the model is that as receptor stimulation of one kind increases, and 'like' ligand emission increases, secretion of the 'non-like' ligand of the second pathway decreases[11].

The 'Wnt' pathway is apparently most important at the very beginning of multicellulars[13]. So far, no Wnt genes have been described in unicellular eukaryotes, or from cellular slime moulds, or from choanoflaggelates. It may be presumed that the appearance of Wnt genes itself was linked to the origin and evolution of multicellular animals from single-celled ancestors[11,13]. The relatively rapid evolution of an original Wnt pathway into similar versions, involving different transcription factors, may be surmised to have occurred about the time of interest, the PreCambrian.

The steady-state version of the model of Appendix A, which will occupy us here, consists of only two (dimensionless) morphogen densities, say, Φ_1 and Φ_2. These two are proportional to the two activated receptor densities. The full nonlinear steady-state version of the model is numerically solved in axial symmetry. This numerical solution, coupling the pattern to geometrical changes[11,17], following growth from the original blastula, leads to a gastrula with animal and vegetal pole eventually in contact. After contact between epithelial sheets of the embryo occurs, the linear steady state version of the model is subsequently employed, and solutions are no longer axially symmetric.

As the total area of the middle surface of the (closed) epithelial sheet increases, the model shows that the amplitudes of the morphogen densities also increase, up to a maximum. Such maxima are determined either by regulatory genes and switches, or attainment of an (impossible) negative morphogen value due to the nonlinearities. These are cutoffs, after which both morphogens decay to zero. After achieving maximum amplitude, a given pattern decays, growth again takes place, to be followed in the next cycle by a more complex pattern.

The linear version of the model is, strictly speaking, only applicable in the small amplitude regime. However, after induction and contact of the two sheets, the linear version will be taken to indicate the general form of the pattern even after it has risen from small amplitude. This tact is adopted largely because the numerical simulation of a more extended model, one with geometry coupled to morphogen in two dimensions, is very much more involved numerically when axial symmetry no longer can be invoked. There are interesting points to be made, and sufficiently convincingly, with the simpler linear version.

The general rule is that a pattern forms when the area is sufficient, as dictated by the two length parameters involved. The area in question is on the middle surface bisecting an epithelial sheet. Growth occurs, the total epithelial area increases, and both morphogen amplitudes rise, up to a maximum. Then the amplitude of this pattern decays, to be followed by further growth. Growth takes place until the next and more complex pattern can be accommodated, one of lower symmetry than the previous pattern. Of the possible allowed patterns, the one that will begin to rise from small amplitude in each cycle will be determined by the geometry of the region, and by conditions on the boundary region. Coupling of pattern to geometry is then a crucial element in the discussions to follow. Morphogen, via the genes, directs geometry, and geometry in turn directs pattern in ongoing growth[11,17].

As the simplest example, no pattern is allowed on a blastula whose radius is below a certain value, determined by the model parameters. As the sphere radius grows, at a certain critical radius a pattern of small amplitude begins to emerge. The upper hemisphere begins to display a steady state morphogen pattern Φ_1, and the lower hemisphere displays morphogen Φ_2 as seen in Fig. 1. Next, Φ_2 increasingly causes cell shape changes as the amplitude of the morphogen increases, and as growth proceeds. The epithelial cells in the lower hemisphere change so that the basal cell surfaces increase relative to the apical surfaces, with perhaps the opposite happening to cells of the upper hemisphere. A gastrulation process then occurs[11]. As the overall animal surface area increases, so also do morphogen amplitudes, and the overlap region between the two morphogens densities sharpens due to the rising effect of ever-present nonlinearities. Such a coupled geometry-morphogen model has been carried out numerically for this simplest situation of gastrulation[11,17]. The shape of the gastrula depends as expected on the form assumed for the nonlinearities, but invagination is the general result. Figure 1 illustrates the morphogen densities, for two specified total areas A, as a function of the coordinate u. Figure 2 shows the resultant gastrula surfaces, for three different total surface areas, for a specific assumed nonlinearity.

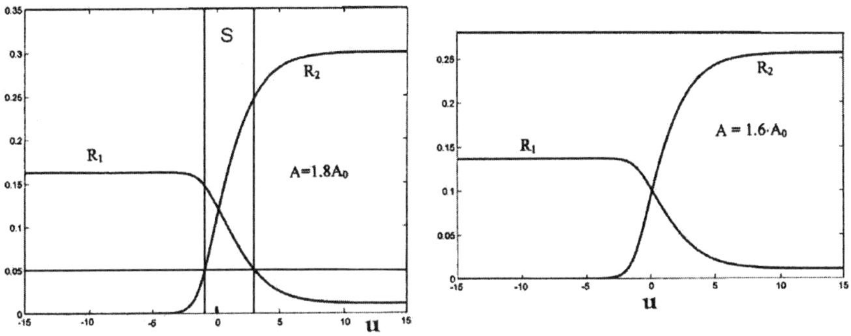

Figure 1. Two numerically obtained morphogen solutions are shown, for the two interacting morphogen densities $R_{1,2}$, and for two different total areas as indicated. There are three regions shown, one left-most corresponding to morphogen density $R_1/R_0 = \Phi_1$, and the right-most corresponding to the morphogen density $R_2/R_0 = \Phi_2/f$. Between these two regions is a third region, for each total area A, between coordinates u_1 and u_2 where the two morphogens are competing and unable to effect differentiation. This is designated by s as a stem cell region, and is designated by a solid annulus centered at the blastopore lip on the corresponding axially symmetric shapes shown in Fig. 2.

The coupling of morphogens to geometry accounts for the most significant nonlinearities, via the Gauss' equation of Sec. 3. The signaling pathways affect cell shape changes only indirectly, of course; activated genes downstream of the activated cell-surface receptor produce (three known) proteins that affect the (three) filaments that determine cell architecture, i.e., the cytoskeleton.

Of great interest is the region between two morphogen densities (e.g., Wnt and Hedgehog (Hh)). There is also a third (often relatively smaller) region delineated between the two morphogens. This region is designated by an s in Fig. 1. The region of overlap is where both morphogens are of about the same density or effectiveness. It may be supposed that this is a region where both morphogens compete roughly equally, a spatial region for which neither morphogen effects differentiation. Since cells in this region are not to be determined, or differentiated by either morphogen density (e.g., by Wnt or Hh) then it seems reasonable that they be designated as 'stem cells', and denoted in Fig. 1 by s. Then control of stem cell proliferation falls to the cooperative effort of the two adjacent morphogens, such control assuring that stem cell proliferation is toward definite adaptive ends of the organism. Somatic stem cells are probably the locus of tumor initiation, and Wnt and Hh pathways function in the normal regulation of stem cell number in at least some tissues. Expansion of the somatic stem cell population may be the first step in the formation of at least some types of cancer[14].

Two well studied equations of mathematical physics are the Helmholtz and the Laplace equations. These are the linear versions of the model (Appendix A) used in Secs. 4 and 5 below, and are

$$\nabla^2 (\Phi_1 - \Phi_2) + k^2 (\Phi_1 - \Phi_2) = 0, \tag{1}$$

and

$$\nabla^2 \left(\frac{\Phi_1}{k_1^2} + \frac{\Phi_2}{k_2^2} \right) = 0. \tag{2}$$

Here $k^2 = k_1^2 + k_2^2$, where k is an inverse length.

3. Coupling Pattern to Cell Shape

This section gives a brief overview of aspects of the coupling of the pattern to cell shape changes. The 'pattern' is viewed as existing on a given geometry of a middle surface bisecting the epithelial sheet of variable thickness. The pattern is thought of as averaged over the sheet thickness at any point. However, coupling of pattern to geometry requires a changing pattern in response to a changing geometry. In turn, the geometry is changed by the changing pattern.

In all that follows, only the average cell shape is considered at any point on the middle surface, so that several (say 5-10) cells are the unit cell[11,17]. It is not in any case usually possible to follow the individual cells in a cell sheet, as they slip and slide around each other.

The pattern is necessarily affected by the surface shape, or the 'geometry'. The Laplacian operator (entering virtually all pattern models) contains the geometry. This can most easily be illustrated by displaying this Laplacian operator (more properly the Laplace-Beltrami operator) in 'conformal' coordinates. The convenient conformal coordinates divide any surface into infinitesimally small squares, where each small area dA at coordinate (u, v) is given by

$$dA = g(u, v) du dv \tag{3}$$

The Laplacian is then[11,17],

$$\nabla^2 = \frac{1}{g(u,v)} \left(\frac{\partial^2}{\partial u^2} + \frac{\partial^2}{\partial v^2} \right) \tag{4}$$

Then it is clear that in order to close the set of equations one must provide an equation for the geometry factor $g(u, v)$. This equation has been given long ago as the highly nonlinear Gauss equation

$$\nabla^2 \ln g = -2K \tag{5}$$

Here K is the Gauss curvature, and is now to be considered a function of the morphogen densities and their derivatives.

Two curvatures are needed to specify any surface, and these are taken to be the Gauss curvature K, and the mean curvature H. How are these to be specified as functions of the morphogens? It has been argued in a recent paper[17] that H and K are most reasonably related by

$$K = H^2 - \lambda_2^2 (\nabla(\Phi_1 - \Phi_2))^2 \tag{6}$$

The numerical parameter λ_2^2 is ~ 1 in numerical work. The fact that K always satisfies $K \leq H^2$ is incorporated as well as the requirement that K is a scalar invariant. Also to be noted is that K takes on its required spherical form of $K = H^2$ at the two points on the symmetry axis. Appearance of a negative contribution to K requires the appearance of the gradient term since a "twist" occurs[17] in the cell shape, and a direction must be specified.

The mean curvature H is assumed to take the form[17]

$$H = \sqrt{\frac{4\pi}{Area}} \left(1 + \lambda_1 \left(\Phi_1 - \Phi_2\right)\right) \tag{7}$$

Constricting apical cell area relative to basal area gives a decrease in H, turning H negative in the region , and vice versa. The square root factor is required so that when morphogens are zero, and the geometry is that of a sphere of radius r, Area $= 4\pi r^2$, and $H = 1/r$ as required. In numerical simulations, $\lambda_1 \sim 5 - 10$.

A two parameter geometry has now been introduced. Suitable invagination of an original spherical surface occurs after the radius has increased to an appropriate size allowing the two morphogens to begin to form, with their respective maxima centered at the animal and vegetal poles. As growth occurs, giving larger total area (Area), then the morphogen amplitudes also increase, leading to progressively greater invagination, simulating a proto gastrula.

Six first order equations have been integrated to give the morphogens and surface shapes shown in Figs. 1 and 2. Of note is the small number of parameters required.

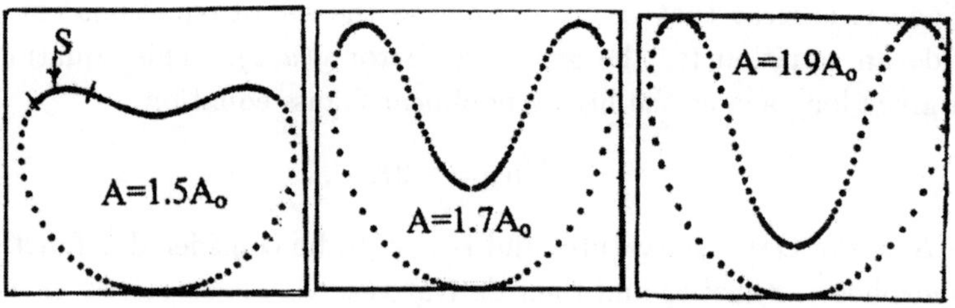

Figure 2. Three axially symmetric, numerically computed surfaces (two corresponding to the morphogen solutions of Fig. 1 are shown. A_0 denotes the largest sphere radius before invagination ('gastrulation') begins. A set of six first order nonlinear equations has been integrated for various total areas.

4. On Urcnidaria

The main groups of bilateral animals are deuterostomes and protostomes. Their last common ancestor is called Urbilateria[15]. Cnidarians (such as sea anemones and corals) and Poriferans (sponges) split off before bilaterians.

The discussion from this point continues where Fig. 2 leaves off. The two epithelial sheets are assumed to come into contact, animal pole contacting vegetal pole. The new boundary conditions set up when contact occurs between the epithelial surfaces is now the question of interest. Consider the imminent contact along the axis of Fig. 2, where the endoderm is shown as about to come into contact with the ectoderm. Two very different scenarios are to be considered. In the first, the surfaces come into contact, new boundary conditions are set up, but the two surfaces do not interpenetrate. The second scenario is a situation when the surfaces come into contact, cell death occurs in the region of contact, the surface reforms and a new opening is produced at this contact point. In the first case, the surface remains topologically equivalent to a sphere; while in the second, the topology is abruptly changed from a topological sphere to a topological torus, or donut. These two scenarios correspond to the genesis of the two sister groups, cnidaria and bilateria, now referred to as Urcnidaria and Urbilateria[15]. We ask what the linear ISP model says of the further evolution from the crucial point of contact in each of the two cases. The equations hereafter employed are the steady state linear equations, namely Eqs. (1) and (2).

Cnidaria typically have a third layer of (mostly) non-cellular mesoglia between two distinct tissue layers. These diploblastic animals have a mouth, which doubles as an anus, and typically a ring of tentacles around the mouth. As boundary condition, it is reasonably supposed that the morphogen densities at the point of contact are equal to the same constant, that $\Phi_1 = \Phi_2$ in this region, and these values now become frozen in, or fixed there.

Appendix B gives the solution of the model when this boundary surface shape is coupled to morphogens and vice versa. Rather than continuing the numerical computation resulting in the shapes shown in Fig. 2 leading up to this point, a simpler approximate tact is taken. The geometry is assumed to be cylindrical, described by two parameters, cylinder height (or length) L, and radius R, and the linear model is used. In spite of such apparently drastic simplifying assumptions, it is believed that interesting key aspects emerge, aspects expected to remain in evidence in the more complex model.

The two-dimensional cylindrical coordinates z and ϕ always reside on the apical/basal bisector of the epithelial sheet, i.e., in the surface. The coordinates are, as always, purely imaginary constructs of the observer, and consequently may be drawn arbitrarily. The pattern, on the other hand, is the property of the animal. It may be wished that the coordinates are chosen cleverly or conveniently in order to make the discussion as transparent as possible.

Imagine a cylinder of length $2L$ and radius R, and initially closed at both ends by circular discs. Next imagine that the top half of the (elastic) cylinder is tucked into the bottom half, forming an inner and outer surface, with the two end discs now (almost) in contact. This gives a caricature of the shapes of Fig. 2, and further ignores the sheet thickness.

The solution of Eqs. (1) and (2) with the boundary condition mentioned may easily be obtained analytically, and is given in Appendix B. The values taken on the two disc end pieces is that $\Phi_1 = \Phi_2 = C$ =constant. The solution for Φ_1 and Φ_2 as a function of z, for any particular radial angle is a dying exponential, with oscillating maximum amplitude in the head region at $z \sim 0$. In Fig. 3, red and blue represent the two morphogen densities. Fig. 3 imagines that the tucked cylindrical shape has been un-tucked, slit down one side and laid out flat, so that both inner ($z < 0$) and outer ($z > 0$) layers are shown. Endoderm has previously been differentiated from ectoderm during gastrulation and before contact occurred, so that the solution shown in Fig. 3 is only the new next pattern superposed on the previous.

Figure 3 shows the small amplitude solution. An exponential decay of both morphogens down the body column occurs, with highest values of each at the head region near $z = 0$. The red and blue patterns alternate around the z axis, with $2n$ (shown is $n = 4$) between regions down the length. The coordinates have been taken as cylindrical, z and ϕ, where $-L \leq z \leq +L$, and $-\pi \leq \phi \leq \pi$.

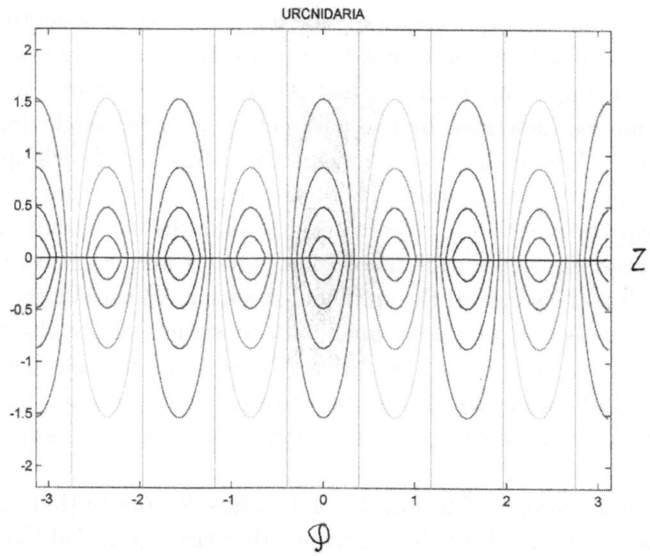

Figure 3. Shown is the solution to the linear equations of Appendix A, Eqs. (A.6) and (A.7), or Eqs. (1) and (2). The axial z and angular ϕ coordinates are indicated. The boundary condition is that both morphogens equal the same constant C at $z = \pm$L at the two flat-disc cylinder ends, which are in contact. The different colors represent the two morphogen densities. The eight straight vertical lines separating the morphogens indicate approximate stem cell regions. The figure is to be viewed as if it were first folded over on itself at the coordinate $z = 0$ so as to have the coordinates $z = +2.2$ and $z = -2.2$ coincide. Next, the angles $\phi = +\pi$ and $\phi = -\pi$ are also made to coincide, representing a cylinder within a cylinder with one open end, a Urcnidaria.

5. On Urbilateria

Obtaining insight into the origin and evolution of segmentation is central to understanding the body plan of major animal groups such as arthropods, annelids, and vertebrates[19].

The second case of interest occurs when the two epithelial surfaces of Fig. 2 make contact along the axis of symmetry, and a new topology is created: a donut topology. The new boundary condition required is that the pattern must be periodic in both coordinates. Again simplifying to a cylindrical caricature of the actual geometry (and again utilizing cylindrical coordinates z and ϕ) allows an analytical solution, given in Appendix C. Now the flat circular discs originally closing the two ends of the cylinder disappear. In this case, the relation between radius R and length L required by the ISP model is simply, from Eqs. (1) and (2) or Appendix C,

$$1 = (\pi m/kL)^2 + (n/kR)^2 \tag{8}$$

The integer m arises from the periodicity in the z direction along the donut axis, while the integer n is associated with the angular variable ϕ.

Imagine a growing animal, and that initially $kR < 1$. From Eq. (8) this necessarily means that $n = 0$, $m = 1$, and $kL = \pi$. When $kR = 1$, n cannot equal 1, since that would require $m = 0$. So the radius R will keep increasing until $kR = 2^{1/2}$, at which point n can become unity, and kL can increase to $(2^{1/2}\pi)$, with $m = 1$. Now a small amplitude pattern with $m = 1$ and $n = 1$ begins to emerge, superseding or overlaying the previous $m = 1$, $n = 0$ pattern. The value $n = 1$ establishes bilateral symmetry, here initially with only one segment, $m = 1$. The morphogen amplitudes are equal at two angles along the z axis, 180° apart.

The radius kR may stabilize near the value $2^{1/2}$. It will be most adaptive for the length to grow very much relative to the radius; natural selection apparently prefers a long digestive tract, with $kL \approx 2^{1/2}\pi m$, $m = m_0 >> 1$, also providing for relatively rapid locomotion. Bilaterality also provides directional, stereo sensors. A bilaterally symmetric, segmented animal with a through gut is now on hand via the simple model.

Further growth may produce the situation of $n = 2$ and $m >> 1$. What is to be especially noted is that lines denoting values along the axis of nearly vanishing morphogen intersect with similar circular ones around the axis, as indicated in Figs. 4 and 5. These lines specify locations where the morphogens are about equal, and thus correspond to stem cell regions. It is proposed that the two pairs of points, the intersections of axial and radial lines, and associated with each segment, provide natural loci for later appendage outgrowth and development, as directed by *hox*, *dll*, and *pax*-6 regulatory genes and associated switches. Pairs of points situated ventrally may designate location of limbs, e.g., legs. Later in evolution, as Hox cluster number increases beyond unity and segments become specialized, dorsal pairs may designate gill positions, or antennae, eye, or later, wing loci. It is interesting to note in this context the Cambrian Hallucigenia[8], an animal with dorsal protrusions corresponding (but not homologous) to its (seven) ventral limb pairs.

Figure 4 displays the solution of the model as given in Appendix C, showing the two morphogens on a rectangle, labeled by distance z along the axis and angle ϕ around the axis. As in the case of the Urcnidaria, the cylinder within cylinder shape has been un-tucked into a single cylinder, with no ends, then slit down the z axis and laid out flat in Fig. 4. Ectoderm ($z < 0$) and endoderm ($z > 0$) are both now shown, and it is remembered that their differentiation has occurred previously in a pattern cycle not now shown.

A segment width according to Eq. (4.1) is $\Delta(kL) \approx 2^{1/2}\pi$, and the corresponding radius will be $kR \approx 2^{1/2}$ (or greater) for a bilateral animal. Then it is expected (as observed) that segment width ΔL will invariably be less than the segment circumference $C = 2\pi R$, or $\Delta L < C$. There does not seem off hand to be a compelling reason to rule out larger segments on the basis of selection; for example, a segment width several times the circumference, (with perhaps each segment supported by a number of leg pairs) might well be imagined. But such do not appear.

Next consider a proto-centipede, the most basal Cambrian ancestor of the centipede. Figure 5 illustrates the growth of such a (possible) animal on the basis of the present model. (The case of $n = 2$ is shown, although $n = 1$ is also possible). The cylinder is shown in side view. At first there is a single segment, as in Fig. 5a, divided into two parts labeled X_1 and Y_1 to denote the two morphogens and to label the first growth cycle.

This first segment is considered as special, determined as different from the rest to follow. Anterior (say, X_1) and posterior (say, Y_1), are differentiated, and deuterostomes are then differentiated from protostomes. One Hox cluster is already present at this early stage. It is supposed in Fig. 5 that the radius has grown to such an extent that $kR \approx 2^{3/2}$ (or perhaps even somewhat greater), so that limb or appendage positions are both designated points. Such points are designated by a) the intersection of the circular stem cell regions periodically placed along the axis separating the two morphogen densities, and b) the four thin linear regions running along the z axis. This provides two symmetrically placed pairs for each segment, one dorsal and one ventral.

A double segment periodicity apparently underlies segment generation

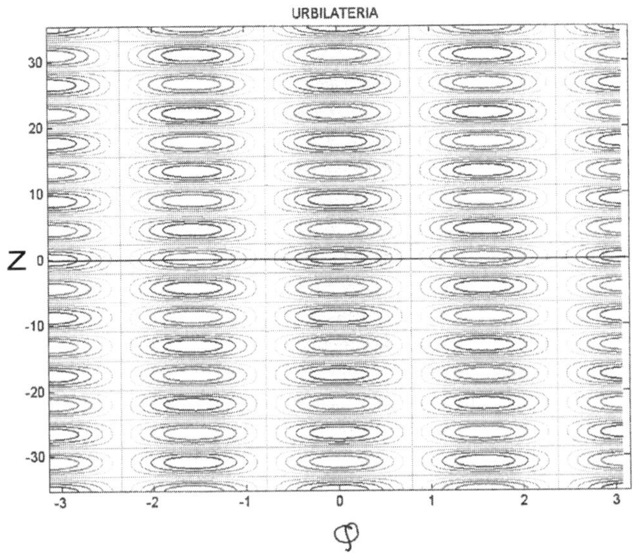

Figure 4. The coordinates z and ϕ indicate the length along the cylinder axis and around the circumference, respectively. Values z ¿ 0 represent the endoderm, while z ¡ 0 represents the ectoderm. The figure is to be thought of as folded and connected much as in Fig. 3, i.e., first connect $z = 33$ to $z = -33$, and then connect $\phi = +\pi$ to $\phi = -\pi$. Each segment consists of the combination of two morphogens separated by a (smaller) stem cell region between morphogens. Fig. 4 shows only four segments ($m = 4$). Also shown are four vertical lines designating stem cell regions between two different morphogens. Fig. 4 then corresponds to the solutions of Eqs. (1) and (2) with $m = 4$ and $n = 2$. Such a model always gives an odd number of leg pairs when all ventral loci have legs.

in centipede development. While the number of leg-bearing segments varies extensively, between 15 and 191, yet it remarkably always gives rise to an odd number of leg pairs[21,22].

The next growth cycle inserts a new segment between X_1 and Y_1 as shown in Fig. 5b, the cylinder again shown in side view. Stem cell regions (thin and linear) are indicated as two dashed lines, one dorsal set and one ventral. The segment is labeled by the integer 2 to indicate that it came about in the second growth cycle. The half-segment Y_1 is special if growth next occurs only at the intersection of Y_1 with X_2, and so on for all subsequent adjacent X's. Subsequent axial growth is controlled between all other X/Y pairs, except for the (X_m/Y_1) pair. This gives

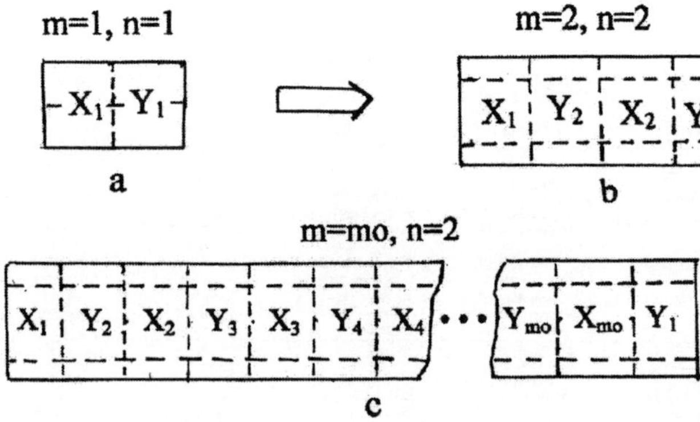

Figure 5. The 3000(+) species of centipede all have an odd number of leg pairs. Fig. 5a shows a side view of the beginning (ectoderm) shape of an animal based on the model equation solutions, with doubly periodic boundary conditions. The two initial morphogen densities are indicated as X_1 and Y_1. In the next growth cycle, shown in Fig. 5b in a cylinder side view, a double segment is added between the first original (double) segment by stem cell growth. The X's and Y's always alternate, and the sub numbers indicate the growth cycle. Further growth is shown in Fig. 5c. The dashed lines indicate the case when radial growth ($kR \geq 2^{3/2}$) has led to the four regions (two are shown) between unlike morphogens, and indicated in Fig. 4 by vertical lines, always an odd number.

segment Y_3/X_3, and so on, giving posterior growth, as often occurs. This is indicated in a side view in Fig. 5c. It is not necessary on the basis of the model that axial growth occurs only between a X_n/Y_1 pair. Axial growth may occur in principle between all segments for n small enough, for example in the embryonic state of certain centipede species, but may revert to only posterior growth along the axis in the adult.

Diverse bilaterian taxa, including representative lophotorchozoa, ecdysozoa and deuterosomes, share aspects of a developmental process where repeated pattern elements are added posteriorly during development. This process of terminal addition suggests that it derived from a shared ancestral mode of development. Modifications of the process of terminal addition of repeated elements apparently occurred in the early Paleozoic radiation of Bilateria[18,19,20].

The stem cells associated with more anterior segments presumably also divide, but do not contribute much to axial growth, but rather migrate into the interior of the animal, where they give rise to mesodermal tissue, such as muscle and nerve.

Consider now that limbs (say, lobopods) are generated at each designated ventral point pair indicated in Fig. 5c. It is easy to see that, no matter how many segments may exist, only an odd number of pairs ever occur, as observed in all (3000+) Centipede species. This occurs independent of the growth algorithm above giving posterior addition of repeated elements[20,21,22,23].

At the other extreme from segmented animals, it may be relevant to point to the $m = 1$ and $n >> 1$ (e.g., esp. $n = 5$) case of Eq. (4.1). This shape resembles the 'pineapple-slice' shaped Burgess Shale animal. Also, the larval stage in usual echinoderm development is bilateral. The initial bilateral larval stage changes into a radially symmetric adult. So Eq. (4.1) with $n >> 1$ can be only (at best) applicable to the adult echinoderm stage. The adult is an animal with a through gut and radial symmetry, without segments, consistent with the $m = 0$ (or $m = 1$) and $n >> 1$ (often $n = 5$ in starfish) parameter set. In present day echinoderms, a deuterostome, development proceeds by way of pouching of the archenteron producing coelomic mesoderm during gastrulation. Complications such as the creation of mesodermal structures in general are not considered in this work, already apparent from the simplicity of the shapes of Fig. 2.

It is of further interest to note that the same pattern generation mechanism as discussed above has been studied in plant patterning[16]. It has been shown that the number four in spiral phyllotaxis (plant patterning) occurs very rarely relative to the numbers two, three, five, eight, etc. These integers refer to the number of leafs or outgrowths on a (cylindrical) stem when there is only one leaf per level, and when the repeat leaf (the one directly above the first) is not counted.

Appendix A.

This appendix outlines the mathematical model[11,17] of two interacting signaling pathways. There are two simple elements of the model. Any model containing these two simple elements will produce interesting patterning, so that the addition of further 'bells and whistles' will not alter the basic concept.

Attention is focused on a small cluster of cells, approximately five-ten, when use of such terms as ligand density and receptor density has meaning. The cells are to be thought of as comprising a closed epithelial surface, so that the densities of the model have dimensions of number/area. Variation of the morphogens (the R's or L's) along the apical-basal cell direction is not considered, or rather thought of as being an averaged value in this dimension.

First of all, each such cell, or rather cell cluster, produces ligand which increases as like kind receptor activation increases. Morphogen R_1, an activated receptor, stimulates production of L_1; otherwise the process would be limited to a purely local one in the absence of like-ligand production, with the particular cell in question then acting as a sink.

The second key element in the model is that activation of a pathway acts to inactivate the other; as R_1 increases, the level of ligand production L_2 is decreased, and similarly for R_2.

The equations representing such a process are then able to be written at once, and are

$$\frac{\partial}{\partial t}L_1 = D_1\nabla^2 L_1 + \alpha R_1 - \beta R_2 + NL \tag{A.1}$$

$$\frac{\partial}{\partial t}L_2 = D_2\nabla^2 L_2 + \beta R_2 - \alpha R_1 + NL \tag{A.2}$$

$$\frac{\partial}{\partial t}R_1 = C_1\bar{R}_1 L_1 - \mu R_1 \tag{A.3}$$

$$\frac{\partial}{\partial t}R_2 = C_2\bar{R}_2 L_2 - \nu R_2 \tag{A.4}$$

The first two terms in Eqs. (A.1) and (A.2) represent in the usual way diffusion of the ligands in the extracellular space. All parameters in the model (e.g., α, β, D_1, C_1, μ, ν) have positive values, as do also, of course, the densities L_1, L_2, R_1 and R_2. The terms αR_1 in Eq. (A.1) and βR_2 in Eq. (A.2) represent the production of like ligand by the corresponding activated receptor. These same terms are used to represent the fact that activation of receptors of density R_2 deactivate or turn off production of free ligands of density L_1, and vice versa.

The transmembrane receptors, which reside in the lateral cell plasma membrane, are relatively immobile. The respective activated densities decay at rates μ and ν, and this decay returns the receptors to their inactive state. Two first terms on the right side of Eqs. (A.3) and (A.4) say that there is a positive rate of change of R_1 or R_2 proportional to both the density of empty receptor sites ($\bar{R}_1$, $\bar{R}_2$) and also to the density of free ligands at the particular local cell site. The density of empty sites may be obtained from the expression

$$R_1 + \bar{R}_1 = R_0 + \eta R_1\,, \quad (R_0 = \mathit{const.})$$

where the last term on the r.h.s. expresses the possibility that the total number of receptors of each type (e.g., '1') increases with activation of that same type receptor, and new (empty) receptors are thus added. Then the empty receptor site density may be written

$$\bar{R}_1 = R_0(1 - \varepsilon_1 R_1)\,, \quad (0 < \varepsilon_1 \leq 1\,, \varepsilon_1 \equiv (1-\eta)/R_0) \qquad \text{(A.5)}$$

and similarly for type 2. The values $\varepsilon = 1$ (and $\eta = 0$), implies that there is no receptor augmentation $\sim R_1$, while $\varepsilon \quad \sim 0$ implies either that there is a new empty receptor created for (almost) every one occupied, or that there are very many more empty sites than occupied ones. When Eq. (A.5) and the analogous equation for type 2 is used in Eqs. (A.3) and (A.4) to eliminate the unoccupied site densities, the model then comprises four coupled equations for four unknowns. The coupling from epithelial shape to morphogen, and back is discussed elsewhere[11,17], and in Sec.3.

The small amplitude, time independent ($\partial/\partial t = 0$) version of Eqs. (A.1)–(A.4) are simply the Helmholtz and Laplace equations

$$\nabla^2(R_1 - fR_2) + k^2(R_1 - fR_2) = 0 \qquad \text{(A.6)}$$

and

$$\nabla^2(R_1/k_1^2 + fR_2/k_2^2) = 0 \qquad \text{(A.7)}$$

The definitions $k^2 = k_1^2 + k_2^2$, $f = \beta/\alpha$, $k_1^2 = \alpha C_1 R_0/(D_1\mu)$, and $k_2^2 = \beta C_2 R_0/(D_2\nu)$ have been used. The k's are the inverse lengths of the model.

Several forms may serve to model the non-linear (N.L) terms on the r.h.s. of Eqs. (A.1) and (A.2). The simplest, and the one used in present simulations is to let

$$R_1 - (\beta/\alpha)R_2 \rightarrow (R_1 - (\beta/\alpha)R_2)/(1 + ((R_1 - (\beta/\alpha)R_2)/c)^2)$$

The constant c is ~ 1. Others forms lead to other nearby surface shapes for the invagination.

The normalized morphogens of the present work are taken as

$$\Phi_1 = R_1/R_0\,, \quad \Phi_2 = (\beta/\alpha)R_2/R_0 \qquad \text{(A.8)}$$

A region of high activation of active receptor density of one morphogen implies low activation of the second. The term NL on the r.h.s. of Eqs. (A.1) and (A.2) indicate that there are expected to be nonlinearities; saturation effects set in for large enough values of either.

Appendix B.

This appendix examines the model of Appendix A with a view to obtaining a solution for an Urcnidaria in cylindrical geometry. This employs a particular boundary condition, as discussed in the text, Sec. 4, namely that both morphogens Φ_1 and Φ_2 are equal to the same constant at the cylinder disc ends. Both of these ends are in contact, and are at the opposite end from the mouth region, $z \sim 0$. The linear equations are (from Eqs. (A.6), (A.7) and (A.8), or Eqs. (1) and (2).

$$\nabla^2(\Phi_1 - \Phi_2) + k^2(\Phi_1 - \Phi_2) = 0 \tag{B.1}$$

$$\nabla^2(\frac{\Phi_1}{k_1^2} - \frac{\Phi_2}{k_2^2}) = 0 \tag{B.2}$$

The constant parameters k, k_1, k_2 and $f \equiv \beta/\alpha \sim 1$ have been defined in Appendix A. The solution to the two equations (B.1) and (B.2) for which both Φ_1 and Φ_2 are constant $= C$ at $z = \pm L$ (and $D = constant$) is

$$\Phi_1 = C + Dk_1^2 \exp(-\alpha|z|/L). \cos(n\phi) \tag{B.3}$$

and

$$\Phi_2 = C - Dk_2^2 \exp(-\alpha|z|/L). \cos(n\phi) \tag{B.4}$$

There is a periodicity around the mouth region, with Φ_1 alternating with Φ_2, and an exponential decay of each morphogen down the z axis.

Stem cell regions are established when the two morphogens are about equal, and the ability of each morphogen to effect cell determination is nullified by the other. This occurs when the angle $\phi \approx (2j+1)\pi/(2n)$, ($j =$ integer), and the Cosine term is nearly zero.

Equation (B.1) dictates that the relation between radius kR and length kL in the urcnidaria case is that

$$1 = (n/kR)^2 - (\alpha/kL)^2 \tag{B.5}$$

The parameter $\alpha \sim 1 \rightarrow 2$, so that Φ_1 and $\Phi_2 \rightarrow C$ as $|z| \rightarrow L$ as required. Then $n \sim kR$, for $R/L << 1$, but n is larger than kR when $R \approx L$, or perhaps much larger, as occurs in sea anemones.

Appendix C.

Here the solution to Eqs. (B.1) and (B.2) are given, again in the case of cylindrical symmetry, but now when the topology is that of a torus, or donut. All solutions must be periodic in both coordinates, and this gives

$$\Phi_1 = C + Dk_1^2 \cos(\pi mz/L). \cos(n\phi) \tag{C.1}$$

and

$$\Phi_2 = C - Dk_2^2 \cos(\pi mz/L). \cos(n\phi) \tag{C.2}$$

Here two integers are defined, m and n, and the solution is periodic for $z \to z + 2L$ as well as $\phi \to \phi + 2\pi$. The condition from Eqs. (1) and (2) relating R, L, m and n for the small amplitude solution of Eqs. (C.1) and (C.2) is

$$1 = (\pi m/kL)^2 + (n/kR)^2 \qquad \text{(C.3)}$$

What is clear is that while kR remains below unity, n must remain zero. So it is supposed that $kL = \pi$ and $m = 1$ at the start of the first cycle of growth while $kR < 1$. The length kL can continue to increase as m increases, with $n = 0$. As kR passes unity and becomes $\approx 2^{1/2}$, n now can become unity, and kL also grows to become $kL \approx 2^{1/2}\pi m_0$, with $m = m_0 > 1$. At this point, a through-gut, bilateral animal has emerged. Length kL may continue to increase while keeping $kR \approx 2^{1/2}$. A double segment periodicity then underlies segment generation[20,21,22,23] for this bilaterally symmetric, through-gut animal.

The next growth cycle may occur after a further increase in radius kR, until $kR \approx 2^{3/2}$, when now $n = 2$. Length kL may continue to increase so that $kL \approx 2^{1/2}\pi m$, where $m >> 1$. A new pattern of small amplitude begins to emerge at each growth cycle, at each area sufficient so that Eq. (C.3) can be satisfied. There are now four narrow linear regions running down the axis where $\cos(n\phi) \approx 0$, $n = 2$, where both morphogens are nearly equal to the same constant C. These are indicated in Fig. 4b by dashed lines. Along with the linear regions, there are also a series of (small) circular regions along the z axis where the two morphogens are also nearly equal, as in Fig. 5. Such are denoted as stem cell regions, regions where the ability of either morphogen Φ_1 or Φ_2 to effect cell differentiation is canceled by the (nearly) same density of the other. The intersections of these two sets of stem cell regions provide positional information for limb or appendage formation.

References

1. S. B.Carroll, J. K. Grenier, S. D. Weatherbee, *From DNA to Diversity: Molecular Genetics and the Evolution of Animal Design,* (Blackwell Science, Malden, MA., 2001)
2. E. H. Davidson, *General Regulatory Systems: Development and Evolution,* (Academic Press, San Diego, Ca., 2001).
3. A. S. Wilkins, *The Evolution of Developmental Pathways*, (Sinauer Assoc., MA, 2002).
4. W. Arthur, "The emerging conceptual framework of evolutionary developmental biology", Nature, **415,** 757-764 (2002).

5. G. Webster and B. Goodwin, *Form and Transformation: Generative and Relational Principles in Biology,* (Cambridge Univ. Press, 1996).
6. J. Maynard Smith, J. Burian, S. Kauffman, P. Alberch, J. Campbell, B. Goodwin, R. Lande, D. Raup, and L. Wolpert, Developmental constraints and evolution, Q. Rev. Biol., **60**, 265, (1985).
7. Pat Willmer, *Invertebrate Relationships: Patterns in Animal Development,* (Cambridge Univ. Press, New York, Cambridge, 1990).
8. J. Valentine, *On the Origin of Phyla,* (Univ. of Chicago Press, 2004).
9. J. D. Murray, *Mathematical Biology,* (Springer-Verlag, Berlin, 1990).
10. I. Salazar-Ciudad, J. Jernvali and S. Newman, "Mechanisms of pattern formation in development and evolution", Development **130**, 2027-2037 (2003). This review contains numerous relevant references.
11. F. W. Cummings, "A model of morphogenesis", Physica A, **339**, 531-547 (2004).
12. D. Panakova, H. Sprong, E. Marois, C. Thiele & S. Eaton, "Lipoprotein particles are required for Hedgehog and Wingless signalling", Nature **435,** 58-61 (2005).
13. A. Kusserow, K. Pang, C. Sturm, M. Hrouda, J. Lentfer, H.A. Schmidt, U.Technau, A. von Haeseler, B. Hobmeyer, M.Q. Martindale, & T. Holstein, "Unexpected complexity of the Wnt gene family in a sea anemone", Nature, **433**, 156-160 (2005).
14. J. Taipale & P. Beachy, "The Hedgehog and Wnt signalling pathways in cancer" Nature, **411**, 349-354 (2001), and references therein.
15. E. De Robertis, and Y. Sasai, Nature **380**, 37-40 (1996).
16. F.W. Cummings & J.C. Strickland, "A model of phyllotaxis", J. Theor. Biol. **192**, 531-544 (1998).
17. F. W. Cummings, "Interaction of morphogens with geometry", Physica A **355**, 427-438 (2005).
18. D. Jacobs, N. Hughes, C. Winchell, "Terminal addition and the evolution of bilaterian form", SICB (www.sicb.org/meetings/2005).
19. E. De Robertis, "The ancestry of segmentation", Nature **387**, 25 (1997).
20. A. Minelli and S. Bortoletto, "Myriapod metamerism and arthropod segmentation", Biol. J. Linn. Soc. **33**, 323 (1988).
21. W. Arthur and A. Chipman, "The centipede Strigamia maritima: what it can tell us about the development and evolution of segmentation", BioEssays **27,** 653 (2005).
22. A. Chipman, W. Arthur and M. Akam, "A double segment periodicity underlies segment generation in centipede development", Curr. Biol., **14,** R557 (2004).
23. C. Kettle, J. Johnstone, T. Jowett, H. Arthur and W. Arthur, "The pattern of segment formation, as revealed by engrailed expression, in a centipede with a variable number of segments", Evolution and Development **5,** 198 (2003).

A SOFTWARE TOOL TO MODEL GENETIC REGULATORY NETWORKS: APPLICATIONS TO SEGMENTAL PATTERNING IN *DROSOPHILA**

FILIPA ALVES AND RUI DILÃO
Nonlinear Dynamics Group, Instituto Superior Técnico
Av. Rovisco Pais, 1049-001 Lisbon, Portugal
E-mail: filipa@sd.ist.utl.pt; rui@sd.ist.utl.pt

Based on the bimolecular mass action law and the derived mass conservation laws, we have proposed a mathematical framework in order to describe the regulation of gene expression in prokaryotes (F. Alves, R. Dilão, C. R. Biologies 328 (2005)). Within this framework, a genetic regulatory network is described by a system of ordinary differential equations. As the number of equations increases exponentially with the number of genes and interactions in the network, we have developed a software package to write symbolically the associated system of differential equations. With this software package, *GeNetSim*, it is possible to follow in time the concentration of proteins and genes of the network. We show some applications of *GeNetSim* to the modeling of a genetic regulatory network involved in the segmentation of the embryo of *Drosophila*.

1. Introduction

For many organisms, our knowledge about gene expression and its regulation is growing. Theoretical models and quantitative tools contribute to the organization and analysis of the information available on gene expression patterns and regulatory interactions. To access the available information computational tools have been developed using different approaches (for reviews, see, for example, Price et al., 2004; Hirano, et al., 2004; Huang, 2004; Anderle et al., 2003; Altschul et al., 1994). Many software tools are designed to organize and retrieve information from the existing gene expression databases (see, for example, Kosman et al., 1998; Janssens et al., 2005), or are focused on the reverse-engineering of the regulatory network

*This work has been partially supported by Fundação para a Ciência e a Tecnologia (Portugal), under the framework of POCTI project C/FIS/13161/1998, and by the grant PRAXIS XXI/BD/18379/98.

associated with a given pattern of gene expression (Kumar et al., 2002; Xing et al., 2005). To predict the expression patterns of specific genes, other tools use the information on the regulatory interactions given by experiments and have been developed in close relation with specific theoretical models of regulation of gene expression (Meir, et al., 2002; von Dassow et al., 2000; Jaeger et al., 2004a; Jaeger et al., 2004b).

One of the difficulties in modeling a genetic regulatory network is related with the large number of equations and parameters describing the system. Often, it is necessary to test different network configurations, parameter values and other model conditions. Writing a new system of equations each time a modification is introduced becomes time-consuming and it is almost impossible not to make mistakes. The system of differential equations, often non-linear, generally has to be solved numerically with efficient methods. Our aim here is to present a computationally efficient tool that performs these tasks automatically, enabling a straightforward analysis of different network scenarios.

The software package GeNetSim has been written in *Mathematica* and is based on a mass action law theoretical framework for the description of the regulation of gene expression in prokaryotes. This approach has been proposed by Alves and Dilão (2005), and in section 2 we make an overview of this technique. We analyse a simple prototype of a genetic regulatory network and we derive the equations describing the concentrations of all the proteins and genes in the system. In section 3, describe all the commands available in this software package. In section 4, we show an application to the segmental patterning in Drosophila.

The main motivation for the development of *GeNetSim* was its application to the modeling of the genetic network involved in the segmentation of the *Drosophila* embryo (Alves and Dilão, 2006). However, this software tool is very flexible, and is easily adapted to other model formulations and contexts. The software package is freely distributed by the authors, or can be downloaded from "http://sd.ist.utl.pt".

2. A model for the regulation of gene expression

The simplest model of gene expression and protein production without any regulatory mechanism can be described by the following kinetic diagram,

$$\begin{array}{l} \mathrm{a} \xrightarrow{p_A} \mathrm{a} + \mathrm{A} \\ \mathrm{A} \xrightarrow{d_A} \end{array} \tag{1}$$

where a and A represent the gene and the protein, respectively, p_A is the production rate of protein A and d_A is its degradation rate.

In this simplified model for protein production, we assume that mRNA concentration is constant during transcription. We also assume that the concentrations of nucleotides, aminoacids and catalysts involved in transcription and translation are not limiting factors. As we are concerned with the regulation of protein production, in the following we will use this simplified model to represent both transcriptional and translational mechanisms. Under these conditions, the concentrations of catalysts, nucleotides and aminoacids will affect the rate constant p_A.

If the collisional interactions that initiate these processes (transcription, translation and protein production) occur randomly in the media, the time evolution of the several components is described by the mass action law (for details see Alves and Dilão (2005)). In this case, the equations for the time evolution of the concentrations are,

$$\begin{aligned} \frac{\mathrm{dA}}{\mathrm{dt}} &= p_A \mathrm{a} - d_A \mathrm{A} \\ \frac{\mathrm{da}}{\mathrm{dt}} &= 0 \end{aligned} \tag{2}$$

where we use the same notation to represent a specific substance and its concentration. The general solution of Eq. (2) is,

$$\begin{aligned} \mathrm{A}(t) &= (\mathrm{A}(0) - \tfrac{p_A}{d_A}\mathrm{a}(0))e^{-dt} + \tfrac{p_A}{d_A}\mathrm{a}(0) \\ \mathrm{a}(t) &= \mathrm{a}(0) \end{aligned} \tag{3}$$

By (3), in the limit $t \to \infty$, the concentration of protein attains the equilibrium value $\mathrm{A}^* = p_A \mathrm{a}(0)/d_A$.

In this model for protein free production, the use of the mass action law relies on the hypothesis that DNA, mRNA and protein molecules are well distributed inside the cell and that the collisional interaction mechanisms are essentially random. The rate constants p_A and d_A are effective velocity constants multiplied by catalysts, nucleotides and aminoacids concentrations.

In the following and in order to maintain models as simple as possible, we will always avoid the explicit introduction of other species than genes and proteins, being gene concentrations always conserved quantities. For a discussion about the role of conserved quantities in these models see Alves and Dilão (2005).

To model any genetic regulatory network where transcription factors account for activation and repression of gene expression, we associate binding sites to genes, where transcription factors (transcriptional activators and repressors) can bind in order to promote or repress protein production, Fig. 1. This approach follows the operon model of Jacob and Monod (1961). In the following, we consider only the case where different transcription factors have different binding sites.

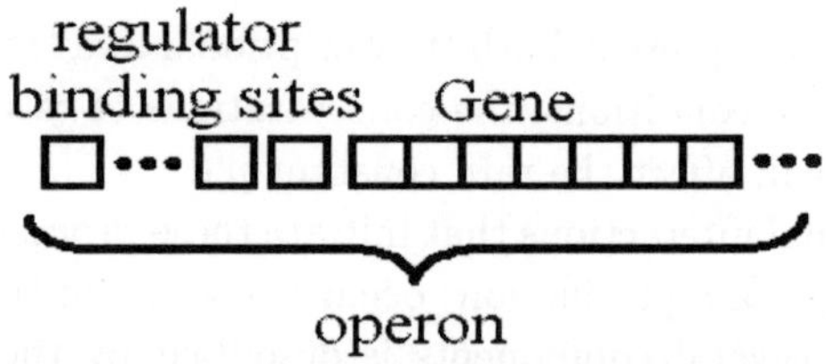

Figure 1. Jacob and Monod operon model for the regulation of protein production in bacteria. The transcriptional regulators bind to the binding sites of the gene and transcription is activated or repressed. In our approach we consider a different binding site for each transcription factor. (Adapted from Alves and Dilão, 2005).

To show how this simplified model for protein production can be applied to a genetic regulatory networok, we take as an example the regulatory network described by the graph in Fig. 2. In this network, protein A activates the expression of the genes b, c and d, leading to the production of the proteins B, C and D, respectively. Protein B represses the production of proteins C and D, and proteins C and D repress the production of protein B. In the following, proteins will be represented by capital letters, and genes by small letters. In the graph in Fig. 2 genes are not represented, only proteins.

In the genetic regulatory network of Fig. 2, gene c has two transcription factors regulating its expression: one activator protein A and one repressor protein B. As we assume that the gene c has an independent regulatory binding site for each of these proteins, the gene c can be in one of four possible bound states: $\mathrm{C}_{0,0}$, $\mathrm{C}_{0,A}$, $\mathrm{C}_{B,0}$ and $\mathrm{C}_{B,A}$. For example, $\mathrm{C}_{0,0}$ means that the two binding sites of the gene c are free, and $\mathrm{C}_{B,0}$ means that the repressor protein B is bound to the gene c.

We say that the gene is activated or is in a "on" state, leading to the consequent protein synthesis, if the activator binding site is occupied and the repressor binding site is free ($\mathrm{C}_{0,A}$). The other states of the gene are repressed states or, $\mathrm{C}_{0,0}$, $\mathrm{C}_{B,0}$ and $\mathrm{C}_{B,A}$ are "off" states of the gene c. In general, denoting by n the number of binding sites of a gene, the number

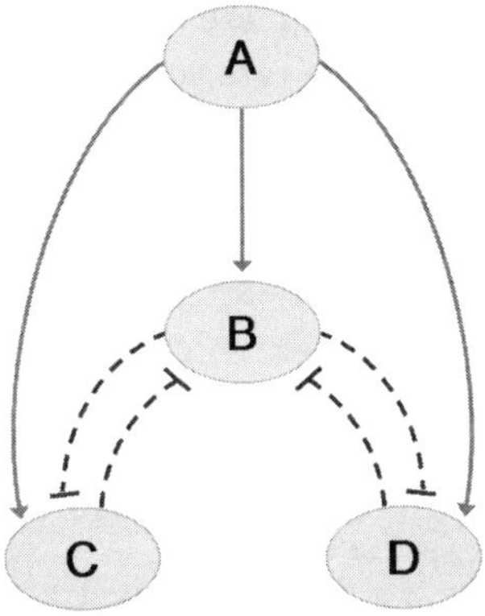

Figure 2. Genetic regulatory network with four genes A, B, C and D. An arrow connecting two genes means that the product of the first gene regulates the transcription of the second gene. The green solid arrows represent activation and the red dashed lines represent repression interactions.

of possible binding states for each gene is 2^n. Clearly gene b has 8 different binding states and genes c and d have 4 different binding states each.

The network of interactions represented by the graph of Fig. 2 are described by the following set of bimolecular mechanisms:

$$
\begin{array}{lll}
\mathrm{B}_{0,0,0} + \mathrm{A} \underset{b_{-a}}{\overset{b_a}{\rightleftharpoons}} \mathrm{B}_{0,0,A} & \mathrm{C}_{0,0} + \mathrm{A} \underset{c_{-a}}{\overset{c_a}{\rightleftharpoons}} \mathrm{C}_{0,A} & \mathrm{D}_{0,0} + \mathrm{A} \underset{d_{-a}}{\overset{d_a}{\rightleftharpoons}} \mathrm{D}_{0,A} \\
\mathrm{B}_{0,0,0} + \mathrm{C} \underset{b_{-c}}{\overset{b_c}{\rightleftharpoons}} \mathrm{B}_{0,C,0} & \mathrm{C}_{0,0} + \mathrm{B} \underset{c_{-b}}{\overset{c_b}{\rightleftharpoons}} \mathrm{C}_{B,0} & \mathrm{D}_{0,0} + \mathrm{B} \underset{d_{-b}}{\overset{d_b}{\rightleftharpoons}} \mathrm{D}_{B,0} \\
\mathrm{B}_{0,0,0} + \mathrm{D} \underset{b_{-d}}{\overset{b_d}{\rightleftharpoons}} \mathrm{B}_{D,0,0} & \mathrm{C}_{0,A} + \mathrm{B} \underset{c_{-b}}{\overset{c_b}{\rightleftharpoons}} \mathrm{C}_{B,A} & \mathrm{D}_{0,A} + \mathrm{B} \underset{d_{-b}}{\overset{d_b}{\rightleftharpoons}} \mathrm{D}_{B,A} \\
\mathrm{B}_{0,0,A} + \mathrm{C} \underset{b_{-c}}{\overset{b_c}{\rightleftharpoons}} \mathrm{B}_{0,C,A} & \mathrm{C}_{B,0} + \mathrm{A} \underset{c_{-a}}{\overset{c_a}{\rightleftharpoons}} \mathrm{C}_{B,A} & \mathrm{D}_{B,0} + \mathrm{A} \underset{d_{-a}}{\overset{d_a}{\rightleftharpoons}} \mathrm{D}_{B,A} \\
\mathrm{B}_{0,0,A} + \mathrm{D} \underset{b_{-d}}{\overset{b_d}{\rightleftharpoons}} \mathrm{B}_{D,0,A} & \mathrm{C}_{0,A} \xrightarrow{p_C} \mathrm{C}_{0,A} + \mathrm{C} & \mathrm{D}_{0,A} \xrightarrow{p_D} \mathrm{D}_{0,A} + \mathrm{D} \\
\mathrm{B}_{0,C,0} + \mathrm{A} \underset{b_{-a}}{\overset{b_a}{\rightleftharpoons}} \mathrm{B}_{0,C,A} & \mathrm{C} \xrightarrow{d_C} & \mathrm{D} \xrightarrow{d_D} \\
\mathrm{B}_{0,C,0} + \mathrm{D} \underset{b_{-d}}{\overset{b_d}{\rightleftharpoons}} \mathrm{B}_{D,C,0} & & \\
\mathrm{B}_{D,0,0} + \mathrm{A} \underset{b_{-a}}{\overset{b_a}{\rightleftharpoons}} \mathrm{B}_{D,0,A} & & \\
\mathrm{B}_{D,0,0} + \mathrm{C} \underset{b_{-c}}{\overset{b_c}{\rightleftharpoons}} \mathrm{B}_{D,C,0} & & \\
\mathrm{B}_{0,C,A} + \mathrm{D} \underset{b_{-d}}{\overset{b_d}{\rightleftharpoons}} \mathrm{B}_{D,C,A} & & \\
\mathrm{B}_{D,0,A} + \mathrm{C} \underset{b_{-c}}{\overset{b_c}{\rightleftharpoons}} \mathrm{B}_{D,C,A} & & \\
\mathrm{B}_{D,C,0} + \mathrm{A} \underset{b_{-a}}{\overset{b_a}{\rightleftharpoons}} \mathrm{B}_{D,C,A} & & \\
\mathrm{B}_{0,0,A} \xrightarrow{p_B} \mathrm{B}_{0,0,A} + \mathrm{B} & & \\
\mathrm{B} \xrightarrow{d_B} & &
\end{array}
\tag{4}
$$

where the kinetic constants b_a, b_c, b_d are the binding affinities of the proteins A, C and D to the gene b regulatory binding sites, and b_{-a}, b_{-c}, b_{-d} are the constants for unbinding the same proteins from the regulatory sites. The constants p_B and d_B are the production and degradation rates of the protein B. Analogously p_A and d_A are the rate constants for protein A; c_a, c_b, c_{-a}, c_{-b}, p_C and d_C are the rate constants for protein C and d_a, d_b, d_{-a}, d_{-b}, p_D and d_D are the rate constants for protein D.

Applying the mass action law to the diagram (4), we obtain a system of differential equations describing the time evolution of the concentrations of all possible gene states and proteins,

$$
\begin{aligned}
\frac{d\mathrm{B}_{0,0,0}}{dt} &= -b_a\mathrm{B}_{0,0,0}\cdot\mathrm{A} - b_c\mathrm{B}_{0,0,0}\cdot\mathrm{C} - b_d\mathrm{B}_{0,0,0}\cdot\mathrm{D} + b_{-a}\mathrm{B}_{0,0,A}\\
&+ b_{-c}\mathrm{B}_{0,C,0} + b_{-d}\mathrm{B}_{D,0,0}\\
\frac{d\mathrm{B}_{0,0,A}}{dt} &= -b_{-a}\mathrm{B}_{0,0,A} - b_c\mathrm{B}_{0,0,A}\cdot\mathrm{C} - b_d\mathrm{B}_{0,0,A}\cdot\mathrm{D} + b_a\mathrm{B}_{0,0,0}\cdot\mathrm{A}\\
&+ b_{-c}\mathrm{B}_{0,C,A} + b_{-d}\mathrm{B}_{D,0,A}\\
\frac{d\mathrm{B}_{0,C,0}}{dt} &= -b_a\mathrm{B}_{0,C,0}\cdot\mathrm{A} - b_{-c}\mathrm{B}_{0,C,0} - b_d\mathrm{B}_{0,C,0}\cdot\mathrm{D} + b_c\mathrm{B}_{0,0,0}\cdot\mathrm{C}\\
&+ b_{-a}\mathrm{B}_{0,C,A} + b_{-d}\mathrm{B}_{D,C,0}\\
\frac{d\mathrm{B}_{0,C,A}}{dt} &= -b_{-a}\mathrm{B}_{0,C,A} - b_{-c}\mathrm{B}_{0,C,A} - b_d\mathrm{B}_{0,C,A}\cdot\mathrm{D} + b_c\mathrm{B}_{0,0,A}\cdot\mathrm{C}\\
&+ b_a\mathrm{B}_{0,C,0}\cdot\mathrm{A} + b_{-d}\mathrm{B}_{D,C,A}\\
\frac{d\mathrm{B}_{D,0,0}}{dt} &= -b_a\mathrm{B}_{D,0,0}\cdot\mathrm{A} - b_c\mathrm{B}_{D,0,0}\cdot\mathrm{C} - b_{-d}\mathrm{B}_{D,0,0} + b_d\mathrm{B}_{0,0,0}\cdot\mathrm{D}\\
&+ b_{-a}\mathrm{B}_{D,0,A} + b_{-c}\mathrm{B}_{D,C,0}\\
\frac{d\mathrm{B}_{D,0,A}}{dt} &= -b_{-a}\mathrm{B}_{D,0,A} - b_c\mathrm{B}_{D,0,A}\cdot\mathrm{C} - b_{-d}\mathrm{B}_{D,0,A} + b_d\mathrm{B}_{0,0,A}\cdot\mathrm{D}\\
&+ b_a\mathrm{B}_{D,0,0}\cdot\mathrm{A} + b_{-c}\mathrm{B}_{D,C,A}\\
\frac{d\mathrm{B}_{D,C,0}}{dt} &= -b_a\mathrm{B}_{D,C,0}\cdot\mathrm{A} - b_{-c}\mathrm{B}_{D,C,0} - b_{-d}\mathrm{B}_{D,C,0} + b_d\mathrm{B}_{0,C,0}\cdot\mathrm{D}\\
&+ b_c\mathrm{B}_{D,0,0}\cdot\mathrm{C} + b_a\mathrm{B}_{D,C,A}\\
\frac{d\mathrm{B}_{D,C,A}}{dt} &= -b_a\mathrm{B}_{D,C,A} - b_{-c}\mathrm{B}_{D,C,A} - b_{-d}\mathrm{B}_{D,C,A} + b_d\mathrm{B}_{0,C,A}\cdot\mathrm{D}\\
&+ b_c\mathrm{B}_{D,0,A}\cdot\mathrm{C} + b_a\mathrm{B}_{D,C,0}\cdot\mathrm{A}\\
\frac{d\mathrm{C}_{0,0}}{dt} &= -c_a\mathrm{C}_{0,0}\cdot\mathrm{A} - c_b\mathrm{C}_{0,0}\cdot\mathrm{B} + c_{-a}\mathrm{C}_{0,A} + c_{-b}\mathrm{C}_{B,0}\\
\frac{d\mathrm{C}_{0,A}}{dt} &= -c_{-a}\mathrm{C}_{0,A} - c_b\mathrm{C}_{0,A}\cdot\mathrm{B} + c_a\mathrm{C}_{0,0}\cdot\mathrm{A} + c_{-b}\mathrm{C}_{B,A}\\
\frac{d\mathrm{C}_{B,0}}{dt} &= -c_a\mathrm{C}_{B,0}\cdot\mathrm{A} - c_{-b}\mathrm{C}_{B,0} + c_b\mathrm{C}_{0,0}\cdot\mathrm{B} + c_{-a}\mathrm{C}_{B,A}\\
\frac{d\mathrm{C}_{B,A}}{dt} &= -c_{-a}\mathrm{C}_{B,A} - c_{-b}\mathrm{C}_{B,A} + c_b\mathrm{C}_{0,A}\cdot\mathrm{B} + c_a\mathrm{C}_{B,0}\cdot\mathrm{A}\\
\frac{d\mathrm{D}_{0,0}}{dt} &= -d_a\mathrm{D}_{0,0}\cdot\mathrm{A} - d_b\mathrm{D}_{0,0}\cdot\mathrm{B} + d_{-a}\mathrm{D}_{0,A} + d_{-b}\mathrm{D}_{B,0}\\
\frac{d\mathrm{D}_{0,A}}{dt} &= -d_{-a}\mathrm{D}_{0,A} - d_b\mathrm{D}_{0,A}\cdot\mathrm{B} + d_a\mathrm{D}_{0,0}\cdot\mathrm{A} + d_{-b}\mathrm{D}_{B,A}\\
\frac{d\mathrm{D}_{B,0}}{dt} &= -d_a\mathrm{D}_{B,0}\cdot\mathrm{A} - d_{-b}\mathrm{D}_{B,0} + d_b\mathrm{D}_{0,0}\cdot\mathrm{B} + d_{-a}\mathrm{D}_{B,A}\\
\frac{d\mathrm{D}_{B,A}}{dt} &= -d_{-a}\mathrm{D}_{B,A} - d_{-b}\mathrm{D}_{B,A} + d_b\mathrm{D}_{0,A}\cdot\mathrm{B} + d_a\mathrm{D}_{B,0}\cdot\mathrm{A}\\
\frac{d\mathrm{B}}{dt} &= p_B\mathrm{B}_{0,0,A} - d_B\mathrm{B} + c_{-b}(\mathrm{C}_{B,0} + \mathrm{C}_{B,A}) + d_{-b}(\mathrm{D}_{B,0} + \mathrm{D}_{B,A})\\
&- \mathrm{B}(c_b(\mathrm{C}_{0,0} + \mathrm{C}_{0,A}) + d_b(\mathrm{D}_{0,0} + \mathrm{D}_{0,A}))\\
\frac{d\mathrm{C}}{dt} &= p_C\mathrm{C}_{0,A} - d_C\mathrm{C} + b_{-c}(\mathrm{B}_{0,C,0} + \mathrm{B}_{0,C,A} + \mathrm{B}_{D,C,0} + \mathrm{B}_{D,C,A})\\
&- b_c\mathrm{C}(\mathrm{B}_{0,0,0} + \mathrm{B}_{0,0,A} + \mathrm{B}_{D,0,0} + \mathrm{B}_{D,0,A})\\
\frac{d\mathrm{D}}{dt} &= p_D\mathrm{D}_{0,A} - d_D\mathrm{D} + b_{-d}(\mathrm{B}_{D,0,0} + \mathrm{B}_{D,0,A} + \mathrm{B}_{D,C,0} + \mathrm{B}_{D,C,A})\\
&- b_d\mathrm{D}(\mathrm{B}_{0,0,0} + \mathrm{B}_{0,0,A} + \mathrm{B}_{0,C,0} + \mathrm{B}_{0,C,A})
\end{aligned}
\tag{5}
$$

However, neither all the equations in (5) are independent. By a standard analysis (Alves and Dilão, 2005), we have the conservation laws,

$$\begin{array}{ll} \mathrm{B}_{0,0,0} + \mathrm{B}_{0,0,A} + \mathrm{B}_{0,C,0} + \mathrm{B}_{0,C,A} + \mathrm{B}_{D,0,0} + \mathrm{B}_{D,0,A} + \mathrm{B}_{D,C,0} + \mathrm{B}_{D,C,A} & = g_1 \\ \mathrm{C}_{0,0} + \mathrm{C}_{0,A} + \mathrm{C}_{B,0} + \mathrm{C}_{B,A} & = g_2 \\ \mathrm{D}_{0,0} + \mathrm{D}_{0,A} + \mathrm{D}_{B,0} + \mathrm{D}_{B,A} & = g_3 \end{array} \tag{6}$$

Threfore, the genetic regulatory network described by the graph of Fig. 2 has a dynamics described by the system of Eq. (5), together with the associated parameters and the conservation laws (6).

3. The *GeNetSim* package

By applying the previously described strategy, the number of equations describing protein production through a genetic regulatory network increase exponentially with the number of genes and interactions considered in the network. The kinetic mechanisms and the differential equations involved become difficult to write extensively without errors. Moreover, each modification introduced into the graph of the regulatory network implies rewriting everything again. To overcome these problems, we have developed a *Mathematica* package, *GeNetSim*, that performs most of these tasks automatically, writing the differential equations associated to any genetic regulatory network.

Initially, *GeNetSim* has been developed to model the gene network involved in the segmentation of the embryo of *Drosophila*. Some of the assumptions and simplifications used are related to this primary application.

The structure of the network of Fig. 2 is codified in *GeNetSim* as shown in Fig. 3.

In the string ***maternal***, we define the set of proteins for which the initial concentration is different from zero. By convention, the names of the proteins should have at least two characters.

In the string ***genenet***, we define the set of proteins of the genetic network to be analized. The proteins included in ***maternal*** should also appear in ***genenet*** if their concentration is not assumed constant during the analyzed integration time. *GeNetSim* writes and solves numerically a set of equations as in (5), describing the concentration variations for the proteins and all the gene states involved.

For each protein YY, we define the set of proteins that have regulatory interactions with YY. This information is contained in the string ***data[YY]***. For each protein YY in ***genenet***, we define as ***ONstates[YY]*** the set of gene states that promote the production of the protein YY. The order of gene states in ***ONstates[YY]*** follows the same order of the transcription factors defined in ***data[YY]***. The value "0" means that the

Genetic Regulatory Network Input

```
maternal = {AA};
genenet = {BB, CC, DD};
data[BB] = {DD, CC, AA};
ONstates[BB] = {{0, 0, 1}};
data[CC] = {BB, AA};
ONstates[CC] = {{0, 1}};
data[DD] = {BB, AA};
ONstates[DD] = {{0, 1}};
extrasubst = {};
```

```
egglength = 0;
tmin = 0;
tmax = 100;
MASS = 0;
txtfilepath =
  "C : /Documents and Settings / rui / My
     Documents / FilipaAlves / GeNetSim / Prog / GeNetSim.dat ";
```

Figure 3. General *Mathematica* input cell for the genetic network of Fig. 2. In the input cell we codify the structure of the genetic regulatory network to be modeled, as well as other general conditions for the simulation. We can also use this cell to insert extra substitution rules in the main system of differential equations.

respective protein is not bound to the DNA binding site, and "1" means that in that specific gene state the corresponding protein is bound to the DNA binding site. For example, in the case of Fig. 2, protein BB has only one ONstate = {0,0,1}". Therefore, protein BB will only be expressed when the protein AA is bound to the DNA and proteins CC and DD are unbound.

GeNetSim also allows the introduction of *Mathematica* substitution rules in the main system of differential equations. In ***extrasubst***, we define for each extra molecule a function that regulates its concentration. This feature allows the user to define extra molecules that may interact with the genetic regulatory network defined in ***genenet***. For example, ***extrasubst***={YY[t]→ ToExpression[2 ZZ[t]]}.

The entries in ***data[YY]*** must have been defined in, at least, one of the categories: ***maternal***, ***genenet*** or ***extrasubst***.

Numerical integration of the protein concentrations will be done between

tmin and ***tmax***, respectively. GeNetSim can generate a text file including all the kinetic mechanisms and differential equations of the model. In ***txtfilepath***, we define the output file. The integer constant ***egglength*** is related with the number of initial conditions that will be used in the determinations of the protein concentrations. If ***egglength***$= n$, with $n \geq 0$, we have $n + 1$ independent initial conditions.

With ***MASS=1***, the protein equations will be fully written, including the interactions with the DNA binding sites, according to the mass-action law. However, we can simplify further and consider that the concentration of proteins doesn't vary significantly in consequence of their binding to the DNA. With ***MASS=0***, the protein differential equations will be written according to this simplification.

GeNetSim automatically generates all the constant names, based on the protein names defined in the strings ***data***. ***PRODbb*** and ***DEGbb*** define, respectively, the production and degradation rates for the protein BB. The constant ***bbaa*** is the binding affinity of the protein AA to the respective binding site associated with the regulation of protein BB expression. The constant ***unbind*** is reaction rate for the unbinding of the proteins from their binding sites. In this version of *GeNetSim*, the value of ***unbind*** was assumed to be equal for all proteins. In Fig. 4, we show the input of the constant values for the genetic regulatory network of Fig. 2.

```
PRODbb = 1.; DEGbb = 1.;
bbaa = 1.; bbcc = 10.; bbdd = 10.;
PRODcc = 1.; DEGcc = 1.;
ccaa = 0.1; ccbb = 10.;
PRODdd = 1.; DEGdd = 1.;
ddaa = 0.05; ddbb = 10.;
unbind = 0.1;
```

Figure 4. Constant values input cell. Here we define the values for the production and degradation rates of all the proteins in *"genenet"*, and also the binding affinities for all the proteins involved in regulatory interactions. The names of the constants must be defined as explained in the text, otherwise they will not be recognized.

Initial conditions should be defined for all the proteins declared in ***maternal***. For the proteins not included in this set, the initial concentration is, by default, set to zero. The initial spatial distribution of the proteins in ***maternal*** can be defined by any function. For each protein YY in the ***maternal*** set, the local concentration ***iniYY[s]*** must be defined for all the integer values of s between 0 and ***egglength***. These local concentrations of the ***maternal*** proteins will be used by *GeNetSim* as initial conditions for the simulation.

For example, in the genetic regulatory network of Fig. 2, a *Mathematica* command for the initial conditions of protein AA is,

$$For[i = 0, i \leq egglength, i++, iniAA[i] = 10./(1. + i)]; \tag{7}$$

The initial conditions for the gene states are automatically set by *GeNetSim.* By default, for $t = 0$, the concentration of the gene states with all the binding sites free has the value 100 and the concentration of all the other gene states have the value 0.

To start a simulation, the *GeNetSim* package must be loaded, and the genetic regulatory network input, constant values and initial conditions sections must be defined. At this point, the function ***BuildNet*** can be called to build the differential equations and the kinetic diagrams of the model of the genetic regulatory network.

After ***BuildNet*** is evaluated, the function ***SolveNet*** is called to solve the differential equations numerically for all the given initial conditions.

After the evaluation of ***SolveNet***, ***YY[s][t]*** is the concentration of the protein YY for the initial condition number $s+1$ at time t. For the gene states of the genetic regulatory network of Fig. 2, **BB001[s][t]** represents the concentration of the gene state $BB_{0,0,A}$ for the initial condition number $s+1$ at time t. The same rule is followed for all the proteins and gene states in the model.

For example, in Fig. 5, we show the time evolution of protein concentrations for the genetic regulation network of Fig. 2, with ***egglength***$= 0$ and $iniAA[i] = 10.0$. Each graph has been generated with the *Mathematica* command,

$$ListPlot[Table[\{t, AA[0][t][[1]]\}, \{t, 0, tmax\}], PlotJoined \rightarrow True]$$

With the choice ***egglength***$= 100$ and the initial conditions as defined in (7), the equilibrium values of the protein concentrations as a function of the initial conditions is represented in Fig. 6. Note that the equilibrium values of the protein concentrations are not continuous functions of the initial conditions. This is a consequence of the role of the conservation law in the solutions of the protein and gene differential equations. As we shall see in the next section, this effect explains the existence of protein gradients along the antero-posterior axis of *Drosophila* without requiring diffusion effects.

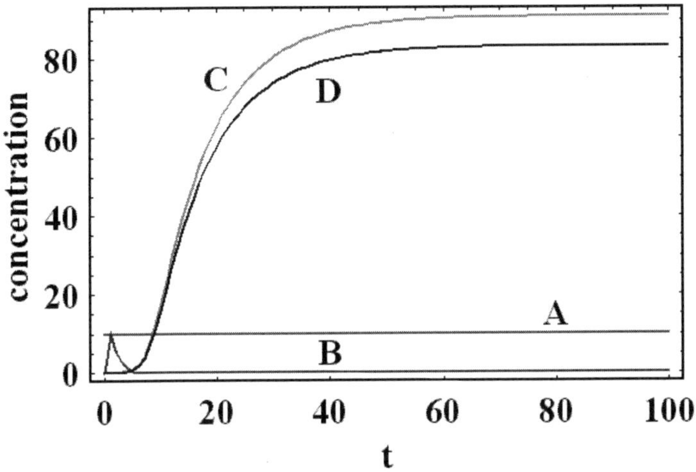

Figure 5. Temporal evolution of protein concentrations for the genetic regulation network of Fig. 2, with ***egglength***$= 0$ and $iniAA[i] = 10.0$.

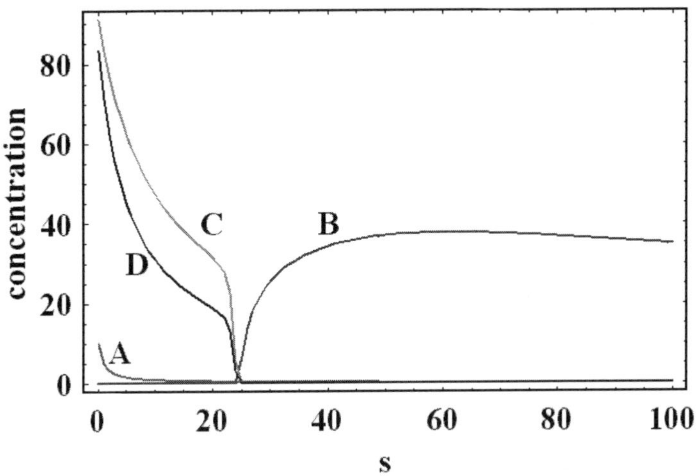

Figure 6. Equilibrium values of the protein concentrations as a function of initial data index parameter s, for the genetic regulatory network of Fig. 2. The initial condition for the protein AA has been defined in (7).

The following additional functions are defined in the *GeNetSim* package:

- ***FICHEIRO***
  Generates the text file ***txtfilepath*** including all the kinetic mechanisms and differential equations of the model.
- ***ALLGENES***
  Gives a set of all the genes (or proteins) included in the model.
- ***ALLSTATES[YY]***
  Displays a list of all the possible states of the gene YY
- ***ONSTATES[YY]***
  Displays a list of the ON states of the gene YY
- ***FULLKINETICS[YY]***
  Generates a list of all the possible unidireccional kinetic mechanisms that involve each state of the gene YY, with the respective constant rates.
- ***FULLKINETICSREV[YY]***
  Generates a list of all the possible unidireccional kinetic mechanisms that involve each state of the gene YY and shows the reverse reaction of each mechanism and the respective constant rates.
- ***KINETICS[YY]***
  Generates a non-redundat list of all the reversible kinetic mechanisms that involve each state of the gene YY, with the respective constant rates.
- ***GENEEQUATIONS[YY]***
  Displays a list of differential equations, one for each state of gene YY.
- ***PROTEINEQUATION[YY]***
  Displays the differential equation that describes the variation of the concentration of protein YY in time.
- ***ALLVARIABLES***
  Shows the set of all the variables in the model.
- ***ALLCONSTANTS***
  Shows a list of the names and values of the constants in the model.
- ***CONSTANT[YY]***
  Shows a list of names and values of all the constants directly involved in the regulation of YY expression.
- ***WRITEPARNAMES***
  Generates a set with the names of all the constants in the model.
- ***WRITEPARVALUES***
  Generates a set with the values of all the constants in the model.
- ***NUMPAR***
  Shows the number of parameters in the model

- ***WRITEARG***
  Displays the set of arguments used in the simulation
- ***WRITEVAR***
  Displays the set of variables used in the simulation
- ***PROTEINVALUES[t]***
  Generates a list of values for the concentration of all the proteins at time t, for all the spatial points between 0 and *egglength*.
- ***TESTCONSERVATIONLAW[YY]***
  Shows a test of the conservation law for the sum of all states of gene YY.
- ***MAXIMUM[YY]***
  Displays the maximum values for the concentration of protein YY, in the spatial and temporal intervals of the simulation. The values are displayed in the form $YY[s][t] = maximumvalue$, where s is space and t is time.
- ***ALLPLOTS[t]***
  Draws a line plot for the spatial distribution of each protein, at time t.
- ***TIMESEQ[YY]***
  Draws a sequence of plots for the spatial distribution of protein YY, one for each time step of the simulation.

4. Segmental patterning in *Drosophila*

In the *Drosophila* egg, the first positional coordinates are set before fertilization. Maternal mRNA molecules are placed at the poles of the oocyte by the mother's ovary cells, defining the antero-posterior axis of the embryo. Fertilization triggers the translation of these maternal mRNA to proteins that regulate the expression of the zygotic genes.

After fertilization, the first 13 nuclear divisions occur without the organization of cellular membranes, giving rise to a syncytial blastoderm. The cytoplasmic membranes only become completely formed three hours after fertilization, in the interphase of the 14th mitotic cycle, just before the onset of gastrulation.

Each of the zygotic genes is transcribed in certain regions of the embryo syncytial blastoderm, and the produced proteins act as transcription factors that regulate the expression of other zygotic genes. The zygotic genes involved in this cascade are divided into three main families: gap, pair-rule, and segment polarity genes. The gap genes are the first to be transcribed zygotically. The proteins resulting from their expression define broad domains along the embryo antero-posterior axis. The gap gene expression is regulated by maternal and gap proteins.

Proteins of maternal origin establish gradients along the antero-posterior axis of the embryo. The mechanisms underlying the establishment of these gradients seem to be associated with diffusion processes occurring in the syncytial blastoderm (Nüsslein-Volhard, 1992).

At the level of gap genes expression, Alves and Dilão (2006) have proposed that segmental patterning along the antero-posterios axis of the embryo of *Drosophila* could be explained, without diffusion, by the genetic regulatory network depicted in Fig. 7.

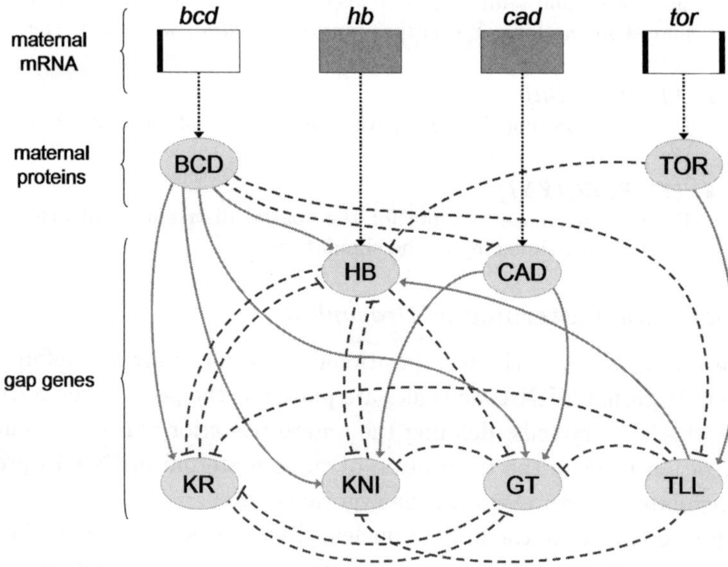

Figure 7. Genetic regulatory network associated with the segmental patterning of gap genes along the antero-posterior axis of the embryo of *Drosophila*.

We model the segmental patterning along the antero-posterior axis of the embryo of *Drosophila* in two stages. In the first stage, mRNAs of maternal origin are translated into proteins that diffuse and degrade within the syncytium. This process is described by a linear system of reaction-diffusion partial differential equations (Alves and R. Dilão, 2006). For each protein of maternal origin, these equations have a non-uniform steady solution along the embryo given by,

$$
\begin{aligned}
\mathrm{P}(x) = & \frac{a\mathrm{A}}{d}\frac{(L_2 - L_1)}{L} \\
& + 2\frac{a\mathrm{A}}{d}\sum_{n=1}^{\infty}\frac{1}{n\pi + \frac{D}{d}\frac{n^3\pi^3}{L^2}}\cos(\frac{n\pi x}{L})\left(\sin(\frac{L_2 n\pi}{L}) - \sin(\frac{L_1 n\pi}{L})\right)
\end{aligned} \tag{8}
$$

where P is protein concentration, D is the diffusion coefficient, a and d are the production and degradation rates, and L is the length of the embryo. This solution is obtained assuming that the mRNA of maternal origin has an initial distribution $A(x) = A$ for $L_1 \leq x \leq L_2$, and $A(x) = 0$ otherwise.

The second stage of the model consists on the description of the genetic regulatory network involving the gap genes and the maternal-effect proteins. At this stage, we assume that the proteins of maternal origin are in a steady state. To model the genetic regulatory network, we apply the methodology developed by Alves and Dilão (2005) to describe the regulation of gene expression in prokaryotes. The diffusion of the zygotic proteins is neglected, and the protein synthesis is described by the genetic regulatory network graph of Fig. 7. As a consequence, at the second stage of the model, the protein production is described by ordinary differential equations. As in general, the protein steady states depend on the conservation laws associated with the mass action law (Alves and Dilão, 2005), at the gap genes stage of development, it is possible to obtain a steady non-uniform protein concentrations along the antero-posterior axis of the *Drosophila* embryo.

```
maternal = {BCD, HB, CAD, TOR};
genenet = {HB, CAD, KR, KNI, GT, TLL};
data[HB] = {TOR, TLL, KR, KNI, BCD};
ONstates[HB] = {{0, 0, 0, 0, 1}, {0, 1, 0, 0, 0}, {0, 1, 0, 0, 1}};
data[CAD] = {BCD};
ONstates[CAD] = {{0}};
data[KR] = {TLL, GT, HB, BCD}; (* 0-bcd  1-hb  2-gt  3-tll *)
ONstates[KR] = {{0, 0, 0, 1}};
data[KNI] = {TLL, GT, CAD, HB, BCD};
ONstates[KNI] = {{0, 0, 0, 0, 1}, {0, 0, 1, 0, 0}, {0, 0, 1, 0, 1}};
data[GT] = {TLL, KR, CAD, HB, BCD};
ONstates[GT] = {{0, 0, 0, 0, 1}, {0, 0, 1, 0, 0}, {0, 0, 1, 0, 1}};
data[TLL] = {TOR, BCD};
ONstates[TLL] = {{1, 0}};
```

Figure 8. *Mathematica* input cell associated with the genetic regulatory network graph in Fig. 7.

In Fig. 8, we show the *Mathematica* input cell associated with the graph in Fig. 7. In this case, we consider as ***ONstates*** all the gene states that have at least one activator bound and no repressors.

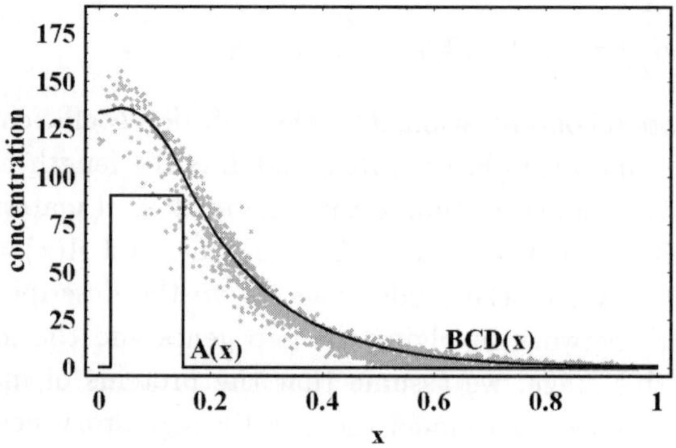

Figure 9. Equilibrium distribution of protein Bicoid along the antero-posterior axis of the *Drosophila* embryo, and calculated from Eq. (8). $A(x)$ is the initial distribution of Bicoid mRNA of maternal origin, and in gray we show experimental data. Parameter values are: $\text{A}(x) = 90$, for $0.02 \le x \le 0.15$, and $\text{A}(x) = 0$, otherwise. $L = 1$, $aA/d = 270$ and $D/d = 0.025$. For example, this combination of parameters can be realized with $D = 10^{-9}$, $a = 1.2 \times 10^{-7}$ and $d = 4 \times 10^{-8}$. (Adapted from Alves and Dilão, 2006).

To calculate the equilibrium distribution of the gap proteins, we take as initial conditions the equilibrium distribution of the proteins of maternal origin calculated from (8). For example, in Fig. 9, we show the equilibrium profile of the Bicoid protein ($BCD(x)$) along the antero-posterior axis of the *Drosophila* embryo, and calculated from Eq. (8). We show also the initial distribution of bicoid mRNA of maternal origin ($A(x)$), and in gray we depict the experimental data (embryo ab7 from the FlyEx database, Myasnikova *et al.*, 2001).

Taking as initial conditions the equilibrium values of the proteins of maternal origin along the antero-posterior axis of the embryo of *Drosophila*, we can run the *GeNetSim* software package in order to obtain the equilibrium distribution of gap proteins. For example, in Fig. 10, we show the gap proteins expression patterns obtained with *GeNetSim* and compared with experimental data.

The development of *GeNetSim* was determinant for building and testing our model of *Drosophila* gap genes segmentation (Alves and Dilão, 2006). As the most time-consuming tasks are performed automatically by the *GeNetSim* software package, it has been possible to rapidly analyze sev-

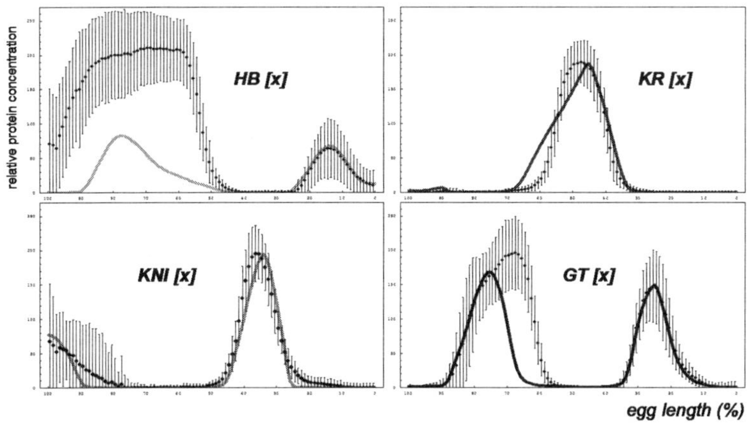

Figure 10. Gap proteins expression patterns in the *wild type* embryo of *Drosophila*, compared with experimental data taken from the FlyEx database, and computed with *GeNetSim*. The experimental integrated data corresponds to the average protein concentrations extracted from the central 10% of y values on the midline of the embryos, in the A-P direction (x coordinate) and registered by the quadratic spline approximation method. The error bars represent standard deviations. After less than 100 time steps the protein gradients have already stabilized. For the comparison with the experimental data, the following proportionality constants were used for the display of the model results: $Hb: 31$; $Kr: 57$; $Kni: 59$; $Gt: 46$. (Adapted from Alves and Dilão, 2006).

eral network configurations, parameter values and other model conditions.

References

Altschul S.F. , Boguski M. S. , Gish W., Wootton J. C. (1994). Issues in searching molecular sequence databases. *Nat Genet* 6(2):119-29.

Alves, F., Dilão, R. (2005) A simple framework to describe the regulation of gene expression in prokaryotes. *C. R. Biologies* 328 (2005) 429444.

Alves, F., Dilão, R. (2006) Modeling segmental patterning in Drosophila: maternal and gap genes. *J. Theor. Biol.*, in press.

Anderle P., Duval M., Draghici S., Kuklin A., Littlejohn T. G., Medrano J. F., Vilanova D., Roberts M. A. (2003). Gene expression databases and data mining. *Biotechniques* Mar; Suppl: 36-44.

Hirano H., Islam N., Kawasaki H. (2004). Technical aspects of functional proteomics in plants. *Phytochemistry* 65 (11): 1487-98.

Huang S. (2004). Back to the biology in systems biology: what can we learn from biomolecular networks? *Brief Funct Genomic Proteomic* 2 (4): 279-97.

Jacob F., Monod, J. 1961. Genetic regulatory mechanisms in the synthesis of proteins, *J. Mol. Biol.* 3 (1961) 318-356.

Jaeger J., Blagov M., Kosman D., Kozlov K. N., Manu, Myasnikova E., Surkova S., Vanario-Alonso C. E., Samsonova M., Sharp D. H., Reinitz J. (2004). Dynamical analysis of regulatory interactions in the gap gene system of Drosophila melanogaster. *Genetics* 167 (4): 1721-37.

Jaeger J., Surkova S., Blagov M., Janssens H., Kosman D., Kozlov K. N., Manu, Myasnikova E., Vanario-Alonso C. E., Samsonova M., Sharp D. H., Reinitz J. (2004). Dynamic control of positional information in the early Drosophila embryo. *Nature* 430 (6997): 368-71.

Janssens H., Kosman D., Vanario-Alonso C. E., Jaeger J., Samsonova M., Reinitz J. (2005). A high-throughput method for quantifying gene expression data from early Drosophila embryos. *Dev. Genes Evol.* 215 (7): 374-81.

Kosman D., Reinitz J., Sharp D. H. (1998). Automated assay of gene expression at cellular resolution. *Pac. Symp. Biocomput.* 3: 6-17.

Kumar S., Jayaraman K., Panchanathan S., Gurunathan R., Marti-Subirana A., Newfeld S. J. (2002). BEST: a novel computational approach for comparing gene expression patterns from early stages of Drosophila melanogaster development. *Genetics* 162 (4): 2037-47.

Meir E., Munro E. M., Odell G. M., Von Dassow G. (2002). Ingeneue: a versatile tool for reconstituting genetic networks, with examples from the segment polarity network. *J. Exp. Zool* . 294 (3): 216-51.

Myasnikova, E., Samsonova, A., Kozlov, K., Samsonova, M., Reinitz, J., 2001. Registration of the expression patterns of *Drosophila* segmentation genes by two independent methods. *Bioinformatics* 17(1), 3-12.

Nüsslein-Volhard, C., 1992. Gradients That Organize Embryo Development. *Scientific American* 275 (2), 54-61.

Price N. D., Reed J. L., Palsson B. O. (2004). Genome-scale models of microbial cells: evaluating the consequences of constraints. *Nat. Rev. Microbiol.* 2 (11): 886-97.

von Dassow G., Meir E., Munro E. M. , Odell G. M. (2000). The segment polarity network is a robust developmental module. *Nature* 406 (6792): 188-92.

Xing B., van der Laan M. J. (2005). A causal inference approach for constructing transcriptional regulatory networks. *Bioinformatics* 05 Aug 30 [Epub ahead of print].

THE MITOCHONDRIAL EVE IN AN EXPONENTIALLY GROWING POPULATION AND A CRITIQUE TO THE OUT OF AFRICA MODEL FOR HUMAN EVOLUTION

ARMANDO G. M. NEVES AND CARLOS H. C. MOREIRA

UFMG, Departamento de Matematica,
Caixa Postal 702, 30123-970 - B. Horizonte - MG, Brazil.
E-mail: aneves@mat.ufmg.br, cmoreira@mat.ufmg.br

We show that a mitochondrial Eve may exist in a population with exponential growth on average, if the growth rate is restricted to a suitable range. As a consequence, neither a population bottleneck, nor a long period of constant population are needed in order to explain the experimentally verified existence of a mitochondrial Eve. Our conclusions are based on a percolation model for mitochondrial DNA inheritance equivalent to the Galton-Watson branching process and are made independent of the particular probability distribution for the number of children by means of an optimization argument over all reasonable distributions.

1. Introduction

The existence of an african mitochondrial Eve[1] is considered as the strongest evidence for the Out of Africa model for human evolution[2]. This model, nowadays dominant among anthropologists, proposes that modern humans, *Homo sapiens*, evolved once in Africa and subsequently colonized the rest of the world replacing, *without mixing with them*, archaic forms such as the *Homo neanderthalensis* and the *Homo erectus*, which had already spread to other continents.

At first geneticists used to take the existence of a mitochondrial Eve as the consequence of a population bottleneck at early times in the history of our species. Avise, Neigel and Arnold[3] proved that a strict bottleneck was not necessary, because extinction of all but one mitochondrial DNA lineages would also probably occur in populations of constant size.

We show that even in an exponentially growing population we may have the necessary lineage extinctions in realistic time and sizeable probability, provided that the average number of children per individual is constrained to a small range. If the average number of children per individual is too small, probably no mithocondrial lineage will survive. If it is too large, probably more than one lineage will survive. If the initial population were around 10,000 individuals[3,4] and the sex ratio fixed at its actual present value[5] of 100 women per 105 men (probability 0.488 of a child being a female), we will show that in a certain sense the range of values for the average number of children compatible with the existence of a mitochondrial Eve is from 2.0492 to 2.0510. This range is obtained without any strong hypothesis on the probability distribution for the number of children per individual.

Our results[6] were obtained by studying mitochondrial DNA inheritance as a percolation model equivalent to the well-known Galton-Watson branching process and an optimization argument over all possible probability distributions for the number of children per individual. Care was also taken in studying the time scale for the lineage extinctions and verifying it is compatible with the time of existence of our species.

Opposed to the Out of Africa model, the Multiregional Evolution hypothesis[7] proposes that modern humans evolved in parallel at many places with occasional genetic exchange among regions in order to maintain uniqueness of the species. In particular, in the multiregional evolution model, modern humans are thought to be descendants of africans as well as of neandertals and other local ancient forms. If a mitochondrial Eve is allowed to exist in a growing population, we do not see why the multiregional evolution hypothesis must be ruled out.

2. Percolation and mitochondrial DNA basics

2.1. *Percolation*

Percolation originally meant the passage of water (or some other fluid) through the soil (or some other porous medium). In this paper we use the same word in the generalized sense probabilists gave to it,[8] i.e. a class of models to describe fluid percolation through a porous medium, in which the medium is described as an infinite set of points (vertices) linked by edges. Edges may be either open or closed to the passage of the fluid according to some statistical law. Typically, but not in this paper, vertices are considered to be the points in the d-dimensional square lattice $\mathbb{Z}^d$ and edges open with probability p and closed with probability $1-p$, independently. Besides fluid percolation, percolation models have also found other applications, such as the spread of diseases in a population or disordered electrical networks.[9]

One is usually interested in calculating the percolation probability, defined as follows. Choose some vertex O to be the origin and some configuration of open and closed edges. Consider the cluster of all vertices connected to the origin by some path of open edges. We say that this configuration percolates if the cluster of vertices connected to O is infinite. The percolation probability is defined as the probability $\theta(p)$ of the percolation event over all configurations.

Percolation models attract so much attention because $\theta(p)$ usually exhibits a "second-order phase transition": there exists a percolation threshold p_c such that $\theta(p)$ is strictly zero for $p < p_c$ and positive for $p > p_c$. They look thus like toy models for Statistical Mechanics and in fact many techniques developed in percolation may be adapted to Statistical Mechanics and vice-versa. But percolation models may be not at all trivial.[9]

As the reader is going to see in a while, the percolation model we shall develop for the existence of a mitochondrial Eve is one in which the medium is a genealogical tree, the edges are parental relationships and the "fluid" is mitochondrial DNA (mtDNA). Before we say more on that, let us also review the basic facts about mtDNA and the mitochondrial Eve.

2.2. *Mitochondrial DNA*

Although most of the genetic information in higher animals is located at cells' nuclei, some DNA may be found in the subcellular organelles called mitochondria. These structures are present in large numbers in nearly every cell and play a key role in metabolism. Mitochondrial DNA is very short; in humans it consists in 16,569 base pairs carrying the information for only 37 genes. despite that, mtDNA of humans and many other species has been the object of recent intensive research,[10] both for its availability (because mitochondria are so numerous) and peculiar inheritance mechanism. Unlike nuclear DNA, which is inherited in equal parts from mother and father, mtDNA is inherited only from the mother. So, in the absence of mutations, the mtDNA of an individual would be identical to the mtDNA of a single ancestor out of his or her 2^n ancestors n generations before, namely the mother of the mother ... of his or her mother.

This simple feature allows one to use mtDNA comparisons among living individuals to look far back in time and draw conclusions about the separation of subpopulations in one species or the speciation process, in which several extant species may descend from one extinct species.[10]

An experiment performed in the late 80's brought press popularity to mtDNA. By examining mtDNA of 147 living humans and taking into account mutations, Cann, Stoneking and Wilson asserted[1] that mtDNA of all living humans could be described as mutations in the mtDNA of a single woman. As we would all be her descendants, this woman was called the *mitochondrial Eve.* By using known mutation rates and exploiting geographical correlations, it could be inferred that the mitochondrial Eve has lived in Africa more or less 200,000 years ago.

The mitochondrial Eve should not be confused with the biblical Eve. Unlike the latter, the mitochondrial Eve is not supposed to be the only woman living at her time. Many other men and women contemporary to her have probably left traces of their nuclear DNA in modern humans, see e.g. Ref. 11 or Ref. 12 for proofs of this in the case of a population with a fixed size. But they left no trace of their mtDNA.

The time and the place in which the mitochondrial Eve lived are considered a strong evidence for the *Out of Africa* model for the origin of our own species.[2] In order to explain why all mtDNA lineages stemming from women coeval to the mitochondrial Eve were extinct, Brown proposed[13] that a severe population bottleneck, in which human population dropped to only a few individuals, must have existed after the mitochondrial Eve. All nowadays living humans would be descendants of these few individuals.

There is a large amount of paleoanthropologic and archaeologic evidence that modern humans coexisted with neandertals in Europe for at least 10,000 years. It is also believed that modern humans coexisted in other parts of the world with other ancient human forms. In particular, supporters of the Out of Africa model assert that there was no mixing among modern and ancient humans, because if some had existed, then we would see its traces in the mtDNA of living humans.

We believe that a bottleneck is a very strong hypothesis, considering that human population, at least in historical times, has been steadily growing, and that achievement of important technological developments in prehistory made it possible for humans to spread all over the world. Avise, Neigel and Arnold[3] argued that a prehistorical population bottleneck is not strictly necessary by showing that stochastic mtDNA lineage extinction can be rapid enough even in *stable-sized* populations.

In the next sections, we will show that a bottleneck is not at all necessary. In fact, we will see that the existence of a mitochondrial Eve can be explained even if the population *grows exponentially*, provided that the growth rate is neither too large, nor too small. If the growth rate of the population is restricted to the correct range, there might have been mixing among descendants of africans and other "ancient" forms and we would still have no sign of that in mtDNA of living humans.

Although our model and formalism are similar to the ones in Ref. 3, we show that one solution discarded there is biologically plausible. We are also able to generalize their methods to arbitrary progeny distributions and to display the model explicitly in the language of an exactly-solvable percolation problem.

3. Percolation model for mtDNA inheritance

The assumptions in our model for mtDNA inheritance are the following:

(A1) Generations are nonoverlapping.
(A2) The numbers of children for each individual are statistically independent and identically distributed random variables assuming value $r \in \{0, 1, 2, \dots\}$ with probability Q_r. The values for the Q_r's are time- and population size-independent. We shall often refer to the *progeny distribution* as the probability distribution for the number of children of individuals.
(A3) A newborn child is a female with probability p and a male with probability $1 - p$. The value for p is also time- and population size-independent.
(A4) There always exist males enough to mate with all females.

Assumption (A1) grants some formal simplicity to the model. (A2) disregards interactions that might come from a number of different sources such as fertility correlations among members of a family, competition (for food supplies, mating partners, etc), cooperation and geographical aspects. One natural attempt to improve the model would be to make the Q_r's dependent on the total population, thus accounting for saturation effects. Although feasible in computer simulations, that would ruin the linearity on which our theoretical analysis relies. The time independence of the Q_r's is also questionable because it disregards the effects of climatic changes. However this assumption may be adequate over an initial period of time long enough to produce most of the lineage extinctions. (A3) adds some generality to the model with respect to the one in Ref. 3, in which $p = 1/2$. Usually p varies from species to species and even for present modern humans,[5] $p \approx 0.488$, strictly less that $1/2$. This fact will be of quantitative relevance in our results. (A4) is assumed, since we do not keep track of the male population. It is a reasonable assumption because males from all concurrent mitochondrial lineages may participate in any single one, without interfering in the mitochondrial inheritance. More on that may be found in a more complete analysis[6] of the model, including men.

Consider now the genealogic tree of an ancestral woman constructed according to the above assumptions. By genealogic tree we mean the graph obtained by considering as vertices the ancestral herself, located at the origin O and all her descendants after an *infinite* number of generations, drawing edges joining each father or mother to their children of either sex. Define as *open* any edge linking a mother to her children and *closed* all other edges. It is clear that the mtDNA lineage of the woman will survive if and only if the configuration of open edges percolates. Therefore mtDNA inheritance may be posed as a problem of edge percolation in a tree graph, in which edges are open with probability p and closed with probability $1-p$. Unfortunately, unlike most percolation models, edges are not statistically independent because any child with both parents in the graph must appear twice, separately linked to each of the parents. An example is shown in Fig. 1(a).

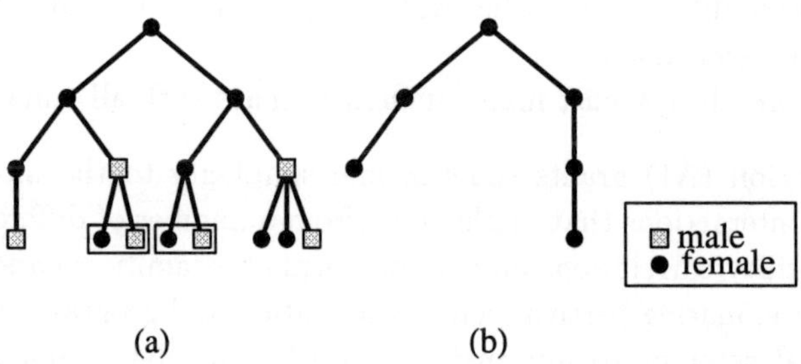

Figure 1. (a) Example of a genealogic tree of the type considered in this paper. The individuals enclosed in the rectangle have both their father and mother in the tree, exemplifying the possibility of statistically dependent edges.
(b) The female genealogic tree corresponding to the complete tree in (a).

In order to overcome the dependence problem, we define the female genealogic tree (FGT) as the tree obtained by deleting all males and their descendants in the complete tree, see Fig. 1(b). In the FGT all edges are open and statistically independent. Percolation in the complete genealogic tree is equivalent to the corresponding FGT being infinite generations long.

If we denote as q_r, $r = 0, 1, 2, \ldots$, the probability that an individual has r children of female sex, then

$$q_r = \sum_{k=r}^{\infty} Q_k \binom{k}{r} p^r (1-p)^{k-r} . \tag{1}$$

The process of growing the FGT under the above assumptions, each female vertex originating r branches with probability q_r, is known as the Galton-Watson process.[14] It was originally introduced in 19th century to study the problem of extinction of family names, with the q_r meaning the probability of a man having r male children.

Let $\theta_n(q_0, q_1, \ldots)$ be probability that the FGT is at least n generations long. We denote by $\overline{\theta}_n \equiv 1 - \theta_n$ the probability that the FGT is long $n-1$ generations or less. The end of the FGT in $n-1$ or less generations may happen either if the woman in its origin has no daughters (with probability q_0), or if she has b daughters, $b = 1, 2, \ldots$, each of them having a female genealogy at most $n-2$ generations long (with probability $q_b \, \overline{\theta}_{n-1}^{\,b}$). Thus,

$$\overline{\theta}_n = S(\overline{\theta}_{n-1}) , \tag{2}$$

where

$$S(x) = q_0 + q_1 x + q_2 x^2 + \ldots \tag{3}$$

is the generating function for the probability distribution of the number of daughters. The initial condition to be used in conjunction with (2) is $\overline{\theta}_1 = q_0$.

The normalization of the q_r's implies that 1 is a fixed point of S. Since S is non-decreasing with all derivatives non-decreasing in $[0, 1]$, it will have a second fixed point $\overline{\theta} \in [0, 1)$ if and only if $S'(1) > 1$. There are no other fixed points in $[0, 1]$. Regarding attractiveness,[15] we may have three different regimes:

(i) If $S'(1) < 1$, then 1 is the only fixed point and it is attractive.
(ii) If $S'(1) > 1$, then 1 is a repulsive fixed point, whereas $\overline{\theta}$ is attractive.
(iii) If $S'(1) = 1$, then 1 is again the only fixed point in $[0, 1]$ and it is *weakly* attractive.

If $\Theta \equiv \lim_{n\to\infty} \theta_n$ is the percolation probability, then

$$\Theta = \begin{cases} 0, \text{ if } \ S'(1) \leq 1 \\ \theta, \text{ if } \ S'(1) > 1 \end{cases} , \tag{4}$$

where $\theta \equiv 1 - \overline{\theta} \in (0,1]$. By writing $S'(1) = \sum_{r=1}^{\infty} r\, q_r$ and using (1) we get that condition $S'(1) > 1$ may be written as $p > p_c$. The percolation threshold p_c is

$$p_c = \frac{1}{\overline{N}}\,, \tag{5}$$

with

$$\overline{N} = \sum_{r=1}^{\infty} r\, Q_r \tag{6}$$

meaning the average number of children (of either sex) per woman. Regime (i) is then the subcritical percolation regime (i.e. $p < p_c$), (ii) is supercritical ($p > p_c$) and (iii) is critical ($p = p_c$).

The value of θ in terms of the q_r's may be obtained by solving $S(\overline{\theta}) = \overline{\theta}$. For $p > p_c$ and p close enough to p_c, we introduce what we shall call *small survival probability approximation.* In this case, an approximate value θ_a for θ is obtained by replacing S by its Taylor polynomial of degree 2 around 1. We find

$$\theta \approx \theta_a = \frac{2(p - p_c)}{p_c\, p^2 \sum_{r=2}^{\infty} r(r-1)\, Q_r}\,. \tag{7}$$

From the non-negativity of S''', it follows that θ_a is actually a lower bound for θ.

The speed of convergence of θ_n to Θ, a useful parameter for simulation purposes, follows from the mean value theorem of Calculus. In regimes (i) and (ii) $\theta_n - \Theta \sim e^{-n/\xi}$ for large n, where, as usual in Statistical Mechanics, the *correlation time*

$$\xi = -1/\ln[S'(1-\Theta)] \tag{8}$$

diverges when $p \to p_c$. In the critical case (iii) exponential convergence is replaced by a much slower power law: $\theta_n \sim \frac{2}{S''(1)\,n}$ for large n.

4. Demographic considerations

Let W be the number of women coeval to the mitochondrial Eve. If different lineages may be considered as independent, then the number r of lineages remaining after n generations is a binomially distributed random variable with

$$\text{Prob}\{r = m\} = \binom{W}{m} \theta_n^m (1-\theta_n)^{W-m}\,. \tag{9}$$

This approach was used in Ref. 3, where the authors concentrated on the probability Π_n for the survival of two or more lineages after n generations. For some specific progeny distributions they found out that in the supercritical regime Π_n tends to a positive value as $n \to \infty$; they considered this to be incompatible with a mitochondrial Eve. The subcritical regime was also discarded because it leads to quick extinction. Instead, the critical regime was selected because with $W = 1,000 \sim 10,000$ it leads to Π_n approaching zero in $n \approx 10^4$ generations. Such values for W and n are well within the range expected by geneticists and paleontologists. For $p = 1/2$, as used in Ref. 3, the critical regime yields $\overline{N} = 2$ and can only account for a stable-sized population.

We shall now argue that for a range of values of $\overline{N}$, the supercritical regime also provides a biologically plausible solution for the existence of a mitochondrial Eve in a growing population. We shall use $W = 5,000$ and $n = 10^4$, as suggested for example in Refs. 3, 4, and, for the sex ratio, $p = 0.488$ as in Ref. 5, assuming that this present value can be extrapolated to the times of early mankind.

From (9) we get that the expected number of surviving lineages after n generations is $W\theta_n$. To be consistent with the existence of a mitochondrial Eve as an event with a not too small probability, this number must not be much smaller or much larger than 1. For illustration purpose, we take it between 1/2 and 2, implying θ_n between $1/(2W)$ and $2/W$. In order to estimate the range of values for $\overline{N}$ consistent with that, we first replace θ_n for θ_a given in (7). This approximation is valid as long as two conditions are fulfilled. First, the number of generations n must be so large that θ_n is close to θ. Finally, the small survival probability approximation must hold in order that θ is close to θ_a. The latter is true, because θ is of order W^{-1}, a small number if $W = 5,000$. The former, $\theta_n \approx \theta$, will be justified soon.

Also, for practical purposes we may truncate the series in the right-hand side of (7) at some order M (limited progeny assumption). In order to calculate the maximum and minimum values of θ_a for given $\overline{N}$, we must respectively minimize or maximize the linear function $\sum_{r=2}^{M} r(r-1)\, Q_r$ under the constraints $\sum_{r=1}^{M} r\, Q_r = \overline{N}$, $\sum_{r=1}^{M} Q_r = 1$ and $0 \leq Q_r \leq 1$ for $r = 0, 1, \ldots, M$. This is an exactly solvable optimization problem (linear programming).

In Fig. 2 we show a plot of the maximum and minimum values for θ_a as functions of $\overline{N}$, taking $M = 10$. The values for $\overline{N}$ consistent with existence of a mitochondrial Eve as a not very rare event are thus such that the maximum θ_a lies over $1/(2W)$, which yields $\overline{N} > 2.0492$, and the minimum

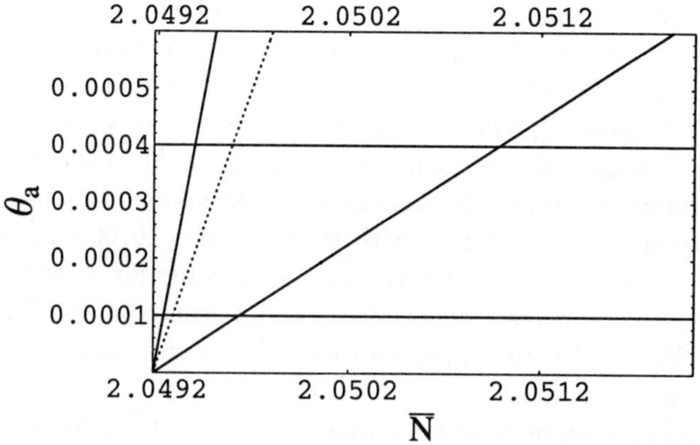

Figure 2. Continuous lines correspond to the maximum and minimum values for θ_a as functions of $\overline{N}$ with $M = 10$. The dotted line is the value of θ as a function of $\overline{N}$ for a Poisson distribution for the progeny. Horizontal lines correspond to the values $1/(2W)$ and $2/W$ for θ_a.

lies below $2/W$, which yields $\overline{N} < 2.0510$.

The value $M = 10$ for the maximum number of children, although reasonable, is arbitrary. However it turns out that the maximum values for θ_a become independent of M for $M \geq 3$ and are obtained with $Q_r \neq 0$ only for $r = 2$ or 3. On the other hand, the minimum values for θ_a do depend on M and tend to 0 as $M \to \infty$. Nonetheless, this minimum value is attained when $Q_r = 0$, $r = 0, 1, 2, \ldots, M - 1$ and $Q_M = \overline{N}/M$, which is quite an artificial probability distribution for the number of children. As imposing realistic conditions on the Q_r's would constrain even more the average number of children consistent with existence of a mitochondrial Eve, we believe that taking $M = 10$ is not a serious limitation in our method. This point is further illustrated in Figs. 2 and 3, in which a Poisson distribution is considered as an example of a realistic distribution for the number of children. Even though more than 10 children are allowed in a Poisson distribution, it follows from Figs. 2 and 3 that the range of values for $\overline{N}$ obtained by using the arbitrary value $M = 10$ is also consistent with a Poisson distribution.

Results plotted in Fig. 2 assume, as already mentioned, that θ_n can be approximated by its limit θ when $n \to \infty$. In other words, the results will be valid provided $n \gg \xi$, where the correlation time ξ is given by (8). In Ref. 6 we prove that in the small survival approximation, ξ depends on the

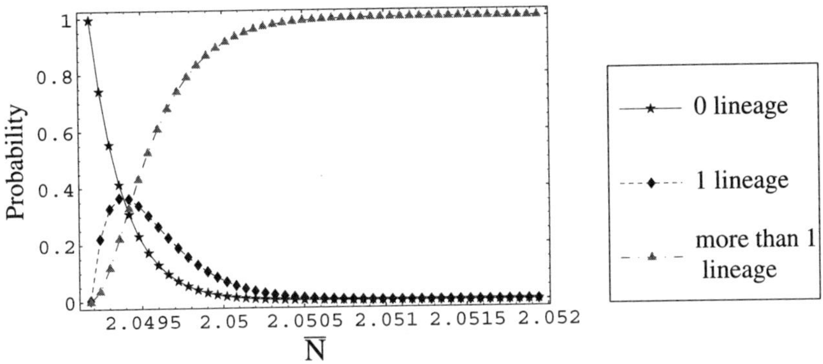

Figure 3. Probability of survival for l mtDNA lineages as a function of the mean number of children $\overline{N}$, for $l = 0$, $l = 1$ and $l > 1$. We considered here that the progeny distribution is the Poisson distribution.

mean of the progeny distribution, but not on its variance, $\xi \approx 1/|p\overline{N} - 1|$. As a consequence, in the range of values for $\overline{N}$ in which we are interested, the graphs of ξ as a function of $\overline{N}$ for various progeny distributions, see Fig. 4, are indistiguishable. In particular, for the largest value $\overline{N} = 2.0510$

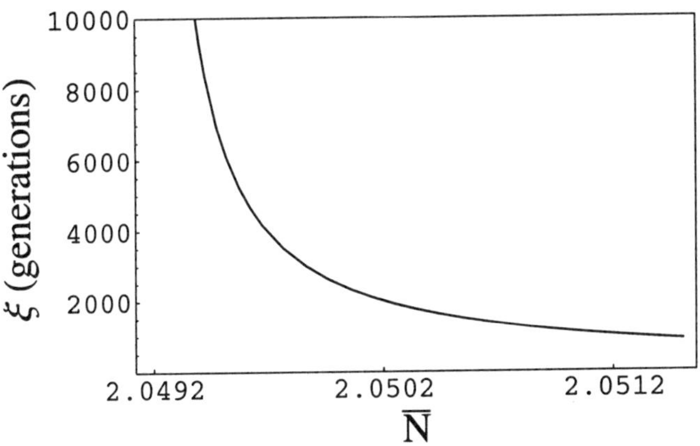

Figure 4. Correlation time ξ as a function of the mean number of children $\overline{N}$. The curve shown is for a Poisson progeny distribution, but the corresponding curve for other distributions is indistinguishable.

compatible with the existence of the mitochondrial Eve, ξ is of the order of

1,100 generations. As geneticists assume that the mitochondrial Eve lived more or less 10,000 generations ago, the approximation of using θ for θ_n is justified for $\overline{N} = 2.0510$. On the other hand, ξ diverges when $\overline{N}$ approaches $1/p$. This means that the range of values for $\overline{N}$ compatible with the Eve should be extended down to $1/p$.

5. Conclusions

We have proposed and solved a percolation model for mtDNA inheritance and lineage extinction compatible with the existence of a mitochondrial Eve in a growing population. According to this model, a mitochondrial Eve is likely only if the mean number of children per individual is constrained to a narrow range.

Although the values in the range we found may seem quite close to 2, so that population looks as if very nearly constant, it is not so. The reader may easily calculate that an initial population of 5,000 women having a fixed number of $2.0510 \cdot 0.488 \approx 1.0009$ female children each will grow to more than 70 million individuals (male and female) in the 10,000 generations separating us from the mitochondrial Eve. Although present human population is 6.4 billion, the roughness of the model and the explosive populational growth in the last few generations may account for the difference.

The exact values for the range of $\overline{N}$ compatible with the mitochondrial Eve depend critically on the sex ratio p and on the number of women W at the time of the mitochondrial Eve, but not on the number of generations since then and on details of the progeny distribution. Of course, our model should be thought of as a simplified version of reality. We think it is a valuable starting point for further work, including simulations for other models[16,17,18].

One tacit assumption in our analysis, supported by biologists, is that all W original mtDNA lineages are equally fit, i.e. there is no natural selection acting on lineage sorting. It also follows from our results, see Fig. 3, that within the range of values for $\overline{N}$ in which a mitochondrial Eve is likely, there is a probability of at least 63% that the number of surviving mtDNA lineages is different from 1. As the probability that this number is 0 is not negligible, we may explain extinction of other hominid species which had existed for some periods, if they had demography similar to our own.

Although we have no argument for deciding between the Out of Africa and Multiregional Evolution models, our result on the possible existence of a mitochondrial Eve in an exponentially growing population shows clearly that a population bottleneck is not necessary in order to produce mtDNA lineage extinctions. So, even if relevant genetic mixing has occurred among early africans and other ancient human forms, e.g. neandertals, a probable outcome of it would still be extinction of all but one mtDNA lineages. We understand thus that the Multiregional Evolution model is compatible with the experimentally verified existence of a mitochondrial Eve.

References

1. Cann, R. L., Stoneking, M., Wilson, A. C. (1987), "Mitochondrial DNA and human evolution", *Nature 325*, 31.
2. Wilson, A. C., Cann, R. L. (1992), "The Recent African Genesis of Humans", *Sci. Am. 266 (4)*, 68.
3. Avise, J. C., Neigel, J. E., Arnold, J. (1984), "Demographic influences on mitochondrial DNA lineage survivorship in animal populations", *J. Molec. Evol. 20*, 99.
4. Takahata, N. (1993), "Allelic Genealogy and Human Evolution", *Mol. Biol. Evol. 10 (1)*, 2.
5. Newell, C. (1988), "Methods and models in Demography", John Wiley and Sons, Chichester.
6. Neves, A. G. M., Moreira, C. H. C. (2005), "Applications of the Galton-Watson process to human DNA evolution and demography", *to appear in Physica A*.
7. Thorne, A. G., Wolpoff, M. H. (1992), "The Multiregional Evolution of Humans", *Sci. Am. 266 (4)*, 76.
8. Broadbent, S. R., Hammersley, J. M. (1957), Percolation processes I. Crystals and mazes, *P. Camb. Philos. Soc. 53* 629.
9. Grimmett G. (1999), "Percolation", 2nd edition, Springer-Verlag, Berlin.
10. Avise J. C. (1986), "Mitochondrial DNA and the evolutionary genetics of higher animals", *Phil. Trans. R. Soc. Lond. B 312*, 325.
11. Derrida, B., Manrubia, S. C., Zanette D. H. (1999), "Statistical Properties of Genealogical Trees", *Phys. Rev. Lett. 82*, 1987.
12. Chang J. T. (1999), "Recent Common Ancestors of All Present-Day Individuals", *Adv. Appl. Probab. 31*, 1002.
13. Brown, W. M. (1980), "Polymorphism in mitochondrial DNA of humans as revealed by restriction endonuclease analysis", *Proc. Natl. Acad. Sci USA 77*, 3605.
14. Harris, T. E. (1989), "The Theory of Branching Processes", Dover, New York.
15. Devaney, R. L. (1992), "A First Course in Chaotic Dynamical Systems", Addison Wesley, Reading.
16. Oliveira, P. M. C., Oliveira, S. M., Radomski, K. P. (2001), "Simulating the mitochondrial DNA inheritance", *Theory Biosci. 120*,77.
17. Oliveira, P. M. C. (2002), "Evolutionary computer simulations", *Physica A 306*, 351.

18. Rohde, D. L. T., Olson, S., Chang, J. T. (2004), "Modelling the recent common ancestry of all living humans", *Nature 431*, 562.

A NEUROCOMPUTATIONAL MODEL OF THE ROLE OF CHOLESTEROL IN THE DEVELOPMENT PROCESS OF ALZHEIMER'S DISEASE*

GIZELLE K. VIANNA, ARTUR EMÍLIO S. REIS AND LUIS ALFREDO V. CARVALHO

COPPE - Universidade Federal do Rio de Janeiro
CP: 6851, CEP: 21945-970, Rio de Janeiro, Brasil.
E-mail: gkupac@gmail.com; artur_reis@yahoo.com.br; MeuCorreioEletronico@Gmail.com

This work presents a mathematical-computational model of the development process of Alzheimer's disease, based on the assumption that cholesterol plays a key role in the formation of hallmark neuropathological lesions that characterize the disease: the senile amyloid plaques and neurofibrillary tangles. The final model, conceived as a system of equations, was implemented as a computer program and, thereafter, two sets of tests were carried out. In the first set of tests, aimed at validating the model, the results obtained from the simulations carried out were qualitatively coherent with in vivo or in vitro experiments found in the consulted literature. In the second set, we performed simulations in order to test a number of hypotheses about the development process of the disease, collected from the literature but yet without experimental confirmation. From the results of these simulations, it was possible to validate those hypotheses and to draw some conclusions about the development process of the disease.

1. Introduction

In silico systems consist of a set of computer programs that provide a useful platform for the testing of hypotheses in biological systems, and may also represent an important tool in the research of specific drugs for the treatment of many diseases. The present work proposes the construction of a mathematical-computational simulation model that represents the physiological alterations resulting from the *Alzheimer's Disease* (AD). It is difficult to achieve a good understanding of the development process of the pathology of AD, since it is generated by a combination of highly connected

*This work has been supported by CAPES.

processes. Therefore, the model built aims at facilitating the comprehension of this disease, that may be considered, due to its characteristics, as an adaptative complex system. Besides attempting to translate the main mechanisms of the disease, the present work is also intended to test some of the hypotheses collected from the literature.

2. The Alzheimer's Disease

During the last century, an important increase in life expectancy has been observed in many countries. The increase in life expectancy of the world population has brought along with it a new and unexpected epidemic: the senile dementia. Of all old-age associated dementias, AD is the most common, affecting almost 7% of persons over 65 years and up to 40% of the persons over 80 years of age. It is estimated that this dementia is responsible almost 70% of all senile dementias, and it represents second most common cause of old-age associated psychological disturbances, after depression[1].

AD consists of a degenerative process, characterized by the occurrence of a series of abnormalities in the brain, selectively affecting neurons of specific regions, such as the cortex and the hippocampus. In AD, microscopic exams reveal a great amount of two physical alterations that characterize the disease and distinguish it from any other diseases: the extracellular senile amyloid plaques (SAP's) and the intracellular neurofibrillary tangles (NT's). SAP's consist of deposits of amyloid β-peptide (Aβ), surrounded by dystrophic axons. NT's contain paired helical filament, composed of hyperphosphorylated forms of Tau protein[2]. Such alterations hinder the functioning of the synapses and the feasibility of neurons, leading to neuronal loss[1].

The complexity of the disease's etiology makes it difficult, or even impossible, to establish a sequence at which events take place, since the development of the disease may occur in different ways in every other individual[3,4]. There are countless hypotheses raised about the mechanisms that may cause the neuronal loss as a consequence of Alzheimer's disease, although none of them stand alone as the sole cause of the disease. Presently, some hypotheses have received greater acceptance, and are being more intensely debated.

Age is the most relevant risk factor for AD. Many physical alterations take place in the brain as an individual grows older: its weight decreases, certain populations of neurons are reduced by cellular demise and the synthesis of neurotransmitters also diminishes. The aging process is yet little understood, but the theory about the oxidative stress conceives it as a result of the accumulation of damages to the tissues caused by free radicals[5]. Some epidemiologic studies also suggest an influence displayed by the cholesterol in the development of AD, and that high plasmatic cholesterol levels increase the risk of developing the disease[6,7,8,9]. Also, there are evidences showing that AD and vascular diseases have some risk factors in common[4]. Furthermore, SAP's, apoptosis, and high levels of free radicals are common events in patients with AD and with coronary artery disease[17].

In short, it can be sustained that the main risk factors for the development of the disease are: the old age, the high level of cholesterol and some genetic components[a]. Therefore, it can be assumed that these risk factors will somehow affect relevant functioning processes of the brain which, once altered, will provide grounds for the appearance of SAP's and NT's and an increase in neuronal loss.

3. The Metabolism of the Cholesterol

There is no consensus about the relationship between the levels of cholesterol in the plasma and in the central nervous system (CNS), in patients with AD. Several researches, conducted under different approaches, have reached divergent conclusions about the effects of cholesterol in the brain. This uncertainty is given to the fact that plasma cholesterol and the brain cholesterol are located in two different depositories, separated by the blood-brain barrier (BBB), that controls the exchange of substances between the fluids of the brain and the blood[10]. The increase of the level of cholesterol in the CNS, as a consequence of the increase of the cholesterol in the plasma, is very light and, as a matter of fact, does not show statistic relevance. However, the level of cholesterol in the frontal cortex demonstrates a more significant increase[11].

The apolipoproteins are lipid carrier molecules that transport cholesterol and fat acids through the circulatory system and through the brain. The apoE is a major apolipoprotein in the CNS, where it coordinates the cholesterol redistribution during the remodeling of the membrane, associated to the plasticity of the synapses[12,13,11]. The gene for the apoE, codified in the chromosome 19, possesses three main alleles $\varepsilon 2$, $\varepsilon 3$ e $\varepsilon 4$, which codify,

---

[a]Among the latter, we will take into consideration only the $\varepsilon 2$, $\varepsilon 3$ e $\varepsilon 4$ alleles, since the focus of the present work is the study of the role of the cholesterol in AD. The study of other genetic factors involved in the pathology of AD remains outside the scope of the present work.

respectively, the apoE isoforms E2, E3, and E4. The allele $\varepsilon 4$ is considered to be a risk factor for AD: the higher the frequency of $\varepsilon 4$, the greater the risk of AD, though the mechanism that leads to the increase of such risk in not known yet[4].

The brain manages to maintain the cholesterol at regular levels. When the level of cholesterol increases to its saturation point, one alternative via of extrusion is turned on. That occurs by the conversion of the cholesterol into a neurotoxic substance, the 24s-hidroxicolesterol (C24S), which results from the altered metabolism of the cholesterol, being the main product of its excess elimination from the brain. When converted into C24S, the cholesterol is capable of crossing the BBB by its own means. In fact, an increase in the levels of this substance in the patients with AD and with vascular dementia has been observed[13,8,14]. Therefore, the production of C24S is directly proportional to the cholesterol in the brain (colCNS). Also, we have considered, for the purpose of building the model, that the conversion into C24S will only occur when there are no apoE available, as it is responsible for cholesterol transportation to the outside of the brain.

In order to avoid the need of inserting a modeling of the metabolism of the cholesterol in the human body, the built model will consider, by simplification, that the amount of cholesterol that enters the CNS (colIN) at a given moment is directly proportional to the cholesterol ingested on a daily basis (colDiet).

Some alterations observed in the functioning of the enzymes of the BBB of animals submitted to the hypercholesterolemia diet have raised the suspicion that the increase of the cholesterol affects somehow the functioning of the barrier[15]. These alterations can be observed close to SAP's, following an increase in the influx of cholesterol[16]. Thus:

$$[colIN] = a_{11} \cdot [colDiet] + b_{11} \cdot [SAP] \tag{1}$$

$$[C24S] = c_{12} \cdot \frac{[colCNS]}{[apoE]} \tag{2}$$

when [apoE] ≥ 0, and:

$$[C24S] = c_{12} \cdot [colCNS] \tag{3}$$

when there is no apoE available in the CNS.

The balance of cholesterol in the brain (colCNS), at each moment, is given by the difference between the cholesterol that entered the brain, via BHE (colIN), minus the cholesterol that exited it, through the same path (colOUT)[b], linked to the apoE or converted into C24S. Thus:

$$[colCNS] = [colIN] - [colOUT] - [C24S] \tag{4}$$

Apparently, the plasma cholesterol levels regulate the apoE expression in the liver and in the brain. Many evidences collected from *in vivo* and *in vitro* experiments show that the hypercholesterolemia increases the expression of the apoE in the brain[8,11]. Therefore, we have considered that the levels of apoE, in the presence of the $\varepsilon 2$, are higher with the occurrence of the $\varepsilon 4$. Consequently, for one given concentration of cholesterol in the CNS, the transport of cholesterol will be more efficient with $\varepsilon 2$ than with $\varepsilon 4$.

The cholesterol distribution in the MPS's of aged rats is similar to the distribution in rats with deficiency of apoE. As the apoE is synthesized in the astrocytes and transported into the neurons, this transportation can be altered with aging[17,18]. The theory of the oxidative stress sustains that the aging process is the result of damages caused by free radicals[5]. Therefore, the variable *age* can be removed from the model and, instead of using it, we will adopt the premise that the levels of free radicals in the CNS (OS) causes physiological alterations.

The synaptic plasmatic membranes (SPM's) consist of two leaflets: the exofacial leaflets (EL) and the citofacial leaflet (CL). They differ in lipid distribution, electrical charge, fluidity, function and location of lipid rafts. Some factors can alter the lipid distribution among these leaflets, such as the chronic ethanol consumption and aging, which practically duplicate the cholesterol level in the EL. However, the total amount of cholesterol found in the membrane is slightly altered. Many functions of the plasma membrane, such as the receptor-effector coupling, the transportation of ions and the translocation of proteins can be influenced by the transbilayer distribution of lipids between the leaflets[18]. Since a greater portion of the cholesterol in the CNS is located in the neuronal membrane, we can state that the amount of cholesterol in the SPM (colSPM) is directly proportional to the cholesterol in the CNS (colCNS).

This value, however, is constantly altered, since a part of the apoE already synthesized is used to remove the excessive cholesterol from the CNS, in the attempt to maintain the ideal level of this steroid in the brain. The apoE binds to the cholesterol in order to export it through the BBB[19]. Since each type of apoE will present different performances, the cholesterol level

---

[b]See Eq. (6).

will be altered according to the rate described by the following equation:

$$\begin{aligned}\frac{d[colSPM]}{dt} = &- b_{14}[colSPM] \cdot [apoE2] - c_{14} \cdot [colSPM] \cdot [apoE3] \\ &- d_{14} \cdot [colSPM] \cdot [apoE4]\end{aligned} \quad (5)$$

Since the cholesterol exits the CNS (colOUT), via BBB, binded to the apoE[20], we shall simplify the calculation of the total amount of cholesterol exported every instant, considering that all of it is originated in the SPM. Therefore, we can calculate the value of this term, which appears in equation (4), through the equation bellow, which corresponds to the same quantity of cholesterol that exited the membrane:

$$\begin{aligned}[colOUT] = \quad & b_{14} \cdot [colSPM] \cdot [apoE2] + c_{14} \cdot [colSPM] \cdot [apoE4] \\ &+ d_{14} \cdot [colSPM] \cdot [apoE3]\end{aligned} \quad (6)$$

It is known that the CL of young brains contains around 85% of the total amount of cholesterol in the plasma membrane[21]. However, with aging, the distribution between the leaflets tends to be equal, accompanying the increasing of cholesterol amount in the SPM. Moreover, it has been argued that the domains of the cholesterol in the membrane, particularly those of the EL, might be altered in AD, becoming vulnerable to the disturbance caused by the Aβ[18]. Indeed, advanced age and the inheritance of apoE $\varepsilon 4$ allele result in duplication of the amount of cholesterol in EL[14]. Again, we shall assume that is not the age, but the accumulation of Aβ and SAP's and the oxidative stress (OE) that affect the distribution of cholesterol between the CL (colCL) and the EL (colEL), resulting in dysfunctions of membrane and cell. Thus:

$$[colCL] = 0,85 \cdot [colSPM] - e_{14} \cdot [SAP] - f_{14} \cdot [A\beta] - g_{14} \cdot [OS] \quad (7)$$

and

$$[colEL] = [colSPM] - [colCL] \quad (8)$$

4. The Formation Senile Amyloid Plaques

The Aβ is a peptide fragment, 39-42 amino acids long, cleaved from the amyloid precursor protein (APP), which forms an insoluble extracellular deposition. A$\beta$is usually generated by the cells and circulates in extracellular fluids in all individuals throughout life, though any factor that affects the balance between its synthesis and its removal will gradually increase its levels[3]. In AD, the protein Aβ is deposited in the senile plaques and also in walls of the cerebral blood vessels[22].

Experiments with transgenic mice placed on a high cholesterol diet have found a strong positive correlation between the levels of Aβ and the levels of both plasma and CNS total cholesterol[23]. *In vitro* studies show that the increase in cholesterol accelerates the formation of deposits of Aβ and that the opposite also occurs; however, an in-depth comprehension of this has not been achieved yet[22].

All the cleavage points of APP are located in the transversal region of the EL. Since rigidity and fluidity of the membranes are affected by the alteration of its cholesterol content, it is possible that alterations in that content will affect the contact between APP and its secretases, as well as the activity of the secretases, thus increasing the production of Aβ[23,22,11,17].

Only a fraction of plasmatic APP is internalized in the cell[27]. In order to simplify the model, we shall suppose that there is an infinite quantity of APP and that it shall not vary. Thus, we may consider that the production rate of Aβ will depend solely on the cholesterol in the EL (colEL) and shall be directly proportional to it. Lipids damaged by oxidation (OS) may also increase the production of the Aβ fragments[8,24]. Gathering this information, we have:

$$\frac{d[A\beta prod]}{dt} = a_{21} \cdot [colEL] + b_{21} \cdot [OS] \tag{9}$$

The apoE acts on the removal of the circulating Aβ, by cerebral capillary sequestration, and transported outside the brain, through the BBB. This transportation can also occur with the Aβ in the free form, without linking to the apoE. It is not known for certain how much of Aβ will be transported in each case[25,13] but, certainly, the efflux of Aβ performed by the apoE is much more rapid than by its own means[10]. We can observe the presence of Aβ in the walls of the main cerebral vessels of individuals as young as 20 years, suggesting that Aβ is drained throughout life along perivascular pathways. Age changes cerebral arteries leading to loss of elasticity, reducing the pulsation of arteries and decreasing the drainage of Aβ[26]. Once again, we will exchange the variable *age*for the level of oxidation in the brain (OS) and the efflux of Aβ (Aβout) will then vary according to the following equation:

$$\begin{aligned}\frac{d[A\beta out]}{dt} &= a_{22} \cdot [A\beta] \cdot [apoE2] + b_{22} \cdot [A\beta] \cdot [apoE3] \\ &\quad + c_{22} \cdot [A\beta] \cdot [apoE4] + d_{22} \cdot [A\beta] - e_{22} \cdot [OS]\end{aligned} \tag{10}$$

Finally, we shall calculate the apoE values, using:

$$\begin{aligned}\frac{d[apoE]}{dt} &= a_{13} \cdot [colCNS] \cdot [\varepsilon 2] + b_{13} \cdot [colCNS] \cdot [\varepsilon 3] \\ &+ c_{13} \cdot [colCNS] \cdot [\varepsilon 4] - b_{14} \cdot [colSPM] \cdot [apoE2] \\ &- c_{14} \cdot [colSPM] \cdot [apoE3] - d_{14} \cdot [colSPM] \cdot [apoE4] \\ &- a_{22} \cdot [A\beta] \cdot [apoE2] - b_{22} \cdot [A\beta] \cdot [apoE3] \\ &- c_{22} \cdot [A\beta] \cdot [apoE4] - d_{13} \cdot [OS]\end{aligned} \quad (11)$$

It is known that the apoE4 possesses less affinity with cholesterol and less speed in its extrusion than the apoE3[19]. It is also less efficient in cholesterol distribution than the apoE2[17]. So, for the same quantity of cholesterol in the CNS, we shall have: $a_{13} > b_{13} > c_{13}$, $b_{14} > c_{14} > d_{14}$ and $a_{22} > b_{22} > c_{22}$.

The Aβ monomers can be degraded by the primary neurons, by the astrocytes and by the microglia, though the incorporation of Aβ into the plasma membranes is inversely proportional to the cholesterol content in the MPS's. The reduction of membrane cholesterol content facilitates the internalization and degradation of Aβ (Aβ degrad). On the other hand, when the cholesterol content is too high, Aβ will not be internalized for degradation, thus increasing the chances of deposition[18]. In mathematical terms:

$$\frac{d[A\beta degrad]}{dt} = a_{23} \cdot \frac{[A\beta]}{[colEL]} \quad (12)$$

The Aβ oligomers, in its turn, may insert itself in the membranes and, once again, cholesterol will be modulating this insertion. Each Aβ oligomer opens itself forming ion channels in the membrane, favoring the influx of calcium ions[27]. The formation rate of those ion channels increases proportionally to the increase of the membrane cholesterol content, though there is a superior limit of saturation. Therefore, we can state that de formation rate of calcium channels obeys the following sigmoid function:

$$\frac{d[Ca^{2+}\ channels]}{dt} = b_{23} \cdot \frac{1}{1 + \varepsilon^{-c_{23}([colSPM]\cdot[A\beta pol]) - d_{23}}} \quad (13)$$

where, b_{23} is the amplitude of the curve, c_{23} is the inclination of the curve, and d_{23} is the inflexion point of the curve.

Aβ tends to bind to the neuronal membrane, maintaining its original and non-toxic α-helical conformation when this membrane presents a low lipid density. When the MPS's cholesterol content increases, the decrease in the Aβ degradation will force its levels to go up. In addition, with the increase of MPS's cholesterol content, Aβ undergoes a transition into a β-sheet conformation. This transformation process acts also as a seed for the toxic amyloid fibril formation[24,21]. In short, we can say that Aβ will aggregate (Aβpol) at a rate directly proportional to the membrane cholesterol level (colMPS) and the balance of Aβ in CNS. Thus:

$$\frac{d[A\beta pol]}{dt} = a_{24} \cdot [colCNS] \cdot [A\beta] \tag{14}$$

Finally, we can calculate the balance of Aβ in the brain, at each instant, following the equation:

$$[A\beta] = [A\beta prod] - [A\beta out] - [A\beta degrad] - [A\beta pol \tag{15}$$

The amyloid deposits are formed by an extracellular aggregation of oligomeric Aβ[22]. Furthermore, many neurotoxic effects of Aβ are measures by free radicals[8]. Thus, we can consider that the rate of formation of the SAP's depends on the quantity of fibril Aβ (Aβpol) and the level of inflammation at the region (OS). Thus:

$$\frac{d[SAP]}{dt} = a_{25} \cdot [A\beta pol] \cdot [OS] \tag{16}$$

5. The Formation of Neurofibrilary Tangles

The Tau protein belongs to a class of proteins of the mammal's brains and is found mostly in axons, being one of the principal components of the microtubules[28]. The microtubules form the cytoskeleton and are indispensable for the maintenance of the neuronal structure and the axonal transport of several substances, such as the neurotransmitters. The role of the Tau in these structures is to stabilize the microtubules, while its rate of phosphorylation will determine its stabilization capacity: the greater the phosphorylation, the lesser is the stability. During the development of the brain, the lack of stability provides greater plasticity, what is crucial for the development and differentiation of the neurons. In the mature brain, however, the instability can determine functional and structural changes, leading to neuronal loss and formation of NT's[29].

Fragments of Tau (ΔTau), generated by caspase-cleavage, are found close to the Aβ deposits, suggesting that intra and extracellular Aβ induces such cleavage. Besides Aβ, other stimuli are capable of activating the caspases, such as oxidative stress. *In vitro*, caspase-cleavage of Tau leads to the neurite, thus contributing for the synaptic deficits and neuronal demise[29]. Thus, we can state that:

$$\frac{d[\Delta Tau]}{dt} = a_{51} \cdot [A\beta] \cdot [Tau] + b_{51} \cdot [OS] \cdot [Tau] \tag{17}$$

After the caspase-cleavage, a change in the conformation of the fragmented Tau (ΔTau)[c] occurs, what catalyzes its filament formation (fTau) in the microtubules. In addition, those fragments act as seed for full-length Tau filament formation[29]. Thus:

$$\frac{d[fTau]}{dt} = a_{53} \cdot [\Delta Tau] + b_{53} \cdot [Tau] \cdot [\Delta Tau] \tag{18}$$

Kinases involved in Tau phosphorylation are regulated by the levels of intracellular calcium. The increase of intracellular levels of this ion may lead to a hyperphosphorylation of Tau[3,?]. Tau also becomes extremely phosphorylized during mitosis[28]. Gathering this data, we have:

$$\begin{aligned}\frac{d[PTau]}{dt} &= a_{54} \cdot [fTau] + b_{54} \cdot [Tau] \cdot [Mit] \\ &+ c_{54} \cdot [Tau] \cdot [Ca^{2+}\ channels]\end{aligned} \tag{19}$$

The phosphorylation of Tau (PTau) can occur as a protective mechanism in order to dissociate it from the microtubules, when it is in the filament formation, what results in destabilization that affects the neuron integrity[29]. Thus, for the disconnected Tau (descTau), we shall have:

$$\frac{d[descTau]}{dt} = a55 \cdot [PTau] \tag{20}$$

The disconnection of Tau favors its accumulation in paired helical filament, which are the main component of NT's[28]. One characteristic occurrence of these filaments is the presence of disconnected and hyperphosphorylized Tau[31,32]. Thus:

$$\frac{d[EN]}{dt} = a_{57} \cdot [descTau] \tag{21}$$

Presently, some researchers affirm that neurons affected by some kind of insult such as toxicity of Aβ and oxidative stress attempt to restart their life cycle. There is, however, a strong connection between the induction of mitosis in mature neurons and neuronal loss. The activation of the life cycle program in mature cells may lead to the cell's apoptosis, what could explain the massive neuronal loss in AD[28,32]. Therefore:

$$\frac{d[Mit]}{dt} = a_{52} \cdot [A\beta] + b_{52} \cdot [OS] \tag{22}$$

All factors leading to neuronal loss in AD (Nloss) that have been considered in the present work can be summarized by the equation bellow:

$$\begin{aligned}\frac{d[Nloss]}{dt} &= a_{61} \cdot [NT] + b_{61} \cdot [SAP] + c_{61} \cdot [Ca^{2+}] \\ &+ d_{61} \cdot [Mit] + e_{61} \cdot [C24s]\end{aligned} \tag{23}$$

[c]The Tau adopts a MC1-immunoreactive conformation.

One simplification of the model was to consider that the level of neuronal loss will characterize AD, since it relates to the cerebral atrophy that occurs in the disease.

6. Results

After the construction of the equations, we have used the Euler's method to solve them simultaneously. One integrating step will correspond to a period of 8 hours, although the nominal value of each step is $1\text{x}10^{-6}$. In total, the simulation of a 100 years period would correspond to 109,500 steps of 8 hours. In order to facilitate visualization of events, we have rounded the number of steps to 100,000, which correspond to 100 years, approximately. The value of time, in the horizontal axis, corresponds to the number of steps executed in the simulation.

The experiments made can be divided in two different sets. The experiments of the first set aimed at validating the model built and elucidating the communication between the several processes of AD. In the second set, we performed simulations aiming at testing the hypotheses about the disease, collected from the consulted literature. The results obtained in each simulation and the conclusions that could be drawn will be reported after the description of each test.

In the tests carried out, a basal cholesterol diet corresponds to a daily consumption of 300mg of cholesterol, and when submitted to a high cholesterol diet that corresponds to a daily consumption of 600mg of cholesterol, that is, twice the amount of an ideal daily diet. Also, we chose to study only the double expression of one type of apoE allele at each simulation. Therefore, there will not be any cases of mixed expression of alleles.

6.1. *Aggregation of Aβ*

Murray[30] raised the hypothesis that, if the process of aggregation of Aβ could be suspended, the disease would not evolve. In the simulation, we tested the system behavior when the aggregation process of Aβ is inhibited. The results are coherent with the previous hypothesis and the following figures illustrate the experiment.

In Fig. 1, we show the production levels of SAP's (PNS's) and NT's (EN's) found in a simulation of daily ingestion of 600mg of cholesterol, in occurrence of the $\varepsilon 4$ allele. In comparison, we can observe that, in Fig. 2, when the aggregation process of Aβ is inhibited, the suppression of the inflammatory process occurs and, consequently, the formation of SAP's is inhibited.

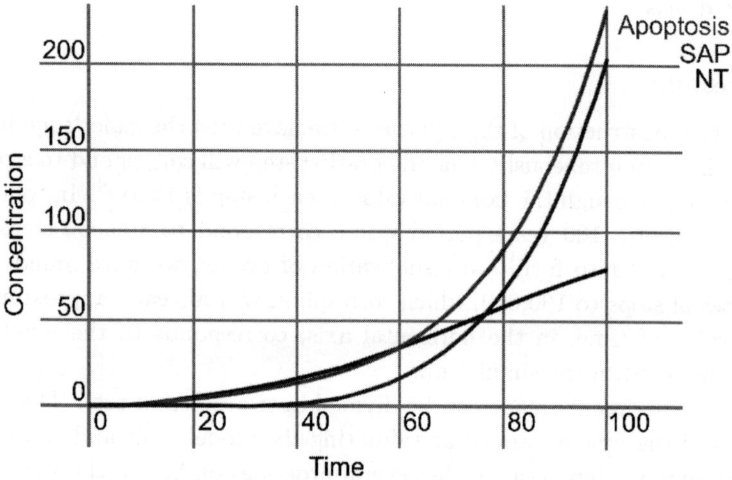

Figure 1. Formation of SAP's and NT's for the $\varepsilon 4$ allele, in a 600mg cholesterol diet and Aβ aggregation.

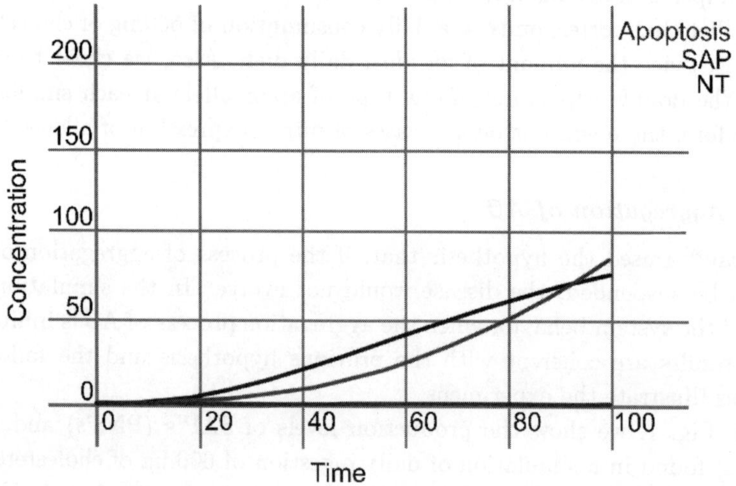

Figure 2. Formation of SAPs and NTs for the $\varepsilon 4$ allele, in a 600mg cholesterol diet, and inhibition of Aβ aggregation.

The formation of NT's, however, is maintained. In fact, the caspase-cleavage of Tau can be activated by the presence of soluble Aβ, and the production of which was not inhibited. Despite not being converted into oligomeric Aβ, the production of Aβ presents a reduction. Coherently, we noticed a light decrease in the formation rate of NT's. It is not difficult to explain the decrease in Aβ production. As SAP's were no longer formed, there will not be any disturbance in the functioning of the BBB by theses structures as well; therefore, the increase of the efflux of cholesterol is avoided. As a consequence, the changes in the distribution of cholesterol between the leaflets of the neuronal membrane will occur much more slowly; hence, the production of amyloid is reduced.

6.2. *Formation of NT's*

In Niemann-Pick's disease, NT's identical to those of AD were found, although SPM's were not encountered. Since individuals with this disease are usually deceased much younger than AD patients, we can suppose that NT's occur before SPM's. This simulation indicates that this phenomenon actually occurs, as shown in Fig. 3, since the formation of NT's happens almost 25 years before SPM's, in the occurrence of $\varepsilon 4$.

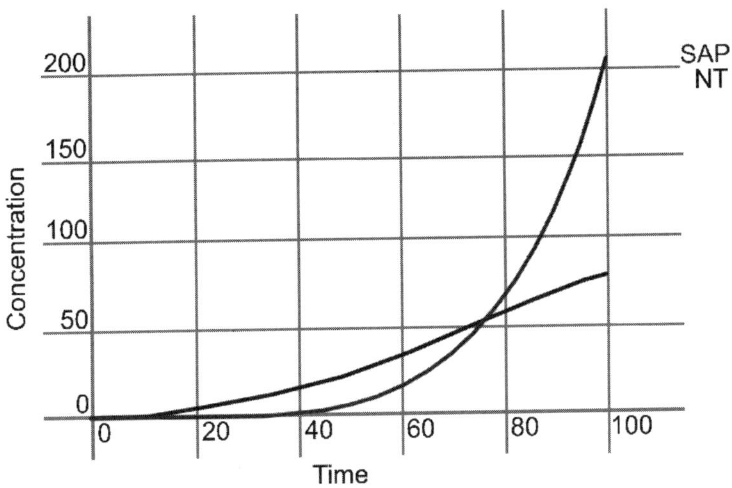

Figure 3. Formation of SPMs and NTs, for the $\varepsilon 4$ allele, in a 600mg cholesterol diet.

6.3. *Inflammation*

There are some positive *feedbacks* in the system that may aggravate the AD pathology. One example is what happens between Aβ and SPM's and the oxidative stress. It is known that Aβ and SPM's initiate an inflammatory process in the brain. On the other hand, the inflammation aggravates the AD development process, while the action of free radicals occurs in the formation of SPM's as much as in the formation of NT's [33]. In another test carried out, we inhibited the occurrence of inflammation. We observed a keen reduction in the development of neuropathological alterations, in diets with daily ingestion of 600mg of cholesterol and the $\varepsilon4$ allele, as shown by the comparison of Fig. 1 and Fig. 4

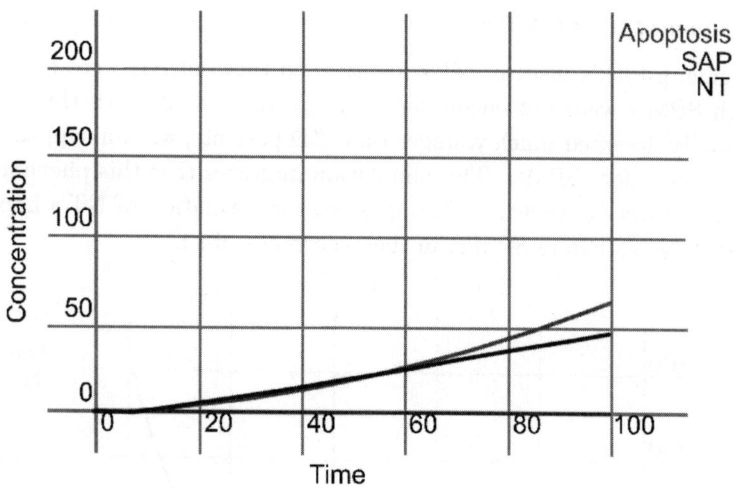

Figure 4. Neuropathological alterations for the $\varepsilon4$ allele, in a 600mg cholesterol diet, without inflammation.

7. Conclusions

The qualitative nature of our results can be considered as quite significant, as they support several *in vivo* or *in vitro* experimental observations, notwithstanding the fact that it was not possible to obtain exact values of many parameters.

The results presented by simulations as well as the results collected from the consulted literature lead to the conclusion that, despite the brain's resistance to dietary lipid composition, the chronic consumption of cholesterol may alter the functioning of certain cerebral proteins and even the structure of neurons. That happens due to the fact that cholesterol alters the composition and certain properties of the neuronal membrane. One consequence of these alterations is the increase of Aβ production and aggregation, at the same time that a reduction of the degradation rate is observed. These factors will contribute to the increase of toxic Aβ formation, leading to the formation of SPM's, what will cause the generation of free radicals, therefore amplifying the oxidative stress. The oxidative stress, in its turn, will oxidize the SPM, creating a positive feedback over Aβ production. Finally, the oxidative stress and the insoluble Aβ will trigger the Tau phosphorylation and, as a consequence, the neuronal structure will destabilize itself, generating NT's.

Inflammation also plays an important role in the development process of AD. In the tests carried out, the inhibition of the inflammatory process resulted in a sharp deceleration of the neurodegenerative process.

References

1. E. R. Kandel, J. H. Schwartz, T. M. Jessel, *Principles of Neural Science*, McGraw-Hill, 4th Edition, USA (2000).
2. M. M. Mesulam, *Annals of the New York Academy of Science*, **924**, 42 (2000).
3. D. J. Selkoe, *Annals of the New York Academy of Science*, **924**, 17 (2000).
4. J. Poirier, *Trends in Molecular Medicine* **9(3)**, 94 (2003).
5. W. F. Ganong, *Review of Medical Physiology*, 16th Edition, Prentice-Hall International Inc., USA (1993).
6. B. Wolozin, et al., *Arch Neurology*, **57**, 1439 (2000) (in YANAGISAWA, 2002).
7. L. M. Refolo et. al., *Neurobiology of Disease*, Academic Press, vol. 8, p.890-899 (2001).
8. M. A. Pappolla, M. A. Smith et. al., *Cholesterol, Oxidative Stress, and Alzheimer's Disease: Expanding the Horizons of Pathogenesis*, In: Smith, M. A.; Perry, G. (Ed) *Serial Review: Causes and Consequences of Oxidative Stress in Alzheimer's Disease*, Free Radical Biology and Medicine, vol. 33, no. 2, p.173-181, Elsevier Science Inc (2002).
9. M. Kivipelto et al., *Neurology*, **56**, 683 (2001).
10. W. Banks et. al., *Annals of the New York Academy of Science*, **826**, 190 (1997).
11. D. Howland et. al., *The Journal of Biological Chemistry*, **273(26)**, 16576, (1998).
12. K. Yanagisawa, *Journal of Neuroscience Research*, **70**, 361 (2002).
13. M. Burns, K. Duff, *Annals of the New York Academy of Science*, **977**, 367 (2002).

14. W. G. Wood, G. P. Eckert, U. Igbavboa, W. E. Müller, *Biochimica et Biophysica Acta*, **1610**, 281 (2003).
15. J. Kalman et al., *Life Sciences*, **68**, 1495 (2001).
16. A. W. Vorbrodt et. al., *Journal of Neurocytology*, **23**, 792 (1994).
17. E. Frears et. al., *Neuroreport*, **10(8)**, 1699 (1999).
18. W. G. Wood et. al., *Neurobiology of Aging*, **23**, 685 (2002).
19. V. N. Trieu, F. M. Uckun, *Biochemical and Biophysical Research Communications*, **268**, 835 (2000).
20. A. R. Koudinov, T. T. Berezov, N. V. Koudinova, *Neuroscience Letters*, **314**, 115 (2001).
21. S. R. Ji, Y. Wu, S. F. Sui, *The Journal of Biological Chemistry*, **277(8)**, 6273 (2001).
22. T. E. Golde, C. B. Eckman, *Drug Discovery Today*, **6(20)**, 1049 (2001).
23. L. M. Refolo et. al., *Neurobiology of Disease*, **7**, 321 (2000).
24. A. Kakio, S. Nishimoto et al., *Biochemistry*, **41**, 7385 (2002).
25. C. L. Martel, *Journal of Neurochemistry*, **69**, 1995 (1997).
26. Weller et. al., *Annals of the New York Academy of Science*, **977**, 162 (2002).
27. N. Arispe, M. Doh, *The FASEB Journal*, **16**, 1526 (2002).
28. S. Illenberger et al., *Molecular Biology of the Cell*, **9**, 1495 (1998).
29. R. A. Rissman, W. W. Poon, M. B. Jones et al., *The Journal of Clinical Investigation*, **114(1)**, 121 (2004).
30. R. K. Murray, *The Biochemical Basis of Some Neuropsychiatry Disorders* in: Murray, R. K., Granner, D. K., Mayes, P. A., et al. *Harper's Biochemistry*, 23 ed., Appleton & Longe, Prentice-Hall International Inc., pp. 750-752 (1993).
31. O. V. Forlenza, W. F. Gattaz, *Revista de Psiquiatra Clínica*, **25(3)**, 114 (1998).
32. R. L. Neve, D. L. McPhie, Y. Chen, *Brain Research*, **886**, 54 (2000).
33. R. Egensperger et al., *Brain Pathology*, **8**, 439 (1998).

THEORETICAL STUDY OF A BIOFILM LIFE CYCLE: GROWTH, NUTRIENT DEPLETION AND DETACHMENT*

GALILEO DOMINGUEZ-ZACARíAS, ERICK LUNA
AND JORGE X. VELASCO-HERNÁNDEZ

Programa de Investigación en Matemáticas Aplicadas y Computación
Instituto Mexicano del Petróleo,
México, DF 07730, México
Email: gdzacari@imp.mx; eluna@imp.mx; velascoj@imp.mx

We present the results of a theoretical investigation, based upon experimental evidence gathered from the existing recent literature, on the interaction of fluid flow and biofilm internal structure, centering on the processes of detachment, spatial structure of microbial communities, and the diffusion and convection of nutrients. The biofilm is viewed as a porous medium with variable porosity, tortuosity and permeability. Our model is a two dimensional approximation to biofilm dynamics and it is capable of reproducing biofilm internal heterogeneities, biofilm surface behavior, nutrient penetration and biofilm critical point of rupture or detachment.

1. Introduction

Biofilm detachment is one of the least understood processes in the life cycle of a biofilm. There is a variety of hypothesis[1] formulated to explain it, among them are we can count the existence of matrix-degrading enzymes, appearance of gas bubbles, nutrient levels and microbial growth, quorum sensing among others.

To understand biofilm physical and chemical characteristics one must consider mutual influences between biofilm structure and processes. Nutrient transport, metabolic residue accumulation, and limited diffusion are affected by biofilm structure. However, as reported by Bayebal et al.[2], this influence is largely hypothetical and awaits elucidation. To this end the development of tools to quantify biofilm structure characteristics and to correlate those with processes is urgently needed. According to several authors[2] biofilm development is associated with : i) structure and function

*This work was supported by grant imp d.00330

that allows survival on surfaces, ii) efficient transport of nutrients to the whole biofilm body.

The analysis of biofilm structure fall into several categories: i) comparative studies, ii) reproducibility of structure, iii) monitoring temporal variations in structure, iv) testing the effect of substances on structure, v) quantifying effect of environmental factors on structure, and vi) parameter estimation of structure characteristics for use in models.

Mai-Prochnow et al.[3] explored the role that cell death plays in the structure of biofilms. They show that some biofilm types have a reproducible pattern of cell death that, in the long term, may play an important role in biofilm dispersion. These authors support the view that cell death inside biofilms may be a widespread phenomenon of importance for biofilm dispersal. However little is known about this mechanisms. One possible explanation has been put forward by Hunt et al.[1] who investigated the role of nutrient starvation in biofilm detachment. According to this author localized nutrient depletion induces starvation in subsets of cells that, given enough time, detach from the biofilm main structure.

Of the many modeling approaches[4–14] available to model the interaction between biofilm structure and the surrounding liquid medium, we have used a hybrid model where conservation equations are coupled to a cellular automaton (CA) that simulates the growth of the bacterial consortia[15].

A biofilm is an assemblage of surface-associated microbial cells enclosed in an extracellular polymeric matrix[16]. Biofilms play important roles in many areas of applied science and technology such as oil enhanced recovery and transportation, pharmaceutical research, food processing, medical devices, antibiotic resistance, electrochemical corrosion, waste water treatment and others[17].

2. The mathematical model

Biofilm growth is modeled with a CA that reproduces cell division, space competition and take in account substrate concentrations. The conservation equations and CA models are coupled through porosity, interface position (biofilm-fluid interface), and concentration and rate consumption of nutrients. The numerical and mathematical details of our model can be consulted in Luna et al.[15]. Biofilm porosity is incorporated into our model through the partial differential equation, which is given by equation 1

$$\frac{\partial}{\partial t}\left[\phi\rho + (1-\phi)\,\rho_b\right] + \nabla\cdot\left(\rho\vec{u}\right) = 0\,, \tag{1}$$

where ρ and ρ_b are the fluid and bacterial densities,respectively, and $\vec{u} = u\hat{x} + v\hat{y}$ is the velocity vector. We suppose Darcy's Law governs the relationship between velocities and pressure. The mass nutrient fraction Y is given considering the balance equation for variable porosity under the hypothesis of low velocities so that dispersion effects can be safely neglected (see equation 2):

$$\phi\frac{\partial Y}{\partial t} + \vec{u}\cdot\nabla Y = \nabla\cdot(\phi D_\phi \nabla Y) + A\,. \tag{2}$$

In the above equation A is the substrate consumption rate, and D_ϕ is the effective molecular diffusion coefficient of nutrients in the fluid.We refer the reader to references to justify the formulas for tortuosity and permeability[18,19,21,15].

Boundary and initial conditions are as follows: initially, there is a given resource concentration inside the biofilm $Y(x, y, t = 0) = Y_{ini}$. The initial biofilm distribution $\Gamma(x, t = 0)$ is given equation 3 as:

$$\Gamma(x, t = 0) = \widetilde{\Gamma}_0 + \widehat{\Gamma}_0 \sin(k_\Gamma x)\ . \tag{3}$$

where, $\widetilde{\Gamma}_0$ is the profile average value, $\widehat{\Gamma}_0$ and k_Γ represent the amplitude and wave number of a small sinusoidal perturbation, respectively. This sinusoidal shape is not that arbitrary as it may appear. Microscopy results show that in biofilms involved in microbiologically induced corrosion or MIC, the bacteria-metal interface, for example, has a periodic character that can be approximated by a profile as the one proposed here. The initial porosity $\phi(x, y, t = 0)$ is calculated as:

$$\phi(x, y, t = 0) = \phi_{\min} + \frac{N_0}{N_T}(1 - \phi_{\min})\ . \tag{4}$$

where, $\phi_{\min}$ is the minimum porosity that the bacterial film can reach, N_0 and N_T are empty and total sites in the arbitrary CA neighborhood used to compute porosity, respectively.

This hypothesis, namely, that bacterial microcolonies are formed by the growth of already attached cells, has been proposed for *Pseudomonas sp.* and *Pseudomonas putida* in an experimental setting[22].

We briefly describe CA basic setting and refer the reader to Luna et al.[15] for further details. Our CA has either of two states at each lattice point: occupied (**1**) by a bacterium or empty (**0**) and also carries information on the nutrient concentration at that point. Colonization or cell division occurs with probability $\frac{N_e R}{4}$ where N_e is the number of occupied nodes in a von Neumann neighborhood of each lattice point, and R is the probability of colonization of unoccupied sites. A substratum minimum concentration $Y_{\min}$, is required also for successful to colonization. When a node is occupied, there is an increment Δa_b in the substratum consumption rate at that point. Mortality due to competition occurs with probability $\frac{N_0 P}{4}$, where N_0 is the number of occupied nodes in a von Neumann neighborhood where P is the probability of death. The node becomes empty if the amount of substratum at that point is below the minimum. The initial condition for the CA was set to a random distribution with 80% occupied lattice points in the area below the fluid bacteria interface $\Gamma(x, t = 0)$. The reader interested in the numerical details of the model implementation is referred to Luna et al.[15] where full information on such issue is provided.

3. Detachment

Studies centered in biofilm disintegration and detachment consider the effect of fluid flow over the biofilm surface[23,24,25,14]. We assume that microbial death within the biofilm weakens attachment strength and mechanical resistance properties, especially at the biofilm-plate interface. *Pseudomonas fluorescens* biofilms form very quickly in well-oxygenated environments but as oxygen is depleted due to growth, disintegration develops[26] may be due to the denaturalization of the extrapolyscacharide substance matrix (EPS). In our model the maximum nutrient concentration exists, for each x, in a neighborhood of $\Gamma(x)$; below this interface, nutrient concentration decreases with depth. If the biofilm grows enough, it increases thickness and therefore, nutrient concentration reaches a critical value at the bottom of the biofilm, near the attachment interface, where bacteria can no longer reproduce or survive. Cell death correlation with nutrient depletion and biofilm growth is a reasonable approximation to the process. As discussed by Lewis[27], biofilm destruction because of nutrient limitation other than oxygen has not been conclusively demonstrated but it is a reasonable mechanism to induce biofilm destruction. Therefore, under this assumption, namely, that nutrient depletion is associated with biofilm destruction, we derive a very simple formula to predict the breaking point of the bacterial film. Biofilm breakage requires that the forces per unit length due to fluid drag (F_d) be greater than a critical value that depends on the thickness and the internal structure of the biofilm

$$|F_d| > \sigma_a \Gamma(x_a) . \tag{5}$$

where Γ is the biofilm thickness and σ_a represents the maximum stress that the biofilm can resist. Full details of this derivation can be consulted in [15].

The cellular automaton describes colonization extinction process through competition where R is the probability of colonization of unoccupied sites (related to bacterial replication), and P is the probability of death due to competition for space and resources. The simulations performed were done for a range of values of P and R , with $\alpha = 0.57$, $\delta = 1 \times 10^{-2}$, $\Omega = 1 \times 10^{-4}$, $\beta = 1 \times 10^{-4}$, $\varepsilon = 0.01$, $\Pi_0 = 0.11$ (σ obtained from [14]), $\Delta a_b = 0.07$, $\Delta a_d = 0.04$, $\frac{Y_{\min}}{Y_f} = 0.5$.

We centered our attention in measurable properties of biofilms. Biofilm growth is one of them. In our model we computed the biofilm mean thickness ($\overline{\gamma}$) as it evolves through changes in parameter space. Figure (1) shows the point at which the biofilm detaches from the plate.

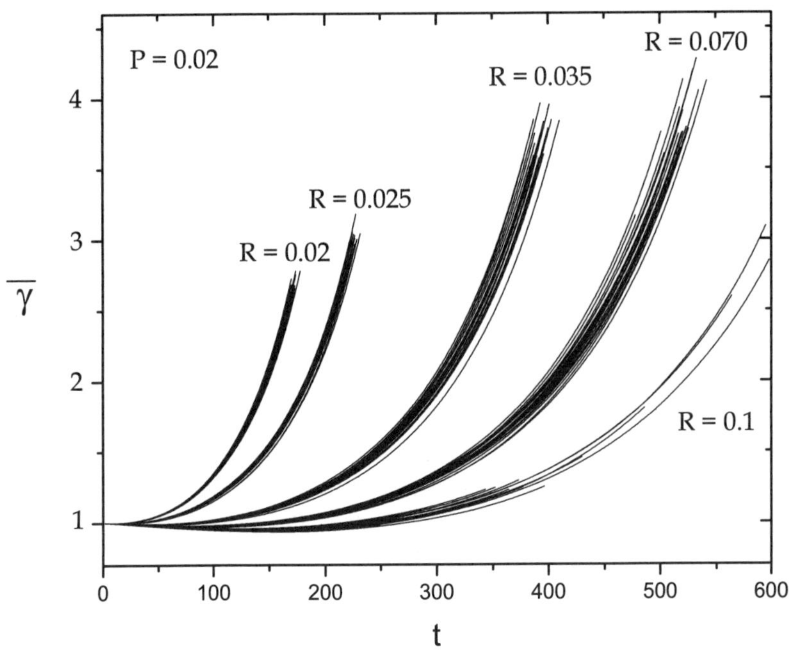

Figure 1. Mean biofilm thickness a a function of time

Other variable of interest is the distribtion of bacterial biomass within

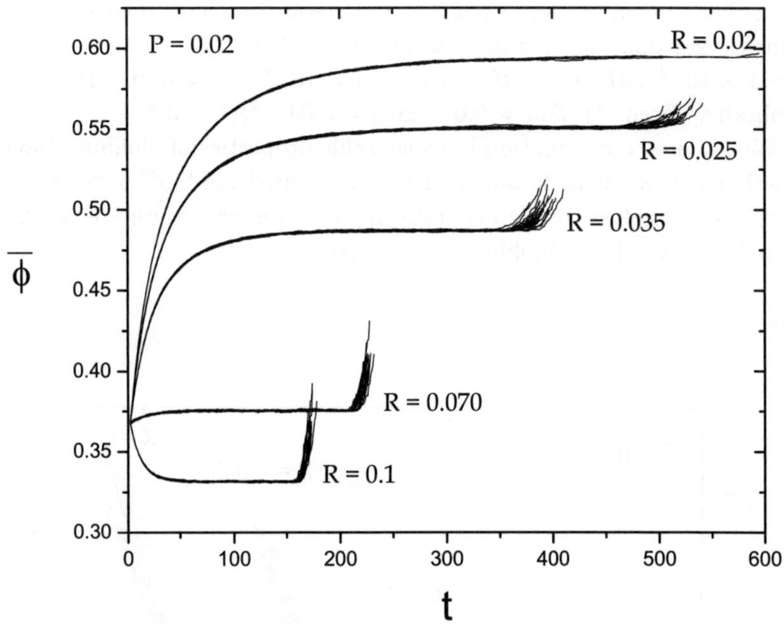

Figure 2. Mean porosity values as a function of time

the biofilm which is given, as an approximation, by mean porosity. Figure 2 illustrates how this variable changes in parameter space. Note that equilibrium mean porosity represented by the platau in the figure is inversely proportional to R. Before biofilm detachment bacterial death increases very quickly at the bottom of the biofilm producing a significative increase in porosity.

Figure 3 shows that our simulations predict that the maximum thickness is roughly the same for all P, implying that competition among bacteria is not a determinant of biofilm growth.

Note that for a given P fixed, as R increases uncertainity for critical thickness becomes narrower implying that variability in thickness before detachment is reduced and it is largely independent of bacterial life history parameters (R and P).

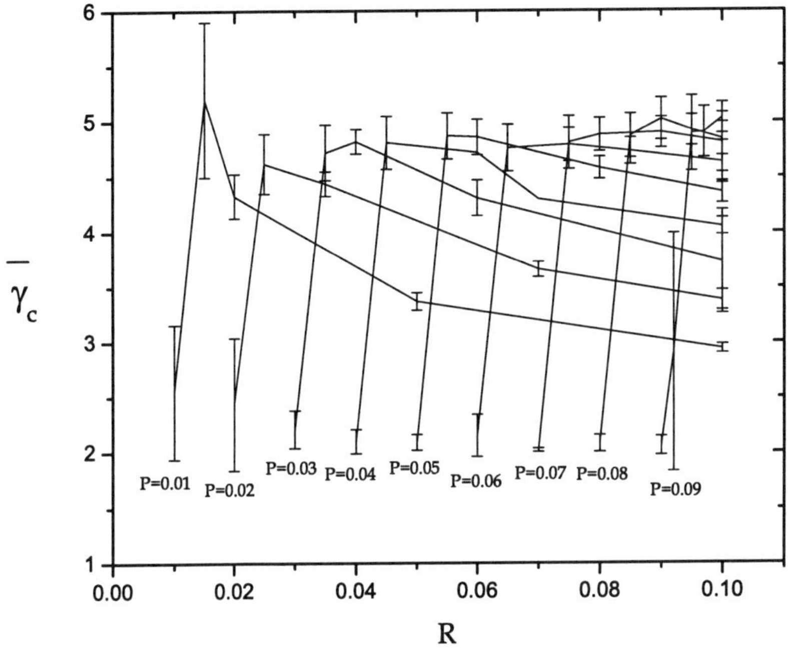

Figure 3. Critical mean thickness as a function of R for different value of P.

We have found that nutrient distribution depends on the biofilm thickness and to a minor degree on the heterogeneities in porosity. It is important to mention that we are modeling biofilms whose metabolism depends on carbon sources present in the fluid that surrounds them. This nutrient therefore must penetrate the biofilm to nourish the component cells.

The mean time to detachment and its standard deviation as functions of R for different values in P are shown in Fig. 4. Here again these statistics were calculated for 20 simulations. This Figure (4) shows that the time needed to reach the breaking condition is inversely proportional R to so that larger colonization ability implies a faster biofilm life cycle. A minimum in the curves is presented when $R = P$ except for $P = 0.01$ where the variance is too large for this behavior to be captured. Note that, roughly, the smaller the death rate P is (competition strength), the longer it takes for the biofilm to detach and that, again, for a fixed P, increases in colonization capabilities (R) decrease the time to detachment.

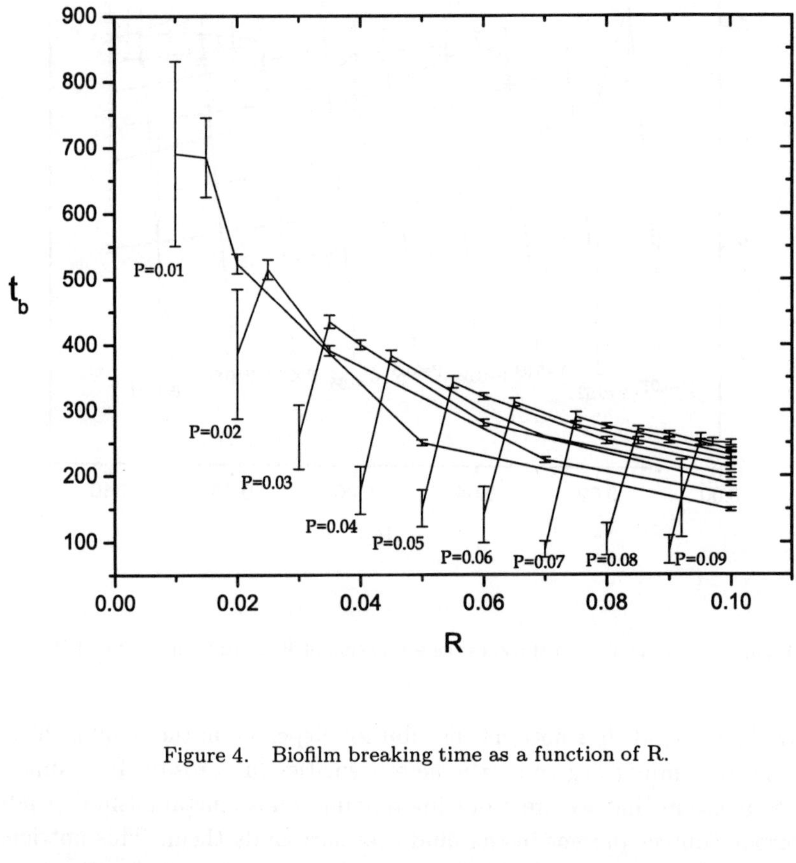

Figure 4. Biofilm breaking time as a function of R.

4. Conclusion

Following the developmental episodes of biofilms postulated by Sauer et al.[28] we have modelled a general biofilm from the irreversible attachment stage through maturation-1 and maturation-2 to the dispersion stage. Our biofilm model passes through a resource limited stage (maturation-1) when diffusion and convection of nutrients become restricted due to biofilm growth, and then reaches a time when it has maximum thickness corresponding to maturation-2 stage of Sauer et al.[28] . In this stage, the P.

aeruginosa biofilm studied by the cited authors, experiences anaerobic or reduced oxygen conditions. In our model this resource limitation occurs too and induces cell death and an increased porosity that, according to our model, leads to the final detachment or dispersion stage. According to Sauer et al.[28] characterize this last stage by the observation of bacterial clusters that move away from the biofilm core leaving hollow structures or voids. Although our model was not constructed to emulate this experimental system it is surprising to see that it predicts exactly this phenomenon right before detachment takes place. Webb et al.[29] report that cell death inside biofilms plays a role in the differentiation and dispersal of subpopulations of surviving biofilm cells but point out that the mechanisms by which voids are produced within the biofilm is unclear.

Our explanation for the process of biofilm development is the formation of cavities or pores that ultimately determine the detachment process. We have also shown that nutrient distribution depends importantly on biofilm thickness and, secondly, on heterogeneities in porosity, tortuosity and permeability, all this parameters related to bacterial distribution within the biofilm. We report that biofilm thickness is inversely proportional to nutrient concentration since transport phenomena are inefficient in providing nutrients to an ever increasing bacterial colony that consumes them. We also have found that there exists a critical thickness for which the nutrient concentration at the bottom interface in not enough to maintain the bacteria alive. At this point, the biofilm begins to lose its mechanical resistance that combined with the fluid drag forces generates the detachment/rupture processes.

We must point out that our model does not consider the effect on biofilm dynamics of the extracellular polymeric substance matrix (EPS) that encloses biofilms. EPS is important in certain types of antibiotic neutralization and that its mechanical and physical characteristics should play an important role in biofilm constitution and transport of cell signaling, nutrient and biocide diffusion. However, it is also known that[27], at least in *Pseudomonas aeruginosa*, the expression of the quorum sensing factor HSL although required for the conformation of a biofilm with typical architecture (channels, mushroom structures, and so forth) is not associated with their ability to resist killing by antibiotics, that is, typical biofilm architecture does not matter that much in this case. Moreover, while studying the properties of antimicrobial resistance in biofilms and detached emboli, Fux et al.[30] report that in large emboli or in stationary state biofilms starvation-induced dormancy is the common underlying mechanism for protection against biocides. These authors conclude that the mechanical disruption of biofilms, with the concomitant disruption of diffusion barriers, is correlated with antibiotic resistance. So although we are aware that not introducing EPS into our model may make it less general, as far as our objectives are concern, it is a reasonable assumption.

In a *S. aureus* biofilm[30] bacteria either detach from the surface where it forms leaving bare spots, or from the main biofilm body (other bacteria) leaving on the surface an attached layer of cells. These same authors report that there is growing evidence that the detachment of biofilms is not only due to the interaction with hydrodynamic forces, a process that is called passive detachment, but actively also in response to population density, changes in substrate concentration and exposure to antimicrobials[30]. Hunt et al.[1] have reported, in a very interesting study, supporting results on the conjecture that nutrient starvation is a cue for detachment, findings aligned in the direction of the conjecture generated with our model. These authors use a mathematical model to explore their hypothesis. The main point here is that biofilm life cycle and the determination of the mechanisms for biofilm detachment require the application of modeling techniques and methodologies. Other recent work in this direction, namely the study of biofilm detachment and associated processes using models has been published by Laspidou and Rittmann[31,32].

Donlan[16] cites work on the physical causes of biofilm detachment where three processes appear to be the most important: erosion or shearing, sloughing, and abrasion. The model presented in this paper addresses the issue of sloughing due to nutrient depletion. As stated by Lewis[27] genes controlling biofilm destruction can be a very important factor for eradication of biofilms. Our model, although in theoretical terms, is able to reproduce breaking/detachment, what amounts to biofilm destruction, under the simple and reasonable hypothesis of postulating a correlation between biofilm destruction and nutrient depletion. We are aware of results that indicate that the mechanisms for detachment modeled here are not universal. For example, Kaplan et al.[33] report that in *Actinobacillus actinomycetemcomitans*, the mechanisms for cell detachment involve the release of indivuduals cells within the biofilm. Nevertheless, there is ample experimental evidence suporting that, at least in vitro, resource limitation and detachment are associated. Telman et al.[34] have published experimental measurements of the effect of increasing liquid shear on the stability and detachment of biofilms. Their results indicate that detachment is a combination of sloughing, the process modeled in this work, and erosion (which we do not consider here).

Also, an important result for our approach is that biofilm life history is important for the detachment process and in this respect the results of Telman et al.[34] refer to the impact of life history in biofilm morphology and sloughing. We have incorporated life history parameters in our bacteria population through parameters R and P. Figures 1, 2, 3 and 4 illustrate how changing life history affects biofilm thickness, biofilm porosity and time to detachment. Finally, we certainly share the point of view that more experimental and computational studies are needed to understand biofilm life cycles, in particular the final stage represented by detachment[1].

Acknowledgments

JXVH completed his part of this work while an International Fellow of the Santa Fe Institute.We thank C.P. Ferreira to help to CA

References

1. S. M. Hunt ,E. Werner, B. Huang ,MA. Hamilton, PS. Stewart, *Applied and Environmental Microbiology* 7418 (2004).
2. H. Beyenal, C. Donovan, et al., *Journal of Microbiological methods* **59(3)**, 395 (2004).
3. A. Mai-Prochnow, F. Evans, D. Dalisay-Saludes, S. Stelzer, et al., *Applied and Environmental Microbiology* **70**, 3232 (2004).
4. S. Pilyugin and P. Waltman, *SIAM J.Appl.Math.* **59**, 1552 (1999).
5. D. A. Jones, H. Smith, *SIAM J.Appl.Math.* **60**, 1576 (2000).
6. E. D. Stemmons, HL. Smith, *SIAM J.Appl.Math.* **61**, 567 (2000).
7. J. Dockery, I. Klapper, *SIAM J.Appl. Math.* **62(3)**, 853 (2001).
8. D. Jones, H. V. Kojouharov, D. Le, HL. Smith, *J.Math.Biol.* DOI. 10.1007/s00285-003-0202-I, (2003).
9. C. Picioreanu, M. C. M. van Loosdrecht, JJ. Heijnen, *Biotechnol. Bioeng.* **57**, 718 (1998).
10. C. Picioreanu, MCM. van Loosdrecht, JJ. Heijnen, *Biotechnol. Bioeng.* **58(1)**, 101 (1998).
11. C. Picioreanu, M. C. M. van Loosdrecht, JJ. Heijnen, *Water Sci. Technol.* **39(7)**, 115 (1999).
12. C. Picioreanu, MCM. van Loosdrecht, JJ. Heijnen, *Biotechnol. Bioeng.* **68**, 355 (2000).
13. C. Picioreanu, MCM. van Loosdrecht, JJ. Heijnen, *Biotechnol. Bioeng.* **69**, 504 (2000).
14. C. Picioreanu, MCM. van Loosdrecht, JJ. Heijnen, *Biotechnol. Bioeng.* **72**, 205 (2001).
15. E. Luna, G. Dominguez-Zacarias, C. Pio Ferreira, JX. Velasco-Hernandez, *Physical Review E* **70**, 061909 (2004).
16. R. M. Donlan, *Biofilms* **8**, 881 (2002).

17. P. S. Stewart, MA. Hamilton, B. Goldstein, BT. Schneider, *Biotechnol.Bioeng.* **49**, 445 (1996).
18. R. Islas-Juarez, M.I. Thesis, Univerisad Nacional Autónoma de México (2003).
19. E. L. Cussler, Cambridge University Press., New York, US (1997).
20. C. N. Davies, *Proc. Inst.Mech. Eng.* **1B**, 185 (1952).
21. P. M. Adler and J-F. Thover, *Appl. Mech. Rev.* **51(9)**, 537 (1998).
22. T. Tolker-Nielsen, UC. Brinch, PC. Ragas, JB. Andersen, CS. Jacobsen, S. Molin, *Journal of Bacteriology* **182**, 6482 (2000).
23. H. Horn, H. Reiff, E. Morgenroth, *Biotechnol. Bioeng.* **81**, 607 (2003).
24. L. Mascari, P. Ymele-Leki, CD. Eggleton, P. Speziale, JM. Ross, *Biotechnol. Bioeng.* **83**, 65 (2003).
25. I. Klapper, C. J. Rupp, R. Cargo, B. Purvedorj, P. Stoodley, *Biotechnol. Bioeng.* **80**, 289 (2002).
26. D. G. Allison, B. Ruiz, C. San Jose, A. Jaspe, p. Gilbert, *FEMS Microbiol.Lett.* **167**, 179 (1998).
27. K. Lewis, *Antimicrobial agents and chemotherapy* **95**, 999 (2001).
28. K. Sauer, A. K. Camper, GD. Ehrlich, JW. Costerton, DG. Davies, *Journal of Bacteriology* **184**, 1140 (2002).
29. J. S. Webb, L. S. Thompson, S. James, T. Charlton, T. Tolker-Nielsen, B. Koch, M. Givskov, S. Kjelleberg, *Journal of Bacteriology* **185**, 4585 (2003).
30. C. A. Fux, S. Wilson, P. Stoodley, *Journal of Bacteriology* **186**, 4486 (2004).
31. C. Laspidou, B. E. Rittmann, *Water Research* **38**, 3349 (2004).
32. C. Laspidou , B. E. Rittmann, *Water Research* **38** 3362 (2004).
33. J. Kaplan, M. F. Meyenhofer, D. H. Fine, *Journal of Bacteriology* **185**, 1399 (2003).
34. U. Telmann, H. Horn, E. Mongenroth, *Water Research* **38**, 3671 (2004).

OPTIMAL CONTROL OF DISTRIBUTED SYSTEMS APPLIED TO THE PROBLEMS OF AMBIENT POLLUTION

SANTINA F. ARANTES AND JAIME E. M. RIVERA

National laboratory for Scientific Computation - LNCC/MCT.
Av. Getúlio Vargas 333
Petropolis, RJ, CEP: 25651-075, Brazil.
E-mail: santina@lncc.br; rivera@lncc.br

We study problems of Optimal Control of Systems of Distributed Parameters. The model consists of a partial differential equation of the parabolic type that carries a solute pollutant in an incompressible fluid, with boundary conditions and initial value, that is, in our model we consider the velocity that the pollutant propagates in an incompressible fluid. The developed mathematical modelling allows to calculate the pollution concentration that is placed in a region of space in such a way that at the time $t = T$, the pollution concentration is the closest as possible to the acceptable maximum concentration in the environment. We will show the existence and uniqueness of the state system and a unique optimal control on a set of permissible functions. We will characterize the optimal control to obtain an optimality system that allows the numerical calculation of the problem and will show the convergence of the method. As an application, we will study the case of the contamination of a river/lagoon by mercury (Hg) in water in movement and still water. The primary aim of this work is to minimize, through the controls, the effect caused by the debris of the pollutant agent.

1. Introduction

Our problem consists of the modelling the concentration of a contaminante/pollutant substance in a determined region. We consider that the concentration varies in the time and that this substance, contaminante/pollutant, is outspread and convected in the region. In our work, the fluid analysed is the water, that we are going to consider as being an incompressible viscous fluid. The domain, given by the rectangle below, whose boundaries are the sides Γ_1, Γ_2, Γ_3 and Γ_4, represents the stretch of a river to be studied. The boundaries Γ_2 and Γ_4 correspond to laterals of the river and Γ_1 and Γ_3 its length. Practical situations, as for example, pollution of air, contaminants in rivers, lagoons, seas or ground have been object of studies of some authors, such as Bedient et al. [2], Couseuil et al. [3] and Schnoor [8]. Some authors, as Banks [1] and Lions [5-6], showed existence and uniqueness of the solution of the system of parabolic equations and the existence of a unique optimal control for the diffusive transient case. In this work we are considering, in addition to the diffusive transient case, the convection of the problem, this means that we are taking in account the velocity of the fluid that in fact is an important datum in the variation of the pollution concentration.

Let Ω be a bounded rectangular region given below with boundary $\Gamma = \Gamma_1 \cup \Gamma_2 \cup \Gamma_3 \cup \Gamma_4$.

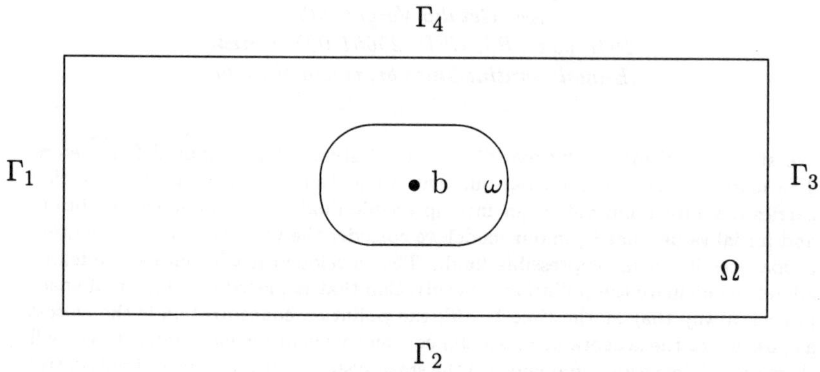

Figure 0.

The state system is

$$\frac{\partial y}{\partial t} - \lambda \Delta y + \beta \cdot \nabla y = f(x,t) \quad \text{in} \quad \Omega \times (0,T) \tag{1}$$

$$y(x,t) = 0 \quad \text{on} \quad \Gamma_1 \times (0,T) \tag{2}$$

$$y(x,t) = g(x,t) \quad \text{on} \quad \Gamma_3 \times (0,T) \tag{3}$$

$$\frac{\partial y}{\partial \nu}(x,t) = h(x,t) \quad \text{on} \quad \Gamma_2 \cup \Gamma_4 \times (0,T) \tag{4}$$

$$y(x,0) = 0 \quad \text{in} \quad \Omega. \tag{5}$$

Where $y(x,t)$ is the pollution concentration in the point $x \in \Omega$ in the instant t, β is the velocity field with $\beta_i \in C^1(\overline{\Omega})$, λ is the diffusion coefficient and $T > 0$ is given. We study the case when the control $v \in L^2(0,T)$ is inserted in the external pollutant source $f(x,t)$ in the form of a pointwise internal control.

The functional J to be minimized on a set of permissible controls $\mathcal{U}_{ad}$ is given by

$$J(v) = \int_\Omega \left| y(x,T;v) - z_d(x) \right|^2 dx + N \int_0^T |v|^2 dt.$$

Here the function $z_d \in L^2(\Omega)$ defines a pattern for the acceptable level of pollution and $N > 0$ is a given constant.

In section 2, We show the existence of a unique solution to system (1-5) and we prove the existence of a unique optimal control u that minimizes the functional J on a set of permissible functions $\mathcal{U}_{ad}$. We get a characterization of the optimal control and from this characterization the optimality system that allows the numerical calculation of the problem. We prove that the optimality system is well posed, that is, that there exists a unique solution which depends continuously on the initial data. To find the numerical solution we create an algorithm based on the uncoupling of the system and we prove that this algorithm is convergent. Finally in section 3, we make the implemention of the numerical solution by using the Finite Elements Method combined with iterative methods, we plot the numerical solution of the problem, analyse the results and get some conclusions.

2. Pointwise Optimal Control

We study the system (1-5), considering

$$f(x,t) = v(t)\delta(x-b),$$

where b determines the point of the pollution entrance.
To facilitate our analysis, we decompose the system (1-5), as follows

$$\begin{aligned}
\frac{\partial y_1}{\partial t} - \lambda\Delta y_1 + \beta\cdot\nabla y_1 &= v(t)\delta(x-b) && \text{in} \quad \Omega\times(0,T)\\
y_1(x,t) &= 0 && \text{on} \quad (\Gamma_1\cup\Gamma_3)\times(0,T)\\
\frac{\partial y_1}{\partial \nu}(x,t) &= 0 && \text{on} \quad (\Gamma_2\cup\Gamma_4)\times(0,T)\\
y_1(x,0) &= 0 && \text{in} \quad \Omega,
\end{aligned} \tag{6}$$

and

$$\begin{aligned}
\frac{\partial y_2}{\partial t} - \lambda\Delta y_2 + \beta\cdot\nabla y_2 &= 0 && \text{in} \quad \Omega\times(0,T)\\
y_2(x,t) &= 0 && \text{on} \quad \Gamma_1\times(0,T)\\
y_2(x,t) &= g(x,t) && \text{on} \quad \Gamma_3\times(0,T)\\
\frac{\partial y_2}{\partial \nu}(x,t) &= h(x,t) && \text{on} \quad (\Gamma_2\cup\Gamma_4)\times(0,T)\\
y_2(x,0) &= 0 && \text{in} \quad \Omega.
\end{aligned} \tag{7}$$

The cost functional J corresponding to these systems is

$$J(v) = \int_\Omega \left| y_1(x,T;v) - \left[\left(z_d - y_2(x,T)\right)\right] \right|^2 dx + N \int_0^T |v|^2 dt.$$

2.1. *Existence and Uniqueness of the Weak Solution of the State System and of Optimal Control*

Given

$$g \in L^2(0,T; H^{\frac{3}{2}}(\Gamma_3)) \quad \text{and} \quad h \in L^2(0,T; H^{\frac{1}{2}}(\Gamma_2 \cup \Gamma_4)).$$

We consider

$$V = \{w \in H^1(\Omega);\ w = 0 \ \text{ in } \Gamma_1 \cup \Gamma_3\}.$$

In V, we define the norm

$$\|v\| = \left(\sum_{i=1}^{n} \int_\Omega \left| \frac{\partial v}{\partial x_i} \right|^{\frac{1}{2}} dx \right)^{\frac{1}{2}}.$$

By using standard methods for Parabolic Partial Differential Equations it is not difficult to show that there exists a unique y_2 weak solution of (7), such that

$$y_2 \in L^2(0,T;V) \cap C^0(0,T;L^2(\Omega)).$$

For the system (6), to show the existence of the weak solution is not simple task because of the Dirac delta. In this case, the functions space where the solution is defined depends on the dimension n of the spatial variable. If $n = 1$, by Sobolev imbedding Theorem we know that the Dirac's delta $\delta \in H^{-1}(\Omega)$. Therefore, for $v \in L^2(0,T)$ there exists a unique y_1 satisfying (6), such that

$$y_1 \in L^2(0,T;V) \quad \text{and} \quad \frac{\partial y_1}{\partial t} \in L^2(0,T;V').$$

Thus, $y_1(T,v)$ makes sense in $L^2(\Omega)$ and the functional J is well defined in $L^2(0,T)$. Instead, for $1 < n \leq 3$ by Sobolev imbedding Theorem we have $H^2(\Omega) \subset C(\overline{\Omega})$ and therefore $\delta \in H^{-2}(\Omega)$. For this case, it is not possible to find a usual weak solution to (6) and thus we prove the existence and uniqueness of ultraweak solution of this system by Transposition Method. The ultraweak formulation is given as follows. Let φ be given in

$L^2(0,T;L^2(\Omega))$ and ϕ solution of

$$\begin{aligned} -\frac{\partial \phi}{\partial t} - \lambda\Delta\phi - \beta\cdot\nabla\phi = \varphi &\quad \text{in} \quad \Omega\times(0,T) \\ \phi(x,t) = 0 &\quad \text{in} \quad (\Gamma_1\cup\Gamma_3)\times(0,T) \\ \frac{\partial\phi}{\partial\nu}(x,t) = 0 &\quad \text{in} \quad (\Gamma_2\cup\Gamma_4)\times(0,T) \\ \phi(x,T) = 0 &\quad \text{in} \quad \Omega. \end{aligned} \tag{8}$$

Definition 2.1. We say that y_1 is ultraweak solution of the system (6) if

$$\int_\Omega\int_0^T y_1\,\varphi\,dxdt = \int_0^T v(t)\phi(b,t)\,dt, \tag{9}$$

for all $\varphi \in L^2(0,T;L^2(\Omega))$ and ϕ solution of (8).

Remark 2.1. Note that this function y_1 satisfying (9), satisfies (6) in the sense of distributions, because for $\varphi \in L^2(0,T;L^2(\Omega))$ and $\phi \in \mathcal{D}(0,T;L^2(\Omega))$ solution of (8), we have

$$\int_0^T\int_\Omega y_1\Big(-\frac{\partial\phi}{\partial t} - \lambda\Delta\phi - \beta\cdot\nabla\phi\Big)dxdt = \int_0^T v(t)\phi(b,t)\,dt. \tag{10}$$

As we are considering the viscous and incompressible fluid, we obtain

$$\beta\big|_{\Gamma_2\cup\Gamma_4} = 0 \quad \text{and} \quad div(\beta) = 0. \tag{11}$$

Therefore, by integrating (10) by parts and using (11), we find

$$\int_0^T\int_\Omega\Big(\frac{\partial y_1}{\partial t} - \lambda\Delta y_1 + \beta\cdot\nabla y_1\Big)\phi dxdt = \int_0^T\int_\Omega v(t)\delta(x-b)\;\phi(x,t)dxdt,$$

for all $\phi \in \mathcal{D}(0,T;L^2(\Omega))$. Thus

$$\frac{\partial y_1}{\partial t} - \lambda\Delta y_1 + \beta\cdot\nabla y_1 = v(t)\,\delta(x-b) \quad \text{in} \quad \mathcal{D}'(0,T;L^2(\Omega)).$$

Theorem 2.1. *If $n \le 3$. Given $v \in L^2(0,t)$, there exists a unique $y_1(v)$ in $L^2(0,T;L^2(\Omega))$ that satisfies*

$$\int_\Omega\int_0^T y_1\,\varphi\,dxdt = \int_0^T v(t)\phi(b,t)\,dt,$$

that is, which is an ultraweak solution of the system (6) *and such that*

$$v \to y_1(v) \;\; \textit{is continuous from} \;\; L^2(0,T) \longrightarrow L^2(0,T;L^2(\Omega)).$$

and

$$t \to y_1(t) \;\; \textit{is continuous from} \;\; [0,T] \longrightarrow H^{-1}(\Omega). \tag{12}$$

Proof. Repeat the prove given in Lions [5]. □

Remark 2.2. When $n > 3$, to get the ultraweak solution of the system (6), by Transposition Method, we need more regularity for the function φ, to secure that $\varphi \in L^2(0,T;H^3(\Omega))$.

Lemma 2.1. *The unique solution, $y(x,t;v)$ of the system (1-5) is an affine application in v, that is,*

$$F(v) = y(x,t;v) - y(x,t;0)$$

is a linear application in v.

It follows from (12) that for $n \geq 2$, there exist functions $v \in L^2(0,T)$ such that $y_1(t;v)$ is not defined in $L^2(\Omega)$, which implies that the functional J is not defined in $L^2(0,T)$. The question, now, is how to define $\mathcal{U}$ such that for $v \in L^2(0,T)$ we have $y_1(.,T;v) \in L^2(\Omega)$.

Definition of the Space $\mathcal{U}$: In the problem (6), for $n = 2,3$ and $\beta = 0$, Lions in the Ref. [7] showed by using Fourier transform that for functions $v \in L^2(0,t)$ we can define

$$\mathcal{U} = \left\{ v \in L^2(0,T);\ \text{such that } y_1(\cdot,T;v) \in L^2(\Omega) \right\}.$$

Such space is characterized as

$$y_1(\cdot,T,v) \in L^2(\Omega) \longleftrightarrow \int_0^T \int_0^T \frac{v(t)v(s)}{(2T-t-s)^{n/2}} dtds < \infty.$$

For our case, by considering $\beta \neq 0$ and writing $y_1 = z + w$, we have

$$\begin{aligned}
&\frac{\partial z}{\partial t} - \lambda \Delta z = v(t)\delta(x-b) && \text{in } \Omega \times (0,T) \\
&z(x,t) = 0 && \text{on } (\Gamma_1 \cup \Gamma_3) \times (0,T) \\
&\frac{\partial z}{\partial \nu}(x,t) = 0 && \text{on } (\Gamma_2 \cup \Gamma_4) \times (0,T) \\
&z(x,0) = 0 && \text{in } \Omega,
\end{aligned} \tag{13}$$

and

$$\begin{aligned}
&\frac{\partial w}{\partial t} - \lambda \Delta w + \beta \cdot \nabla y_1 = 0 && \text{in } \Omega \times (0,T) \\
&w(x,t) = 0 && \text{on } (\Gamma_1 \cup \Gamma_3) \times (0,T) \\
&\frac{\partial w}{\partial \nu}(x,t) = 0 && \text{on } (\Gamma_2 \cup \Gamma_4) \times (0,T) \\
&w(x,0) = 0 && \text{in } \Omega.
\end{aligned} \tag{14}$$

In the problem (13), we use Lions'[7] result. In (14), from theory of parabolic equations, $w \in L^2(0,T;V) \cap C^0([0,T];L^2(\Omega))$.

Thus, we define the space $\mathcal{U}$ of functions of controls as

$$\mathcal{U} = \left\{ v \in L^2(0,T);\ \int_0^T \int_0^T \frac{v(t)v(s)}{(2T-t-s)^{n/2}} dtds < \infty \right\}.$$

With the norm

$$|||v|||_{\mathcal{U}}^2 = \int_0^T |v|^2 dt + \int_0^T \int_0^T \frac{v(t)v(s)}{(2T-t-s)^{n/2}} dtds.$$

Properties of the Space $\mathcal{U}$: (see the Refs. Lions [6] and [7]).

- $\mathcal{U}$ with the above norm is a Hilbert Space.
- $\mathcal{U}$ does not depend on Ω.
- $\mathcal{U}$ does not depend on boundary conditions.
- $C_0^\infty(0,T) \subset \mathcal{U}$.
- $L_0^2(0,T) = \{v \in L^2(0,T); v \equiv 0 \text{ in } (T-\epsilon, T), \epsilon \geq 0\}$ is dense in $\mathcal{U}$, because of the regularizing effect.
- $|||v|||_{\mathcal{U}}$ is equivalent to the norm

$$\|v\|_{\mathcal{U}}^2 = \int_0^T |v|^2 dt + \int_\Omega |y(x,T;v)|^2 dx.$$

The space $\mathcal{D}(0,T)$ is dense in $\mathcal{U}$. Therefore, if $\mathcal{U}'$ denotes the dual space of $\mathcal{U}$, then

$$\mathcal{D}(0,T) \subset \mathcal{U} \subset L^2(0,T) \subset \mathcal{U}' \subset \mathcal{D}'(0,T).$$

Definition of $\mathcal{U}_{ad}$: We consider closed convex subset $\mathcal{U}_{ad}$ of $\mathcal{U}$.

- $\mathcal{U}_{ad} = \mathcal{U}$, or
- $\mathcal{U}_{ad} = \{v \in \mathcal{U};\ v \geq \psi \geq 0 \text{ a.e. in } (0,T)\}$, ψ given in $H^1(0,T)$.

One can show that J is strictly convex, continuous and coercive on $\mathcal{U}_{ad}$. Therefore, by Direct Method of the Calculation of Variations we have the following Theorem.

Theorem 2.2. *There exists a unique element* $\mathbf{u} \in \mathcal{U}_{\mathbf{ad}}$ *that minimizes the functional* J*, where* $\mathbf{u}$ *is the optimal control.*

The pair $(u, y(u))$ is called of the optimal solution for the state system (1-5).

2.2. *Characterization of Optimal Control*

Theorem 2.3. *The optimal control u is characterized by*

$$\int_\Omega \Big[y_1(x,T;u) - \big(z_d - y_2(x,T)\big)\Big]\big[y_1(x,T;v-u)\big]dx + N\int_0^T u(v-u)dt \geq 0, \tag{15}$$

$\forall\, v \in \mathcal{U}_{ad}$ *and* $u \in \mathcal{U}_{ad}$.

Proof. Consider $f(\theta) = J\big((1-\theta)u+\theta v\big)$, $\forall v \in \mathcal{U}_{ad}, u \in \mathcal{U}_{ad}$ and $0 \leq \theta \leq 1$. Note that $f(0) \leq f(\theta)$, $\forall\, \theta \in [0,1]$. Thus, f has a minimum in zero and therefore $f'(0) \geq 0$. By differentiating $f(\theta)$, the result follows. □

The estimate in (15) is not appropriate to numerical analysis, because does not describe u in explicit form. To given an appropriate characterization of u, we introduce the adjoint state.

Adjoint State: The system of the adjoint state is given by

$$\begin{aligned}
&-\frac{\partial q}{\partial t} - \lambda\Delta q - \beta\cdot\nabla q = 0 && \text{in } \Omega\times(0,T)\\
&q(x,t) = 0 && \text{on } \Gamma_1\times(0,T)\\
&q(x,t) = g(x,t) && \text{on } \Gamma_3\times(0,T)\\
&\frac{\partial q}{\partial\nu}(x,t) = h(x,t) && \text{on } \Gamma_2\cup\Gamma_4\times(0,T)\\
&q(x,T) = y_1(x,T;u) - \big[z_d - y_2(x,T)\big] && \text{in } \Omega.
\end{aligned}$$

This system admits a unique solution q satisfying

$$q \in L^2(0,T;V)\cap C^0([0,T];L^2(\Omega)) \quad \text{and} \quad \frac{\partial q}{\partial t} \in L^2(0,T;V').$$

Also here, by decomposing this system, we obtain

$$\begin{aligned}
&-\frac{\partial q_1}{\partial t} - \lambda\Delta q_1 - \beta\cdot\nabla q_1 = 0 && \text{in } \Omega\times(0,T) \qquad (16)\\
&q_1(x,t) = 0 && \text{on } (\Gamma_1\cup\Gamma_3)\times(0,T)\\
&\frac{\partial q_1}{\partial\nu}(x,t) = 0 && \text{on } (\Gamma_2\cup\Gamma_4)\times(0,T)\\
&q_1(x,T) = y_1(x,T;u) - \big[z_d - y_2(x,T)\big] && \text{in } \Omega,
\end{aligned}$$

and

$$\begin{aligned}
&-\frac{\partial q_2}{\partial t}-\lambda\Delta q_2-\beta\cdot\nabla q_2=0 && \text{in } \ \Omega\times(0,T)\\
&q_2(x,t)=0 && \text{on } \ \Gamma_1\times(0,T)\\
&q_2(x,t)=g(x,t) && \text{on } \ \Gamma_3\times(0,T)\\
&\frac{\partial q_2}{\partial\nu}(x,t)=h(x,t) && \text{on } \ (\Gamma_2\cup\Gamma_4)\times(0,T)\\
&q_2(x,T)=0 && \text{in } \ \Omega.
\end{aligned}$$

We have $q_1 \in C^\infty([0,T)\times\Omega)$. Therefore, we can define $q_1(b,t)$ for $t<T$. By using the adjoint state (16), the state system (6) and the equations (11), we arrive to a similar Lions' [6] result.

Theorem 2.4. *One has $q_1(b,t)\in\mathcal{U}'$ and the condition (15) is equivalent to*

$$\int_0^T \Big(q_1(b,t)+Nu\Big)(v-u)dt\geq 0,\quad \forall v\in\mathcal{U}_{ad},\quad u\in\mathcal{U}_{ad}.$$

The integral above denotes the duality between $\mathcal{U}'$ and $\mathcal{U}$.

Remark 2.3. This characterization is better than such in (15), because it is possible to write the equation as a system of the Partial Differential Equation for some convex closed $\mathcal{U}_{ad}$. For example:

- If $\mathcal{U}_{ad}=\mathcal{U}$, we arrive that the optimal control is given by

$$u=-\frac{1}{N}q_1(b,t).$$

- If $\mathcal{U}_{\psi_{ad}} = \{\, v\in\mathcal{U};\ v\geq\psi\geq 0 \ \textit{a.e.} \text{ in } (0,T)\}$, with ψ given in $H^1(0,T)$, we get

$$u_\psi=\psi+\left(\frac{1}{N}q_{1\psi}(b,t)+\psi\right)^- \quad \text{if} \quad \psi\neq 0,$$

  and

$$u_0=\left(\frac{1}{N}q_{1_0}(b,t)\right)^- \quad \text{if} \quad \psi\equiv 0.$$

We will summarize some properties of the optimal control.

Lemma 2.2. *Consider $u_\psi(t)>0$, $t\in(0,T)$ and $(t_1,t_2)\subseteq(0,T)$, then the following properties are satisfied.*

(i) If $\psi(t)\leq u_0\ \forall\ t\in(0,T)$ then $u_\psi\equiv u_0$.

(ii) If $\psi(t) \leq u_0(t)\ \forall\, t \in (t_1, t_2)$ *then* $u_\psi(t) \leq u_0(t)\ \forall\, t \in (t_1, t_2)$.
(iii) If $\psi(t) \geq u_0(t)\ \forall\, t \in (t_1, t_2)$ *then* $u_\psi(t) \geq u_0(t)\ \forall\, t \in (t_1, t_2)$.

A graphic that illustrates this properties, among the optimal controls u_0 when $\psi \equiv 0$ and u_ψ when $\psi \neq 0$ for the following functions ψ:

$$\psi_0 \equiv 0, \ \ \forall\, t \in (0, T),$$

$$\psi_1 = \begin{cases} 0.1 & \text{if } t \in [0, 20000) \\ 0.2 & \text{if } t \in [20000, 40000) \\ 0.3 & \text{if } t \in [40000, 60000) \\ 0.4 & \text{if } t \in [60000, 80000) \\ 0.0 & \text{if } t \in [80000, 100000], \end{cases} \qquad \psi_2 = \begin{cases} 0.0 & \text{if } t \in [0, 20000) \\ 0.3 & \text{if } t \in [20000, 40000) \\ 0.4 & \text{if } t \in [40000, 60000) \\ 0.5 & \text{if } t \in [60000, 80000) \\ 0.0 & \text{if } t \in [80000, 100000] \end{cases}$$

it is the following

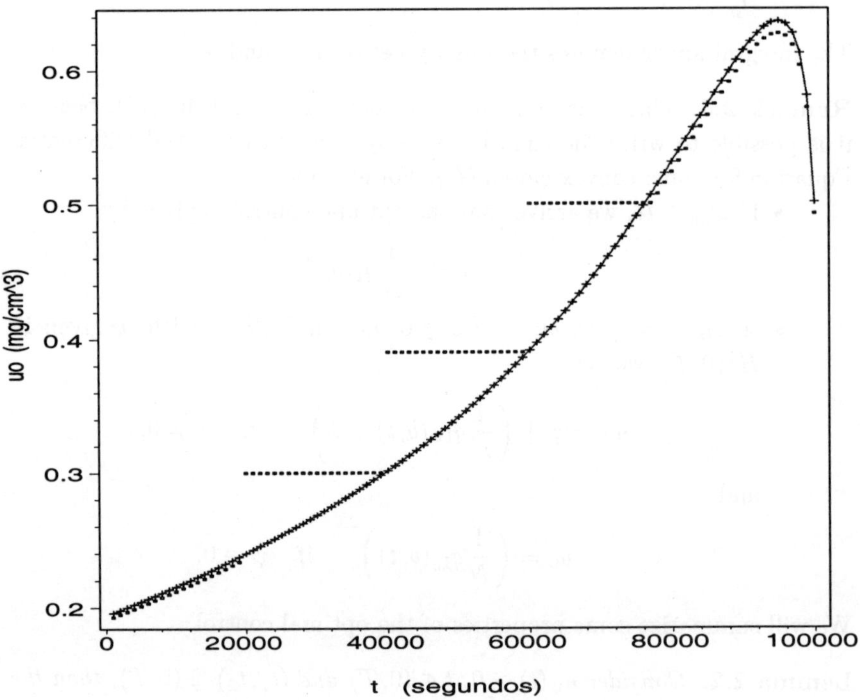

Figure 1. Comparison of u_0 with u_{ψ_1} and u_{ψ_2}
——— u_0, $++++\, u_{\psi_1}$, $\ldots\ldots\, u_{\psi_2}$

For this case, we consider $T = 100000s$, $\beta = (0,0)cm/s$, $\Omega = 1000cm^2$, $z_d = 10mg/cm^3 \times sen\,(\pi x/1000) \times sen\,(\pi y/1000)$, $N = 10$, $\lambda = 1cm^2/s$ and $h = g = 0$.

Optimality System: The optimality system is fundamental for the numerical calculation of the solution of the problem. In this work, we are going to consider the convex and closed subset $\mathcal{U}_{\psi_{ad}}$ given above. We have the following Theorem.

Theorem 2.5. *The optimal control u, that minimizes the functional J is characterized by the unique solution $\{u, y, q\}$ of the optimality system*

$$\begin{aligned}
&\frac{\partial y}{\partial t} - \lambda\Delta y + \beta\cdot\nabla y = \left[\psi + \left(\frac{1}{N}q_1(b,t) + \psi\right)^{-}\right]\delta(x-b) && \Omega\times(0,T)\\
&-\frac{\partial q}{\partial t} - \lambda\Delta q - \beta\cdot\nabla q = 0 && \Omega\times(0,T)\\
&y(x,t) = q(x,t) = 0 && \Gamma_1\times(0,T)\\
&y(x,t) = q(x,t) = g(x,t) && \Gamma_3\times(0,T)\\
&\frac{\partial y}{\partial \nu}(x,t) = \frac{\partial q}{\partial \nu}(x,t) = h(x,t) && (\Gamma_2\cup\Gamma_4)\times(0,T)\\
&y(x,0) = 0 && \Omega\\
&q(x,T) = y_1(x,T;u) - \left[z_d - y_2(x,T)\right] && \Omega. \qquad (17)
\end{aligned}$$

This system can be decomposed

$$\begin{aligned}
&\frac{\partial y_1}{\partial t} - \lambda\Delta y_1 + \beta\cdot\nabla y_1 = \left[\psi + \left(\frac{1}{N}q_1(b,t) + \psi\right)^{-}\right]\delta(x-b) && \Omega\times(0,T)\\
&-\frac{\partial q_1}{\partial t} - \lambda\Delta q_1 - \beta\cdot\nabla q_1 = 0 && \Omega\times(0,T)\\
&y_1(x,t) = q_1(x,t) = 0 && (\Gamma_1\cup\Gamma_3)\times(0,T)\\
&\frac{\partial y_1}{\partial \nu}(x,t) = \frac{\partial q_1}{\partial \nu}(x,t) = 0 && (\Gamma_2\cup\Gamma_4)\times(0,T)\\
&y_1(x,0) = 0 && \Omega\\
&q_1(x,T) = y_1(x,T;u) - \left[z_d - y_2(x,T)\right] && \Omega, \qquad (18)
\end{aligned}$$

and

$$
\begin{aligned}
&\frac{\partial y_2}{\partial t} - \lambda\Delta y_2 + \beta\cdot\nabla y_2 = 0 && \text{in} \quad \Omega\times(0,T)\\
&-\frac{\partial q_2}{\partial t} - \lambda\Delta q_2 - \beta\cdot\nabla q_2 = 0 && \text{in} \quad \Omega\times(0,T)\\
&y_2(x,t) = q_2(x,t) = 0 && \text{on} \quad \Gamma_1\times(0,T)\\
&y_2(x,t) = q_2(x,t) = g(x,t) && \text{on} \quad \Gamma_3\times(0,T)\\
&\frac{\partial y_2}{\partial \nu}(x,t) = \frac{\partial q_2}{\partial \nu}(x,t) = h(x,t) && \text{on} \quad (\Gamma_2\cup\Gamma_4)\times(0,T)\\
&y_2(x,0) = q_2(x,T) = 0 && \text{in} \quad \Omega.
\end{aligned} \tag{19}
$$

Note that the systems (17), (18) and (19) are not Cauchy problems. For the numerical solution of the problem (19) we find no difficulty and it is enough to solve it in y_2. Now, to solve the system (18), due to the coupling term its solution can not be obtained by a direct calculation, to solve it we will uncouple the system (18). What is not simple for, $n = 2, 3$, because $\delta \in H^{-2}(\Omega)$.

2.3. *Approach and Numerical Convergence*

We are going to regularize the system (18). Let ϵ be given with $0<\epsilon<T$. By the regularizing effect of parabolic equations, we rewrite the system (18) in the form

$$
\begin{aligned}
&y_{1t}^{\epsilon} - \lambda\Delta y_1^{\epsilon} + \beta\cdot\nabla y_1^{\epsilon} = \chi_{[0,T-\epsilon]}\left[\psi^{\epsilon} + \left(\frac{1}{N}\int_\Omega \phi_i q_1 dx + \psi^{\epsilon}\right)^{-}\right]\phi_i\\
&-q_{1t}^{\epsilon} - \lambda\Delta q_1^{\epsilon} - \beta\cdot\nabla q_1^{\epsilon} = 0\\
&y_1^{\epsilon}(x,t) = q_1^{\epsilon}(x,t) = 0\\
&\frac{\partial y_1^{\epsilon}}{\partial \nu}(x,t) = \frac{\partial q_1^{\epsilon}}{\partial \nu}(x,t) = 0\\
&y_1^{\epsilon}(x,0) = 0\\
&q_1^{\epsilon}(x,T) = y_1^{\epsilon}(x,T;u) - \left[z_d - y_2(x,T)\right].
\end{aligned} \tag{20}
$$

Where

$$
y_1^{\epsilon}(x,T;u) \longrightarrow y_1(x,T;u) \quad \text{strong in } H^1(\Omega)\cap V, \quad \text{as } \epsilon\longrightarrow 0 \tag{21}
$$

and $\phi_i \longrightarrow \delta(x-b)$ strong in $H^{-2}(\Omega)$, for $\phi_i \in C_0^\infty(\Omega)$. The system above is important, because for $z_d \in H^1(\Omega)$, we have $y_1(\cdot,T;u)$, $q_1(\cdot,T) \in H^1(\Omega)$, we will see this later. To solve the system numerically (20), we use the following algorithm. To simplify notation, we eliminate the index ϵ.

The uncoupled system (20) is given by

$$
\begin{aligned}
&{y_1}_t^n - \lambda \Delta y_1^n + \beta \cdot \nabla y_1^n = \chi_{[0,T-\epsilon]} \left[\psi + \left(\frac{1}{N} \int_\Omega \phi_i q_1^{n-1} dx + \psi \right)^- \right] \phi_i \\
&-{q_1}_t^{n-1} - \lambda \Delta q_1^{n-1} - \beta \cdot \nabla q_1^{n-1} = 0 \\
&y_1^n(x,t) = q^{n-1}(x,t) = 0 \\
&\frac{\partial y_1^n}{\partial \nu}(x,t) = \frac{\partial q^{n-1}}{\partial \nu}(x,t) = 0 \\
&y_1^n(x,0) = 0 \\
&q_1^{n-1}(x,T) = y_1^{n-1}(x,T;u) - \left[z_d - y_2(x,T) \right] \\
&y_1^0(x,T) = \text{ given.}
\end{aligned}
\tag{22}
$$

We are going to show to the convergence of the system (22) for the system (20) and later we will see that when $\epsilon \longrightarrow 0$, the system (20) converges to the system (18).

Theorem 2.6. *The unique solution of system (22) converges to the unique solution of the system (20) provided that N is large enough.*

Proof. Consider $R^n = y_1^n - y_1$ and $S^n = q_1^n - q_1$. From both systems (20) and (22), we have

$$
\begin{aligned}
&R_t^n - \lambda \Delta R^n + \beta \cdot \nabla R^n \\
&\qquad = \chi_{[0,T-\epsilon]} \left[\left(\frac{1}{N} \int_\Omega \phi_i q_1^{n-1} dx + \psi \right)^- - \left(\frac{1}{N} \int_\Omega \phi_i q_1 dx + \psi \right)^- \right] \phi_i \\
&-S_t^{n-1} - \lambda \Delta S^{n-1} - \beta \cdot \nabla S^{n-1} = 0 \\
&R^n(x,t) = S^{n-1}(x,t) = 0 \\
&\frac{\partial R^n}{\partial \nu}(x,t) = \frac{\partial S^{n-1}}{\partial \nu}(x,t) = 0 \\
&R^n(x,0) = 0 \\
&S^{n-1}(x,T) = R^{n-1}(x,T).
\end{aligned}
\tag{23}
$$

Consider $z_d \in H^1(\Omega)$. We will show that

$$
R^n, S^n \to 0 \quad \text{in} \quad L^2(0,T;L^2(\Omega)), \quad \text{as} \quad n \to \infty.
$$

Note that $S^n \in C^\infty([0,T) \times \Omega)$ and by the regularizing effect of parabolic equations, we have

$$
\|S^n\|_{L^2(0,T-\epsilon;H^2(\Omega))} \le c \|R^n(\cdot,T)\|_{L^2(\Omega)}. \tag{24}
$$

We first show that

$$R^n(x,T) \to 0 \quad \text{in} \quad L^2(\Omega), \quad \text{as} \quad n \to \infty. \tag{25}$$

By multiplying the first and the second equations of the previous system, respectively, by S^n and $-R^n$, adding the result of the two new equations, integrating in Ω by parts, using the equations (11) and the inequality $|f^- - g^-| \leq |f - g|$, we get

$$\frac{1}{2}\frac{d}{dt}\int_\Omega R^n S^n dx \leq \chi_{[0,T-\epsilon]}\frac{1}{N}\Big|\int_\Omega \phi_i S^{n-1}dx\Big|\Big|\int_\Omega \phi_i S^n dx\Big|.$$

By integrating from 0 to T and using that $\phi_i \longrightarrow \delta(x-b)$ strong in $H^{-2}(\Omega)$, we have

$$\int_\Omega R^n(x,T)S^n(x,T) \leq \frac{2}{N}\int_0^{T-\epsilon} \big|S^{n-1}(b,t)\big|\,\big|S^n(b,t)\big|dt.$$

By using the final condition of S^n, the Cauchy-Schwartz's and (24) inequalities and the Sobolev imbedding Theorem, we arrive to

$$\int_\Omega |R^n(x,T)|^2 dx \leq \frac{c}{N}\big\|R^{n-1}(x,T)\big\|_{L^2(\Omega)}\big\|R^n(x,T)\big\|_{L^2(\Omega)}.$$

Therefore

$$\int_\Omega |R^n(x,T)|^2 dx \leq \frac{c^2}{N^2}\int_\Omega |R^{n-1}(x,T)|^2 dx.$$

By choosing N such that $\alpha := \frac{c}{N} < 1$, we have

$$\int_\Omega |R^n(x,T)|^2 dx \leq \alpha^{2n}\int_\Omega |R^0(x,T)|^2 dx.$$

Which gives the desired result (25).
Now, we will show that

$$S^n \to 0 \quad \text{in} \quad L^2(0,T;L^2(\Omega)), \quad \text{as} \quad n \to \infty. \tag{26}$$

Observe that by using (11), we find

$$\int_\Omega (S^n \beta \cdot \nabla R^n) dx = -\int_\Omega R^n div(S^n\beta) dx = -\int_\Omega (R^n \beta \cdot \nabla S^n) dx.$$

In particular

$$\int_\Omega (\,\beta \cdot \nabla R^n)R^n dx = 0 \qquad \text{and} \qquad \int_\Omega (\,\beta \cdot \nabla S^n)S^n dx = 0. \tag{27}$$

Thus, by multiplying the second equation of (23) by S^n, integrating over $\Omega \times (0,T)$ by parts, using (27) and the final condition of S^n, we obtain

$$\int_\Omega \int_0^T |\nabla S^n|^2 dx dt \leq \frac{1}{2\lambda} \int_\Omega |S^n(x,T)|^2 dx = \frac{1}{2\lambda} \int_\Omega |R^n(x,T)|^2 dx.$$

Above, by using (25) and the Poincare's inequality, we have (26).
It remains to show that

$$R^n \to 0 \quad \text{in} \quad L^2(0,T;L^2(\Omega)), \quad \text{as} \quad n \to \infty. \tag{28}$$

For this we are going to use the method of transposition, defined in (9). Consider

$$\int_\Omega \int_0^T R^n \, \varphi dx dt = \int_0^{T-\epsilon} \Big[\Big(\frac{1}{N} q_1^{n-1}(b,t) + \psi \Big)^- - \Big(\frac{1}{N} q_1(b,t) + \psi \Big)^- \Big] \phi(b,t) dt, \tag{29}$$

for all $\varphi \in L^2(0,T;L^2(\Omega))$ and ϕ solution of (8).
From system (8), by using multiplicative techniques and relation (27), we arrive to

$$\int_0^T |\phi(b,t)|^2 \, dt \leq C \|\varphi\|^2_{L^2\left(0,T;L^2(\Omega)\right)}. \tag{30}$$

As in the equation (29) φ is arbitrary, by taking $\varphi = R^n$, making use of the Cauchy-Schwartz's, $|f^- - g^-| \leq |f - g|$ and relation (30), we have

$$\begin{aligned} \int_\Omega \int_0^T |R^n|^2 dt dx &\leq \frac{1}{N} \int_0^{T-\epsilon} \big|S^{n-1}(b,t)\big| \, \big|\phi(b,t)\big| dt \\ &\leq \frac{C}{N} \Big(\int_0^T \big|S^{n-1}(b,t)\big| dt \Big)^{\frac{1}{2}} \Big(\int_\Omega \int_0^T \big|R^n\big|^2 dt dx \Big)^{\frac{1}{2}}. \end{aligned}$$

Therefore

$$\int_\Omega \int_0^T |R^n|^2 dt dx \leq \frac{C^2}{N^2} \int_0^T \big|S^{n-1}(b,t)\big| dt.$$

By using the Sobolev imbedding Theorem, (24) and (25), we obtain (28).
The result of the Theorem follows from (26) and (28). □

The Lemma below will be useful in that follows.

Lemma 2.3. *If $z_d \in H^1(\Omega)$, then the solution of system (20) satisfies $y_1^\epsilon(.,T), q_1^\epsilon(.,T) \in H^1(\Omega)$.*

Proof. By using the regularizing effect of the parabolic equations, and reasoning as in Lions[6], our conclusion follows.

Finally, we will show the convergence of the system (20) to the system (18).

Theorem 2.7. *System (20) converges to the system (18), as* $\epsilon \longrightarrow 0$.

Proof. By multiplying the second equation of the system (20) by q^ϵ, integrating over $\Omega \times (0,T)$, using the equations (27) and the Lemma (2.3), we get

$$\int_\Omega \int_0^T |\nabla q^\epsilon|^2 dt dx \leq \frac{1}{2\lambda} \int_\Omega |q^\epsilon(x,T)|^2 dx \leq C. \tag{31}$$

Now, by multiplying the second equation of the system (20) by Δq^ϵ and proceeding as before, we have

$$(\lambda - \delta) \int_\Omega \int_0^T |\Delta q^\epsilon|^2 dt dx \leq \frac{1}{2} \int_\Omega |\nabla q^\epsilon(x,T)|^2 dx + c_\delta \int_\Omega \int_0^T |\beta \cdot \nabla q^\epsilon|^2 dt dx,$$

with δ small enough such that $\lambda - \delta > 0$. From the Lemma (2.3) and of the inequality (31), we have

$$\int_\Omega \int_0^T |\Delta q^\epsilon|^2 dt dx \leq C, \quad \text{as} \quad \epsilon \longrightarrow 0.$$

Thus, q^ϵ is limited in $L^2(0,T; H^2(\Omega) \cap V)$. Therefore, we can extract a subsequence, denoted in the same way, such that

$$q^\epsilon \longrightarrow q' \quad \text{weakly in} \quad L^2(0,T; H^2(\Omega) \cap V).$$

In particular

$$q^\epsilon(\cdot,T) \longrightarrow q'(\cdot,T) \quad \text{weakly in} \quad H^2(\Omega) \cap V.$$

From system (20), convergence (21) and the uniqueness of the weakly limit, it follows $q(x,T) = q'(x,T)$. Moreover, because of the uniqueness of the system (20), we obtain that $q = q'$. From the inequality (31), we have

$$\int_\Omega \int_0^T |\nabla q^\epsilon - \nabla q|^2 dt dx \leq \frac{1}{2\lambda} \int_\Omega |q^\epsilon(x,T) - q(x,T)|^2 dx \longrightarrow 0.$$

Therefore

$$q^\epsilon \longrightarrow q \quad \text{strongly in} \quad L^2(0,T; H^2(\Omega) \cap V). \tag{32}$$

Note also that when ϵ tends to zero,

$$\psi^\epsilon \longrightarrow \psi \quad \text{strongly in} \quad H^1(0,T). \tag{33}$$

Now, by using the ultraweak solution (as defined in (9)), considering

$$\varphi = y^\epsilon - y, \quad v = \psi + \Big(\frac{1}{N} q(b,t) + \psi\Big)^-, \quad v^\epsilon = \psi^\epsilon + \Big(\frac{1}{N} q^\epsilon(b,t) + \psi^\epsilon\Big)^-$$

and taking in account (32) and (33), we arrive at

$$\int_{\Omega}\int_{0}^{T} \left|y^{\epsilon} - y\right|^{2} dxdt = \int_{0}^{T} \left(v^{\epsilon} - v\right) \phi(b,t)\, dt \longrightarrow 0.$$

Thus

$$y^{\epsilon} \longrightarrow y \quad \text{strongly in} \quad L^2\big(0,T;L^2(\Omega)\big). \tag{34}$$

From both convergence (32) and (34), our conclusion follows. □

Remark 2.4. By varying the place $x = b$ of the pollution entrance, the system (1-5) does not change and for each place of the pollution entrance $x = b$, there exists a unique control $u_b \in \mathcal{U}_{ad}$ such that

$$J(u_b) = \inf\left\{J(v);\ v \in \mathcal{U}_{ad}\right\}.$$

As the optimal controls $u_b \in \mathcal{U}_{ad} \subset \mathcal{U}$, it follows that among the optimal controls u_b there exists a unique optimal control, denoted by u_{b_o}, such that

$$J(u_{b_o}) = \inf\left\{J(u_b);\ b \in \Omega\right\}.$$

This means that in Ω there exists a unique point b_o (called strategic point or optimal point) where pollution is, still, less harmful to the environment.

Remark 2.5. If the pollutant source is located, e.g.,

$$f(x,t) = v(t)\phi(x),$$

where $\phi(x) \in L^2(\Omega)$ with compact support in $\omega \subset \Omega$ determines the region of the pollution entrance, then we get more easily the same results of the pointwise source due to function ϕ being more regular.

3. Computational Results, Analysis and Conclusion

By the Brazilian Legislation, the acceptable level z_d of pollution of mercury (Hg) in drinking water is given in the next Table, containing some more constants than the ones used in the graphics.

Diffusion coefficient of Hg	$\lambda = 6 \times 10^{-6} cm^2/s$
Acceptable level of Hg	$z_d = 2\times 10^{-6} mg/cm^3 sen(\pi x/100) sen(\pi y/100)$
Velocity field (cm/s)	$\beta = (a\ sen\ \pi(100-y)/100\ ,\ 0),\ a \geq 0$
Cost constant	$N = 1 \times 10^5$
Final time	$T = 8640000$ s

Due to the fact that Mercury spreads out very slowly, it is enough to choose the small space domain. In case we increase too much the domain, a longer time will be necessary for observing the diffusion of Mercury, besides increasing a lot the computational cost. Thus, we consider a square domain $\Omega = (0, 100) \times (0, 100)cm^2$, where point (0,0) corresponds to $\Gamma_1 \cap \Gamma_2$ (see Figure 0). We use a mesh of 10000 equal-length quadratic elements with $\Delta x = \Delta y = 1\, cm$ and 400 steps in the time with $\Delta t = 21600\, s$ fixed.

We choose the boundary condition $h = 0$ and in the boundary Γ_3 the pollution is given by the function g defined by

$$g(x,t) = \begin{cases} (ay^2 + by + c)\, t & \text{if} \quad y \in [w_i, w_f], \\ 0 & \text{if} \quad y \notin [w_i, w_f]. \end{cases}$$

Where $[w_i, w_f] \subset \Gamma_3$ and the values w_i, w_f, a, b and c are given according to each graphic.

The sequence of functions ϕ_l, of compact support in $\omega \subset \Omega \subset \mathbb{R}^2$, is chosen as being

$$\phi_l(x) = \begin{cases} \frac{1}{l^2} & \text{if} \quad x \in \omega, \\ 0 & \text{if} \quad x \notin \omega. \end{cases}$$

Where ω is a square of the side l centered in the point $b \in \omega \subset \Omega$.

We plot the cases of the internal pollutant source $f = v(t)\phi_l(x)$. By making $l \to 0$, we get the graphic for the pointwise pollutant source $v(t)\delta(x - b)$.

The numerical problem is solved by using stabilized method SUPG of semidiscrete finite elements, see Hughes [4].

3.1. *Computational Results and Analysis*

In the figures (2)-(6), we are considering in the function g defined above $a = b = c = 0$ and the velocity field is zero, these are cases of the still water.

In this case (still water), we note that initially we have a certain amount of pollution u, such that in elapsing time, this amount diminishes so that the pollution concentration in the final time $y(x, T; u)$ does not exceed the allowed value z_d. We still observe through the graphics of the concentration that as the square region ω diminishes, e.g., $l \to 0$, we graphically confirm the convergence of the pollution concentration with source $v(t)\phi_l(x)$ for the pollution concentration with source $v(t)\delta(x-b)$.

In the following figures (5) and (6), we vary the place $x = b$ of the pollution entrance.

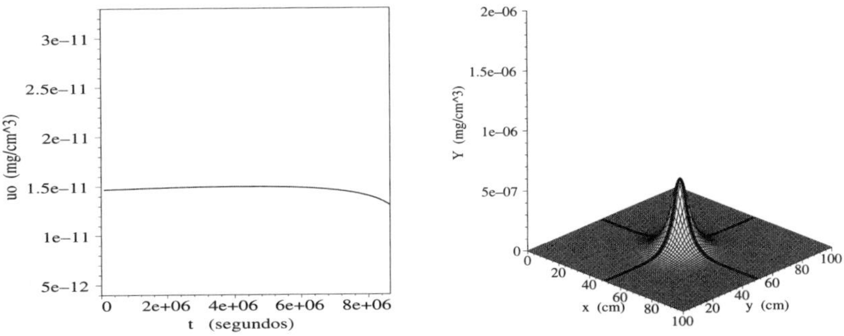

Figure 2. Control u_0 Concentration y, $\omega = (47, 53) \times (47, 53) cm^2$.

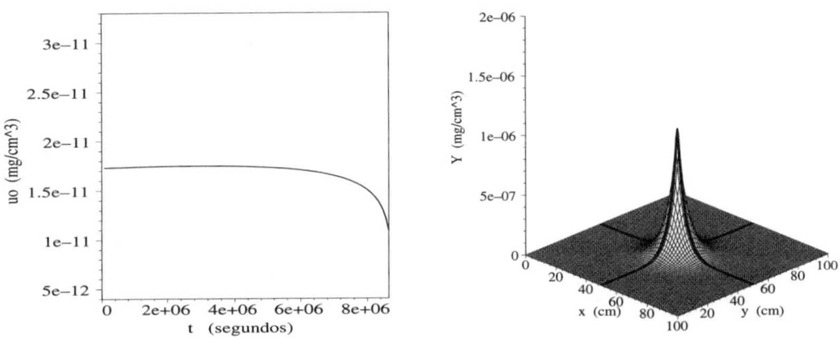

Figure 3. Control u_0 Concentration y, $\omega = (49, 51) \times (49, 51) cm^2$.

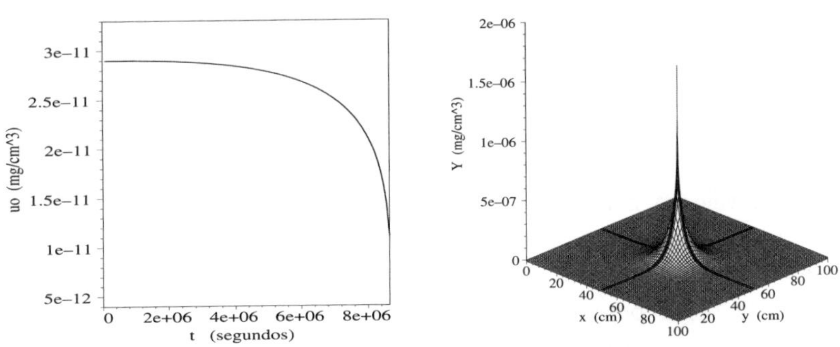

Figure 4. Control u_0 Concentration y, $\omega = (49.9, 50.1) \times (49.9, 50.1) cm^2$.

Note that $l = 2.0cm$ is the same of the Figure 3. We observe that the poured amount of pollution depends on the entrance place of pollutant. We noted that in the central region of the domain, the pollutant has more freedom of spreading, then a greater amount of pollutant can be poured in this place and consequently the pollution concentration at the final time

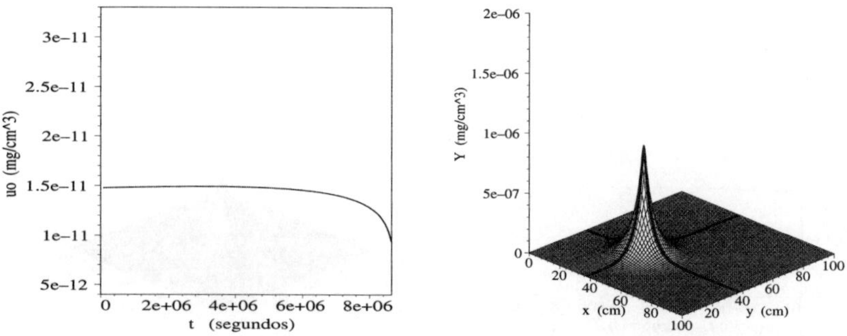

Figure 5. Control u_0 Concentration y, $\omega = (36, 38) \times (36, 38) cm^2$.

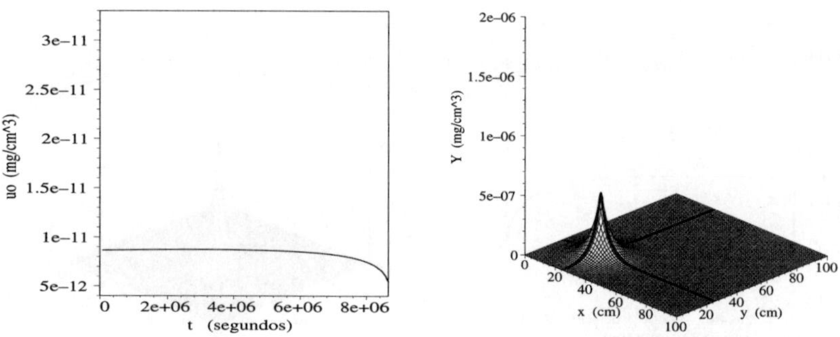

Figure 6. Control u_0 Concentration y, $\omega = (24, 26) \times (24, 26) cm^2$.

reaches a bigger index close to z_d. Now, as the pollutant source moves from the central region, the poured amount of pollution diminishes and consequently the pollution concentration also diminishes.

At last, in figures (7) and (8), we take in the function g, $[w_i, w_f] = [30, 70] cm$, with $a = -0,03 \times 10^{-17}$, $b = 3 \times 10^{-17}$ and $c = -60 \times 10^{-17}$. We consider the velocity field nonzero given by $\beta = (2 \times 10^{-5} cm/s \; sen \; \pi(100 - y)/100 \; , \; 0)$, these are cases where the water is in movement.

Here, we verify that the control u or the amount of the pollution entrance starts being poured in a defined level, is increased and later diminished in elapsing time. For this case (water is in movement), for the pollution concentration we note a similar behavior to the one of the still water when we diminish the square region ω.

3.2. *Conclusion*

Numerically, the solution of the problem agrees with the expected.

A difficulty for the numerical validation of the model is in the fact

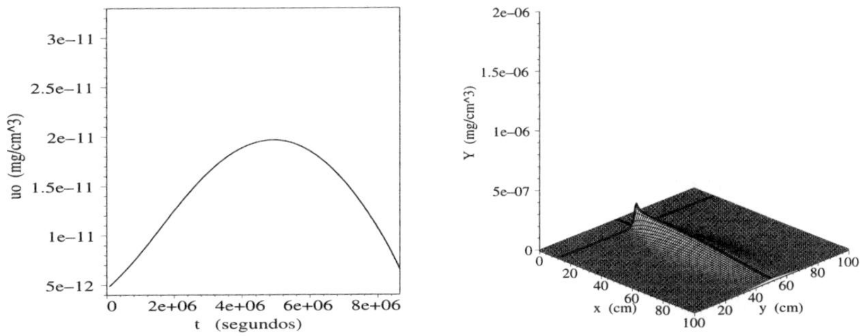

Figure 7. Control u_0 Concentration y, $\omega = (11, 13) \times (49, 51) cm^2$.

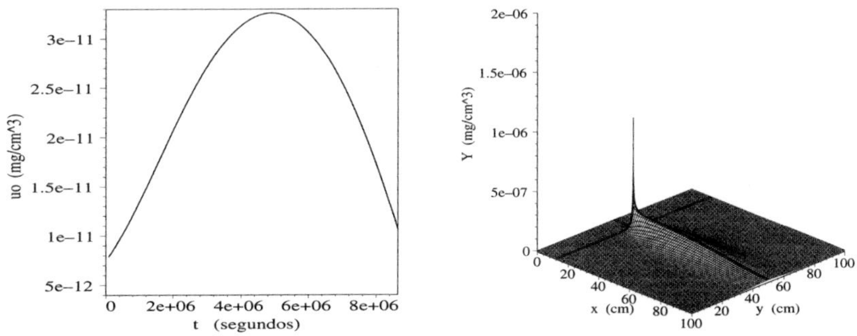

Figure 8. Control u_0 Concentration y, $\omega = (11.9, 12.1) \times (49.9, 50.1) cm^2$.

that the convergence of the method depends on the parameter N, due to the uncoupling method used. We are working to get a new uncoupling independent on this parameter.

In this problem, other situations can be studied. For example, the concerning function z_d, we could consider it dependent on time. Also, can be considered the optimal control in the boundary or the initial data, instead of considering in internal points. In these cases, the mathematical analysis depends on the regularity of the initial data, a problem that can be solved by taking the initial data in more regular spaces and the numerical analysis does not change very much. The care that we must have is not to prescribe the boundary in question. However, if we consider the optimal control in more than one parameter of the model, the problem becomes much more complex and the expected results are not so obvious, due to the difficulty that we have in controlling the mixture of the pollutant agent.

References

1. H.T.Banks, Control and Estimation in Distributed Parameter Systems. *Siam, Philadelphia. (1996)*
2. P. B. Bedient, H. S. Rifai and C. J. Newell, Ground Water Contamination Transport and Remediation. Pretice - Hall PTR, New Jersey (1994).
3. H. X. Couseuil and M. D. Martins, Contaminação de Águas Subterrâneas por Derramamento de Gasolina. Engenharia Satinária e Ambiental, Vol.2, N.2. (1997).
4. Hughes and J.R. Thomas, The finite Element Method-Linear static and Dynamical Finite Element Analysis Dover Publications, INC. Mineola, New York (2000).
5. J.L. Lions, Some Aspects of the Optimal Control of Distributed Parameter Systems. Université de Paris and I.R.I.A., Regional Conference Series in Applied Mathematics (1972).
6. J.L Lions, Function Spaces and Optimal Control of Distributed Systems. Rio de Janeiro-UFRJ (1980).
7. J.L. Lions, Some Methode in The Mathematical Analysis of Systems and Their Control. Science Press - Beijing, China (1981).
8. J.L. Schnoor, Environmental Modeling: Fate and Transport of Pollutants in Water, Air and Soil. John Wiley and Sons, Inc. New York. (1996)

MODELING THE IN VIVO DYNAMICS OF VIRAL INFECTIONS*

RUY M. RIBEIRO
Theoretical Biology and Biophysics,
Los Alamos National Laboratory,
MS K710, Los Alamos, New Mexico 87545, USA
E-mail: ruy@lanl.gov

The understanding of HIV infection has benefited enormously from detailed studies of the dynamics of the virus and immune responses. In particular, mathematical models have been successfully applied to the analysis of antiretroviral treatment. Modeling this treatment as a perturbation on the pre-therapy steady state of viral load revealed that virus and infected cells are cleared rapidly. Indeed the half-life of free HIV is approximately 1 hour, and the half-life of infected cells is about 1 day. I will give an overview of these results in the context of the dynamics of other infections and what it means for the success of treatment and the possibilities to develop a vaccine.

1. Introduction

1.1. *HIV a Global Pandemic*

Human immunodeficiency virus (HIV) infects 39.4 million people worldwide, especially in developing countries of sub-Saharan Africa, with almost 5 million new infections just in 2004 [1]. In 2004, 3.1 million people died of the complications occurring in late stage HIV [1], when acquired immunodeficiency syndrome (AIDS) manifests itself.

Although the number of infected individuals in western countries is relatively low, and it has been stable or declining for some time, the proportion of infected adults in sub-Saharan Africa is staggering, reaching up to 35% of the adult population in countries like Botswana or South Africa [1]. Moreover, the incidence of this infection is growing in large countries like China, India and Russia [1]. The population wide effects of this disease

*This work was done partially under the auspices of the Department of Energy and supported by NIH grants RR18754-02, and RR065555-14.

are devastating, by crippling the main productive age group in the population (young adults). It has reduced the life expectancy in several African countries, inverting a trend for increased quality of life that was verified before the explosion of the epidemic. For example, the life expectancy in Botswana has been reduced from 65 to 33 in the last 15 years, almost exclusively due to the AIDS epidemic. It has also made millions of children orphans, many of them also infected.

The medical treatments that exist, and help make HIV a manageable chronic infection are very expensive, and mostly only available to people living in richer countries, unless the government has a national strategy to fight the disease, such as is the exemplar case in Brazil. Today, it is generally believed that only an inexpensive and efficient vaccine can curb this pandemic. Unfortunately, so far, all efforts to develop such a vaccine have faced difficult scientific problems that have yet to be resolved.

1.2. *Basic Biology of HIV Infection*

HIV is a retrovirus and encodes its genetic material in two RNA positive-strands, linked by hydrogen bonds. These RNA molecules are about 9800 nucleotides long and include three main coding regions: *gag*, *pol* and *env* [2, 3]. These three regions are common to all retroviruses. *env* codes for the surface glycoproteins (gp) that make the virus envelope; *gag* codes for the matrix protein (MA) that lines the virus envelope, as well as the capsid proteins that both protect (CA) and form the core (NC), which contains the genome of the virion; and *pol* codes for the essential viral enzymes - reverse transcriptase (RT), integrase (IN) and protease (PR). Besides these proteins, HIV also expresses others which compose the viral particle (Vif, Vpr), control viral gene expression (Tat, Rev), and regulate host-cell functions (Vpu, Nef) [3].

The main target cell populations for HIV infection are those expressing the membrane surface protein CD4 and an appropriate co-receptor [4]. Two of the targets for infection are T-helper lymphocytes and macrophages, which are essential components of the immune system. HIV infects a cell by a succession of 8 more or less well characterized steps [2]. 1) Binding of the envelope proteins to target cell receptors. 2) Fusion of the envelope with the plasma membrane and release of the capsid into the cytoplasm. 3) Uncoating of the virus by shedding the capsid and partially shedding the nucleocapsid. 4) Reverse transcription of the viral RNA into DNA, with the help of the viral RT enzyme. 5) Integration of the pro-viral DNA into a random locus of the host-cell genome, via specific catalyses by the integrase enzyme. 6) Copying by the cell machinery of the viral DNA into viral RNA for new virions and mRNA for synthesis of the viral proteins. This transcription step may be regulated by viral proteins, such as Tat and Rev. 7) Assembly of the structural proteins (like MA, CA, NC) near the plasma membrane; and splicing by the protease enzyme of the long polyprotein, formed by transcription of the mRNA, into the individual proteins that form the virus particle together with the two strands of RNA. 8) Budding of new virions from the infected cell and the restart of the infection cycle.

As the infection cycles repeat themselves over and over again, the number of CD4+ T-helper cells, one of the main targets of infection, declines steadily [2]. Why this happens is still one of the biggest mysteries of HIV infection. It is likely associated with a hyperactivated immune system for most of the chronic phase of the infection. When the CD4+ T-cell count reaches below 200 cells μl^{-1}, from the normal steady state of 1000 cells μl^{-1}, the patient by definition enters the stage of AIDS (Figure 1). From this point on, and as the CD4+ T-cells decline even further, the patient's immune system has greater and greater difficulty fighting common opportunistic infections, which at higher CD4 counts may have been innocuous. Eventually the person dies from these infections. After primary infection, the viral load stays more or less constant until the late stage of infection when it increases dramatically, in association with morphological disaggregation of the lymph nodes [2].

1.3. *Treatment of HIV Infection*

Most of the drugs used against HIV can be classified in two classes: reverse transcriptase inhibiting (further subdivided into nucleoside and non-nucleoside analogues) and protease inhibiting drugs [5]. As their names indicate, the two classes of drugs are directed at distinct steps of the infection cycle, and lead to different outcomes. The RT-inhibitors (RTI) prevent new infections of uninfected cells, whereas the protease inhibitors (PI) prevent the formation of new infectious viruses, but not infection with virus already present. In addition to these drugs, there is one approved fusion inhibitor in the clinic. There are also new types of drugs being developed targeted at other steps of the lifecycle [6-12]: fusion, proviral DNA integration, and viral gene regulation.

All the drugs available at the moment select for drug resistant mutants when used in monotherapy [5]. Many of the primary re-

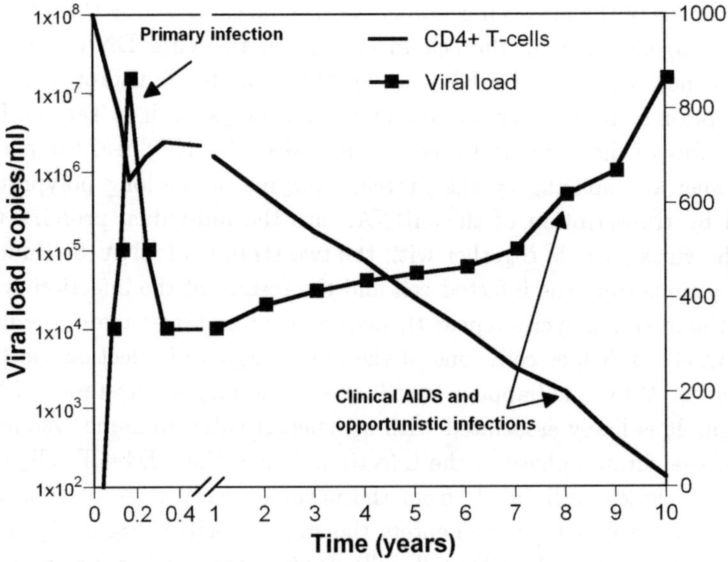

Figure 1. Schematic profile of viral load and CD4+ T-cell count in the peripheral blood of a HIV infected individual. Notice the break in the x-axis. The onset of clinical AIDS occurs when the CD4+ T-cell count drops below 200 cells μl^{-1}.

sistant mutants, those appearing first during therapy, differ from the consensus sequence by only one point mutation [13] (see also http://resdb.lanl.gov/Resist_DB/default.htm). Hence, it is not surprising that they are selected quickly, since it is expected that all viable one-point mutants are present in the viral quasi-species. Even though the fitness of these strains might be lower than the wildtype's fitness, their replication may lead to the emergence of better adapted strains, with additional mutations, that usually bring the viral load back to its initial level. (Indeed, viral load could recover just due to increases in target cells in the presence of less fit virus.) For these reasons, monotherapy with any of these drugs is not recommended, unless there are no alternatives [14]. Therefore, combination therapy is the preferred treatment against HIV, in principle, with at least one drug of each category (PI and RTI) [14]. The hope is that in these situations more than one point mutation will be necessary to confer resistance.

When treatment is initiated, there is a large and rapid decay in circulating virus (viral load), after an initial delay. Over the first week to ten days the viral load can decay several orders of magnitude, as the infection-reinfection cycles are stopped or diminished by the therapy (Figure 2) [15].

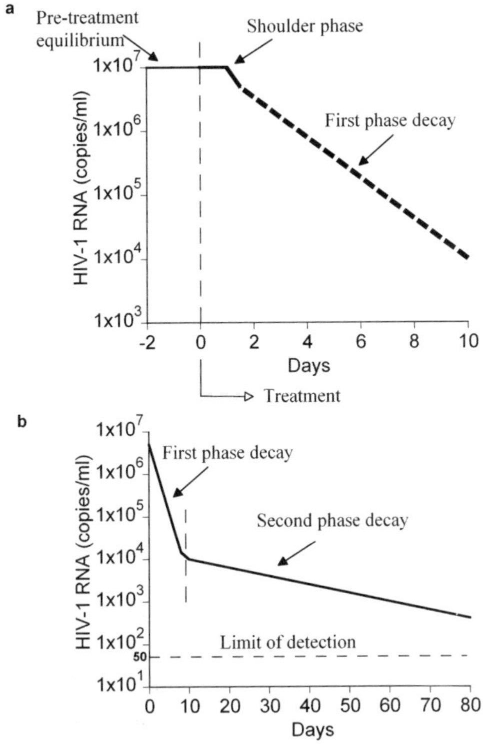

Figure 2. Typical pattern of viral load decay under drug therapy. (a) First 10 days of decay. (b) Long-term decay of the virus.

1.4. *Mathematical Modeling of Viral Infections*

The use of mathematical modeling and computer simulation in the study of infectious disease and, in particular HIV infection, has proven quite fruitful. The study of the dynamics of HIV infection represents a paradigm of success in the application of mathematical models to further our knowledge of disease pathogenesis [15, 16]. With these analyses, it was possible to quantify the rapidity of HIV infection and replication, the rate of virion clearance, the lifespan of productively infected cells [15, 17-19], and predict the impact of treatment and the appearance of drug resistant mutations [20, 21]. Moreover, models have helped clarify controversial issues relating to the mechanism of T cell depletion in HIV infection and motivated new experimental and clinical studies [22].

These same techniques have been used to model other viral infections, most notably hepatitis C virus [23] and hepatitis B virus infections [24]. Although biologically (and medically) these are all very different infections, the mathematical approach to each is similar, and modeling has afforded important insights into the effects of treatment, the mode of action of drug, and the responses to treatment in different patient groups.

Here I will present the modeling approach used to analyze dynamic data from viral infections. These models are driven by the data, and as such represent simple abstractions of a complex biological system. The objective is not to reproduce the complex system behavior in qualitative terms, rather we seek to understand key aspects of the biology, interpret experimental data and make physiological recommendations, which can be used in the clinic.

2. Modeling HIV infection

2.1. *Developing the Model*

In the simplest model, we only take into account the key players of the system. Thus, we have uninfected target cells (T), infected cells (I) and free virus (V). For example, target cells could correspond to CD4+ T-helper cells expressing the right amount/combination of the co-receptor to be susceptible to infection. We assume that these cells are produced at a constant rate λ per day, die at a rate d per cell per day, and can be infected by free virus, according to a simple mass action infection term, i.e., βVT. This generates infected cells, which are lost at a rate δ, larger than d, to reflect viral effects in shortening the infected cell lifespan. Finally, free viruses are produced by infected cells at a constant rate p per cell per day, and are cleared from circulation at a rate c per virus per day [25]. Thus, the differential equations describing this system are

$$\begin{aligned} \frac{dT}{dt} &= \lambda - dT - \beta VT \\ \frac{dI}{dt} &= \beta VT - \delta I \\ \frac{dV}{dt} &= pI - cV \end{aligned} \tag{1}$$

This is clearly a simple set of equations describing a complicated phenomenon. For example, clearance of free viruses can occur by a variety of mechanisms including filtration by the reticuloendothelial system, opsonization, binding to follicular dendritic cells, phagocytosis, etc., but we use only one term to include all these. A few of the assumptions warrant further discussion. First, we don't explicit model the origin of target cells, we simply assume that there is a inexhaustible supply that trickles into the susceptible compartment at a constant rate. Other models have been proposed where these cells are maintained homeostatically, or are slowly exhausted. Infection occurs at a rate given by a mass action term, which assumes a well mixed compartment and random motion. This may not be the case in solid tissue, such as lymph nodes, where most of infection is probably occurring. It is also not clear how virions are produced. Here we assume a continuous production at a constant rate (p), however it is possible that HIV infection is lytic, that is the viruses burst the cell open as they leave the infected cell. If this is the case, it may be more appropriate to model virion production with a term proportional to the death of infected cells such as $N\delta I$, where N is the burst size, i.e. the average number of virions produced when a cell dies.

The reasons to use this simplified description are very practical. First, and foremost, as we will see, this formulation has been quite successful in helping us to interpret data and understand better the process of infection. Two, its simplicity goes hand in hand with the paucity of data available to analyze the model. In most experiments we only have the viral load measured in blood over a short period of time, and with more or less frequent sampling. Three, some of the shortcomings such as assuming a constant supply of target cells are not so important since we use this model to analyze only short term experimental data, when this supply can well be considered inexhaustible. For other types of modeling, such as understanding pathogenesis, *i.e.*, the long term outcome of infection and its progression, this model is clearly inappropriate and many variants have been discussed in the literature.

Upon infection, the model indicates that after a brief transient period, the three populations converge to the stable steady state given by

$$\begin{aligned} T^* &= \tfrac{\delta c}{p\beta} \\ I^* &= \tfrac{\lambda}{\delta} - \tfrac{cd}{p\beta} \\ V^* &= \tfrac{p}{c} I^* \end{aligned} \tag{2}$$

The data show that, although total CD4+ T-cells continually decrease at a rate of up to 10 cells month^{-1} (at least in peripheral blood) [2], viral load is usually stable for long periods of time, of the order of months to years, albeit with fluctuations. Indeed, a remarkable aspect of this infection, and an indication of its dynamical control, is that if an individual stops treatment, the viral load tends to go back to its pre-treatment steady state.

2.2. *Drug therapy: modeling*

The drugs most extensively used today, as mentioned above, block reverse transcription of the virus, thus preventing new infections; or block infectious virion production. The first effect can be modeled by reducing the infection term in equation 1 by a factor dependent on the drug effectiveness, η . Thus, the infection term becomes $(1-\eta)\beta VT$, since only (1-η) infections will be successful, the remaining being blocked by the drug. Likewise, a protease inhibitor reduces the production of free infectious virus by a factor ε. And the corresponding production term becomes $(1-\varepsilon)pI$. Here the details of the mechanism of drug action are important. Protease inhibitors block the maturation of virions after they are assembled. This occurs because the virions are initially produced enclosing a polyprotein, which needs to be cleaved by the protease to finalize virion maturation and render the virions infectious. Viruses are still being produced at rate pI, but the drug blocks maturation of a fraction ε of these virions. Since our assays for viral load do not distinguish between mature and immature virions, the model needs to keep track of both types. We, thus, introduce a new equation for the number of non-infectious virions being produced, and the modified equations are [25]

$$\begin{aligned}
\tfrac{dT}{dt} &= \lambda - dt - (1-\eta)\beta V_I T \\
\tfrac{dI}{dt} &= (1-\eta)\beta V_I T - \delta I \\
\tfrac{dV_I}{dt} &= (1-\varepsilon)pI - cV_I \\
\tfrac{dV_{NI}}{dt} &= \varepsilon pI - cV_{NI}
\end{aligned} \tag{3}$$

Notice that here we assume the same clearance rate for infectious and non-infectious virions. This system of non-linear equations does not have a closed form solution. However, if during the short period of drug treatment that we analyze (3-5 weeks), we assume that the target cell population remains approximately constant, we can linearize the system and easily solve it. If treatment starts at the infected steady state (equations 2), the solution for $V(t) = V_I(t) + V_{NI}(t)$, the quantity measured in the viral load assays, is [26]

$$V(t) = V_0 \frac{(\lambda_2 + \varepsilon c)e^{\lambda_1 t} - (\lambda_1 + \varepsilon c)e^{\lambda_2 t} - \varepsilon\theta e^{-ct}}{\theta(1-\varepsilon)} \tag{4}$$

Where the eigenvalues are c, $\lambda_1 = -1/2(c + \delta + \theta)$, and $\lambda_2 = -1/2(c + \delta - \theta)$, with $\theta = ((c-\delta)^2 + 4c\delta(1-\eta)(1-\varepsilon))^{1/2}$. This solution is valid for $\varepsilon \neq 1$, and a simplified expression is obtained if we solve for the case of a 100% efficient protease inhibitor regimen, *i.e.* $\varepsilon = 1$, as has been done by Perelson *et al.* [27].

$$V(t) = V_0 e^{-ct} + \frac{cV_0}{c-\delta}\left(\frac{(c-\delta\eta)}{c-\delta}(e^{-\delta t} - e^{-ct}) - (1-\eta)\delta t e^{-ct}\right) \qquad (5)$$

2.3. *Drug therapy: data fits*

We can fit the solution of the viral dynamic model (equation 4 above) to data of viral load decay under drug treatment [15] using standard non-linear fitting algorithms, such as Levenberg-Marquardt. There are, however, a couple of practical points to take into consideration. Often, we observe a delay between the start of treatment and the initial decrease in viral load. This can very well be due to the pharmacological delay, which is the time it takes for the drug to act after it is ingested. This is easily incorporated into the model, by making $V(t) = V_0$ for $t < \tau$, and $V(t)$ equal to the solution 4 only after the delay τ, with $t = t - \tau$. Another important issue is that the viral clearance is so rapid, that it is difficult to fit c, and to see its effects at the time scale of usual sampling. Indeed, many times the effects of c and τ are subsumed into the viral load "shoulder" observed at the beginning of treatment (see Figure 2a) [28-30]. However, fixing τ at some small value or neglecting it allows fitting c. The value thus estimated is a lower bound for the clearance of free viruses.

The first experiments where drug therapy data were collected with high frequency were done in 1995 [17, 18]. Since then, as better drug regimens and assays have developed, data accumulated allowing an estimation of both c and δ. The best estimates available today indicate that $c = 23 \pm 11$ day^{-1} [31] and $\delta = 1.0 \pm 0.3$ day^{-1} [32]. As mentioned above c is difficult to measure from treatment experiments, so it is worth pointing out that the current estimate comes from independent experiments involving plasma apheresis. In this type of experiments, plasma is taken from the patient and the virus filtered out, before the plasma is re-infused into the patient [31]. So this intervention also corresponds to a perturbation of the steady state, and very frequent measurements of the viral load (every few minutes), tracking its decay, allow the calculation of the viral clearance rate. Unfortunately, this is a difficult and somewhat risky technique and it was only performed on 4 patients, thus the large standard deviation.

2.4. *The Long Term Decay of Virus*

After the initial very fast decay in viral load, which corresponds to the clearance of free virus and the loss of infected cells, the virus continues to decrease but at a much smaller rate. In Figure 2b, we show an illustration of this continued decay of virus under treatment, past the first couple of weeks. This decay is thought to be due to the loss of other virus producing cells, which are much longer lived than those we call I in equations 1 [33]. These I cells most likely correspond to productively infected CD4+ T-helper cells, which are actively producing viruses.

Several proposals have been put forward regarding the nature of the long-lived infected cells. They could correspond to a population of CD4+ T-helper cells turning over more slowly (chronically infected cells), or to other cell populations like infected macrophages. In addition, modeling has shown that the decay of virus in this phase is compatible with the disappearance of virus trapped on the surface of follicular dendritic cells [34], without the need to invoke other infected cell populations.

Extensions of the basic model have been proposed and used to fit long-term data showing the second phase of decay [33]. These fits indicate that the half-life of cells producing virus in this phase is much longer, about 17 days [19, 35]. If this rate of decay was constant for a long period, it can be shown that 100% efficient treatment would eventually eradicate infection after 2-3 years [19, 35]. Unfortunately, this is not the case, and very minor populations that decay even more slowly exist and become important producers of virus when the viral load is very low, eg. below the detection limit of $\sim$50 copies/ml.

In particular, HIV, being a retrovirus, integrates into the genome of the host cell and can remain silent (not copying itself) for long periods of time. These cells are said to be latently infected, and the best estimates indicate that this compartment decays with a half-life of >6 months [36-40]. This, thus, seems to indicate that viral eradication by antiretroviral therapy alone will be exceedingly difficult.

2.5. *Implications of Viral Dynamics Results*

HIV is a chronic infection that lasts many years until it manifests itself in the form of AIDS. For this reason, it was initially thought that it was a latent infection. That is, the virus persisted without continued replication inside long-lived infected cells, such as is the case for instance in herpes simplex virus [10]. However, the results of mathematical modeling of drug treatment experiments proved otherwise.

If the virus clearance rate is ~ 23 day^{-1}, and if during chronic infection viral load is maintained at $\sim 5 \times 10^5$ copies/ml, then we can calculate that, in the volume of 15 l of extra-cellular fluid in an average adult, the total production (and clearance) of virus reaches $\sim 1.7 \times 10^{11}$ viruses per day. The same calculation can be done to assess the loss and production of CD4+ T-helper cells per day. It comes out to $\sim 3 \times 10^7 \times 1$ T-cells per day [41]. These levels of turnover continue for the 10-year duration of an average infection. Thus, HIV infection is not silent or latent, rather it is better described as a very dynamic equilibrium between continued massive destruction and production of viruses and cells.

Another important consequence of such high turnover, coupled with high mutation rates (3×10^{-5} per base pair per infection cycle [42]), is that the viral population exhibits great variation and diversity. These characteristics play a crucial role in viral diversification and evolution [43, 44]. In fact, HIV exists as a swarm of different mutants, more or less clustered around the consensus or wildtype sequence. This swarm of mutants is observed at any specified moment in time, and thus classical competitive exclusion is very rare. Instead, many viral genotypes, and even phenotypes, co-exist *in vivo* [44-47], not only because they are replication competent, but also because they are continuously being created by mutation from the wildtype and from each other. This ensemble of viral variants actively participating in the evolution of the virus is usually called a quasi-species [48-50]. The importance of the quasispecies is that the viral population, as a whole, can adapt quickly and efficiently to any changes in the environment. In fact, the innumerable different mutants available increase by many fold the likelihood of at least one being able to thrive in the new environment. For example, drug resistance can be quickly acquired upon the start of treatment.

With the mutation and production rates indicated above, we can calculate that for HIV's $\sim$ 9800-base long genome, we have: 9800 base pairs $\times 3 \times 10^{-5}$ mutations per base pair per infection cycle $\times 3 \times 10^7$ replications per day $= \sim 9 \times 10^6$ one-point mutations per day. Similarly, $\sim$265 two-point mutations per day are generated. This last calculation assumes that the mutants are generated by simultaneous mutation at two bases. However, the most likely scenario is that the virus will accumulate mutation in successive replication rounds [21]. Mutants differing by three nucleotides from the wildtype are more likely to be created from existing two-point mutants then directly from the wildtype strain. Thus, we can imagine a network of mutants emerging due to changes from the wildtype (Figure 3) [21]. The exact composition (*i.e.*, frequencies for each mutant) of this quasispecies depends on the relative fitness of each mutant. It can be shown, in a simplified scenario where all intermediate viruses have the same fitness disadvantage s in relation to wildtype (*i.e.*, fitness of wildtype is 1 and of the mutant is 1-s) [21], that the frequency of each mutant relative to the wildtype is approximately $n!(\mu/s)^n$. Here $n!$ indicates the combinatorial number of ways in which the virus can evolve from the wildtype to the n-point mutant. Since this is a relative frequency, it can not be larger than one and when s is very small or n very large the approximations used to derive the formula break down.

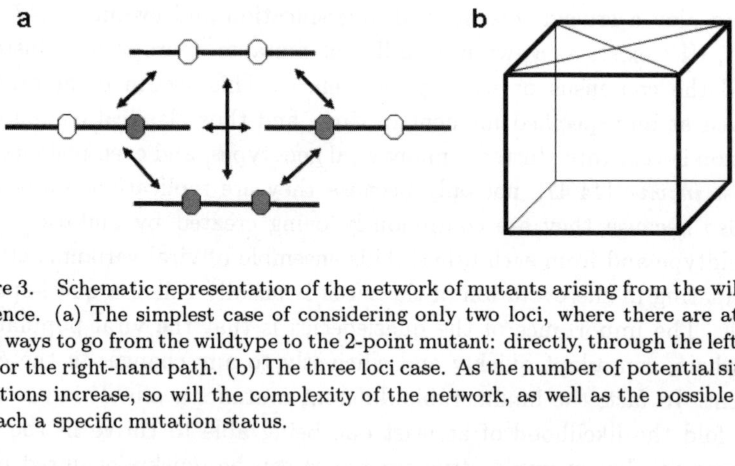

Figure 3. Schematic representation of the network of mutants arising from the wildtype sequence. (a) The simplest case of considering only two loci, where there are at least three ways to go from the wildtype to the 2-point mutant: directly, through the left-hand path or the right-hand path. (b) The three loci case. As the number of potential sites for mutations increase, so will the complexity of the network, as well as the possible paths to reach a specific mutation status.

In practice, this huge diversity means that the virus can evolve very quickly. This is reflected in the huge viral diversity at the population level, which for example makes vaccine development much more difficult [51-53]. And also in the diversity at the individual level, which leads to the appearance of drug resistance for many suboptimal regimens. In addition, this diversity helps the virus escape the vigorous immune response that the host usually develops against it [51].

2.6. *Comparison with Other Infections*

After the initial success in applying mathematical modeling concepts to the study of HIV dynamics, the same ideas were applied to a number of other chronic viral infections [23, 24]. It then became clear that the fast turnover observed in HIV is shared by other apparently latent infections such as hepatitis C (HCV) and hepatitis B (HBV) infections. These infections affect even more people worldwide than HIV, with 200 million people infected with HCV [54] and 400 million with HBV [10].

Mathematical modeling of these infections helped elucidate the viral lifecycles [23], the dynamic nature of the infections [55], the mode of action of different drug regimens [55-57], the relative efficacy of different therapy protocols, the differences in treatment response among diverse infected populations [58, 59], as well as having important health care applications. Modeling showed that the estimated drug efficacy and viral load 24h post-initiation of treatment were very good predictors of final outcome of therapy 52 weeks later [60].

In Table 1, we present a comparison among the three infections in terms of some of the modeling-estimated dynamic parameters. An interesting aspect is that although the initial response, corresponding to the clearance of free viruses is similar in all cases; there is large variation in the decay of infected cells, both comparing the 3 infections and within the hepatitis infections. These differences likely reflect the cytopathicity of the infection, with HIV infection probably leading to a fast death of the infected cells, whereas both HBV and HCV infections do not per se kill the cell (hepatocyte).

Table 1. Comparison of estimated dynamic parameters in three viral infections [55, 61, 62]

	HIV	**HBV**	**HCV**
Virions			
Half-life (h)	< 1	24	3
Daily turnover (%)	> 100	50	> 99
Daily production (virus)	2×10^{11}	2×10^{12}	5×10^{11}
Infected cells			
Half-life (days)	0.7	10 to > 100	3 to 70
Daily turnover (%)	63	1 to 7	10 to 25

3. Conclusions

Mathematical modeling of viral infections, especially HIV, has provided a new and powerful tool to understand the lifecycle, pathogenesis and treatment of these diseases. These models when closely tied with experimental work and validation have helped change therapeutic protocols and define new priorities of research.

The continued interaction between theoretical biologists, from many different backgrounds such as mathematics, physics, computer science and chemistry, and experimental scientists, such as clinicians, virologists and immunologists, has the potential to provide great new insights into the mechanisms of disease and host response to disease.

References

1. UNAIDS/WHO (2004) (UNAIDS/WHO, Geneva).
2. J.A. Levy HIV and the Pathogenesis of AIDS, American Society of Microbiology, Washington, 1998.
3. N. Nathanson, R. Ahmed, F. Gonzlez-Scarano, D. Griffin, K. Holmes, F. Murphy, H. Robinson, Viral Pathogenesis, Lippincot-Raven ublishers, Philadelphia, 1997, pp. 940
4. E.A. Berger, P.M. Murphy, J.M. Farber, Chemokine receptors as HIV-1 coreceptors: roles in viral entry, tropism, and disease, Annu. Rev. Immunol. 17 (1999) 657-700.
5. A.M. Vandamme, K. VanLaethem, E. DeClercq, Managing resistance to anti-HIV drugs - An important consideration for effective disease management, Drugs 57 (1999) 337-361.
6. G. Barbaro, A. Scozzafava, A. Mastrolorenzo, C.T. Supuran, Highly active antiretroviral therapy: current state of the art, new agents and their pharmacological interactions useful for improving therapeutic outcome, Curr. Pharm. Des. 11 (2005) 1805-1843.
7. D. Daelemans, A.M. Vandamme, E. de Clercq, HIV gene regulation as target for anti-HIV chemotherapy, Antiviral Chemistry & Chemotherapy 10 (1999) 1-14.
8. G.A. Donzella, D. Schols, S.W. Lin, J.A. Este, K.A. Nagashima, P.J. Maddon, G.P. Allaway, T.P. Sakmar, G. Henson, E. DeClercq, J.P. Moore, AMD310, a small molecule inhibitor of HIV-1 entry via the CXCR4 co- receptor, Nature Medicine 4 (1998) 72-77.
9. M.J. Endres, S. Jaffer, B. Haggarty, J.D. Turner, B.J. Doranz, P.J. Obrien, D.L. Kolson, J.A. Hoxie, Targeting of HIV- and SIV-infected cells by CD4-chemokine receptor pseudotypes, Science 278 (1997) 1462-1464.
10. B.N. Fields, Fundamental Virology, Lippincont-Raven Publishers, Philadelphia, 1996, pp. 1340
11. Y. Pommier, A.A. Pilon, K. Bajaj, A. Mazumder, N. Neamati, HIV-1 inte-

grase as a target for antiviral drugs, Antiviral Chemistry & Chemotherapy 8 (1997) 463-483.
12. K. Van Vaerenbergh, K. Van Laethem, J. Albert, C.A. Boucher, B. Clotet, M. Floridia, J. Gerstoft, B. Hejdeman, C. Nielsen, C. Pannecouque, L. Perrin, M.F. Pirillo, L. Ruiz, J.C. Schmit, F. Schneider, A. Schoolmeester, R. Schuurman, H.J. Stellbrink, L. Stuyver, J. Van Lunzen, B. Van Remoortel, E. Van Wijngaerden, S. Vella, M. Witvrouw, S. Yerly, E. De Clercq, J. Destmyer, A.M. Vandamme, Prevalence and characteristics of multinucleoside-resistant human immunodeficiency virus type 1 among European patients receiving combinations of nucleoside analogues, Antimicrob. Agents Chemother. 44 (2000) 2109-2117.
13. R.F. Schinazi, B.A. Larder, J.W. Mellors, Mutations in retroviral genes associated with drug resistance, International Antiviral News 5 (1997) 129-142.
14. P.G. Yeni, S.M. Hammer, M.S. Hirsch, M.S. Saag, M. Schechter, C.C. Carpenter, M.A. Fischl, J.M. Gatell, B.G. Gazzard, D.M. Jacobsen, D.A. Katzenstein, J.S. Montaner, D.D. Richman, R.T. Schooley, M.A. Thompson, S. Vella, P.A. Volberding, Treatment for adult HIV infection: 2004 recommendations of the International AIDS Society-USA Panel, JAMA 292 (2004) 251-265.
15. A.S. Perelson, Modelling viral and immune system dynamics, Nat Rev Immunol 2 (2002) 28-36.
16. M.A. Nowak, R.M. May Virus dynamics: Mathematical principles of immunology and virology, Oxford University Press, Oxford, 2000.
17. X. Wei, S.K. Ghosh, M.E. Taylor, V.A. Johnson, E.A. Emini, P. Deutsch, J.D. Lifson, S. Bonhoeffer, M.A. Nowak, B.H. Hahn, M.S. Saag, G.M. Shaw, Viral dynamics in human immunodeficiency virus type 1 infection, Nature 373 (1995) 117-122.
18. D.D. Ho, A.U. Neumann, A.S. Perelson, W. Chen, J.M. Leonard, M. Markowitz, Rapid turnover of plasma virions and CD4 lymphocytes in HIV-1 infection, Nature 373 (1995) 123-126.
19. A.S. Perelson, P. Essunger, Y. Cao, M. Vesanen, A. Hurley, K. Saksela, M. Markowitz, D.D. Ho, Decay characteristics of HIV-1-infected compartments during combination therapy, Nature 387 (1997) 188-191.
20. R.M. Ribeiro, S. Bonhoeffer, Production of resistant HIV mutants during antiretroviral therapy, Proc. Natl. Acad. Sci. U. S. A. 97 (2000) 7681-7686.
21. R.M. Ribeiro, S. Bonhoeffer, M.A. Nowak, The frequency of resistant mutant virus before antiviral therapy, AIDS 12 (1998) 461-465.
22. R.M. Ribeiro, H. Mohri, D.D. Ho, A.S. Perelson, In vivo dynamics of T cell activation, proliferation, and death in HIV-1 infection: why are CD4+ but not CD8+ T cells depleted?, Proc. Natl. Acad. Sci. U. S. A. 99 (2002) 15572-15577.
23. T.J. Layden, J.E. Layden, R.M. Ribeiro, A.S. Perelson, Mathematical modeling of viral kinetics: a tool to understand and optimize therapy, Clin Liver Dis 7 (2003) 163-178.
24. R.M. Ribeiro, A. Lo, A.S. Perelson, Dynamics of hepatitis B virus infection, Microbes Infect 4 (2002) 829-835.
25. R.M. Ribeiro, A.S. Perelson, The Analysis of HIV Dynamics Using Math-

ematical Models, in: Wormser G.P. (Ed. AIDS and Other Manifestations of HIV Infection, Elsevier, San Diego, 2004, p. 905-912.

26. A.S. Perelson, P.W. Nelson, Mathematical analysis of HIV-1 dynamics in vivo, SIAM Review 41 (1999) 3-44.
27. A.S. Perelson, A.U. Neuman, M. Markowitz, J.M. Leonard, D.D. Ho, HIV-1 dynamics in vivo: Virion clearance rate, infected cell lifespan and viral generation time, Science 271 (1996) 1582-1586.
28. A.V. Herz, S. Bonhoeffer, R.M. Anderson, R.M. May, M.A. Nowak, Viral dynamics in vivo: limitations on estimates of intracellular delay and virus decay, Proc. Natl. Acad. Sci. U. S. A. 93 (1996) 7247-7251.
29. P.W. Nelson, J.E. Mittler, A.S. Perelson, Effect of drug efficacy and the eclipse phase of the viral life cycle on estimates of HIV viral dynamic parameters, J. Acquir. Immune Defic. Syndr. 26 (2001) 405-412.
30. P.W. Nelson, J.D. Murray, A.S. Perelson, A model of HIV-1 pathogenesis that includes an intracellular delay, Math. Biosci. 163 (2000) 201-215.
31. B. Ramratnam, S. Bonhoeffer, J. Binley, A. Hurley, L. Zhang, J.E. Mittler, M. Markowitz, J.P. Moore, A.S. Perelson, D.D. Ho, Rapid production and clearance of HIV-1 and hepatitis C virus assessed by large volume plasma apheresis, Lancet 354 (1999) 1782-1785.
32. M. Markowitz, M. Louie, A. Hurley, E. Sun, M. Di Mascio, A.S. Perelson, D.D. Ho, A novel antiviral intervention results in more accurate assessment of human immunodeficiency virus type 1 replication dynamics and T-cell decay in vivo, J. Virol. 77 (2003) 5037-5038.
33. M. Di Mascio, R.M. Ribeiro, M. Markowitz, D.D. Ho, A.S. Perelson, Modeling the long-term control of viremia in HIV-1 infected patients treated with antiretroviral therapy, Math. Biosci. 188 (2004) 47-62.
34. W.S. Hlavacek, N.I. Stilianakis, D.W. Notermans, S.A. Danner, A.S. Perelson, Influence of follicular dendritic cells on decay of HIV during antiretroviral therapy, Proc. Natl. Acad. Sci. U. S. A. 97 (2000) 10966-10971.
35. A.S. Perelson, P. Essunger, D.D. Ho, Dynamics of HIV-1 and CD4+ lymphocytes in vivo, AIDS 11 (1997) S17-24.
36. B. Ramratnam, J.E. Mittler, L. Zhang, D. Boden, A. Hurley, F. Fang, C.A. Macken, A.S. Perelson, M. Markowitz, D.D. Ho, The decay of the latent reservoir of replication-competent HIV-1 is inversely correlated with the extent of residual viral replication during prolonged anti-retroviral therapy, Nat. Med. 6 (2000) 82-85.
37. L. Zhang, B. Ramratnam, K. Tenner-Racz, Y. He, M. Vesanen, S. Lewin, A. Talal, P. Racz, A.S. Perelson, B.T. Korber, M. Markowitz, D.D. Ho, Quantifying residual HIV-1 replication in patients receiving combination antiretroviral therapy, N. Engl. J. Med. 340 (1999) 1605-1613.
38. D. Finzi, J. Blankson, J.D. Siliciano, J.B. Margolick, K. Chadwick, T. Pierson, K. Smith, J. Lisziewicz, F. Lori, C. Flexner, T.C. Quinn, R.E. Chaisson, E. Rosenberg, B. Walker, S. Gange, J. Gallant, R.F. Siliciano, Latent infection of CD4+ T cells provides a mechanism for lifelong persistence of HIV-1, even in patients on effective combination therapy, Nat. Med. 5 (1999) 512-517.
39. M. Di Mascio, G. Dornadula, H. Zhang, J. Sullivan, Y. Xu, J. Kulkosky, R.J.

Pomerantz, A.S. Perelson, In a subset of subjects on highly active antiretroviral therapy, human immunodeficiency virus type 1 RNA in plasma decays from 50 to <5 copies per milliliter, with a half-life of 6 months, J. Virol. 77 (2003) 2271-2275.

40. T.W. Chun, A.S. Fauci, Latent reservoirs of HIV: obstacles to the eradication of virus, Proc. Natl. Acad. Sci. U. S. A. 96 (1999) 10958-10961.
41. T.W. Chun, L. Carruth, D. Finzi, X. Shen, J.A. DiGiuseppe, H. Taylor, M. Hermankova, K. Chadwick, J. Margolick, T.C. Quinn, Y.H. Kuo, R. Brookmeyer, M.A. Zeiger, P. Barditch-Crovo, R.F. Siliciano, Quantification of latent tissue reservoirs and total body viral load in HIV-1 infection, Nature 387 (1997) 183-188.
42. L.M. Mansky, H.M. Temin, Lower in-Vivo Mutation-Rate of Human-Immunodeficiency-Virus Type-1 Than That Predicted From the Fidelity of Purified Reverse- Transcriptase, Journal of Virology 69 (1995) 5087-5094.
43. E. Domingo, E. Martinezsalas, F. Sobrino, J.C. Delatorre, A. Portela, J. Ortin, C. Lopezgalindez, P. Perezbrena, N. Villanueva, R. Najera, S. Vandepol, D. Steinhauer, N. Depolo, J. Holland, The Quasispecies (Extremely Heterogeneous) Nature of Viral-Rna Genome Populations - Biological Relevance - a Review, Gene 40 (1985) 1-8.
44. J.J. Holland, J.C. Delatorre, D.A. Steinhauer, Rna Virus Populations As Quasi-Species, Current Topics in Microbiology and Immunology 176 (1992) 1-20.
45. C.K. Biebricher, Darwinian Selection of Self-Replicating Rna Molecules, Evolutionary Biology 16 (1983) 1-52.
46. E. Domingo, M. Davila, J. Ortin, Nucleotide-Sequence Heterogeneity of the Rna From a Natural- Population of Foot-and-Mouth-Disease Virus, Gene 11 (1980) 333-346.
47. J. Gomez, M. Martell, J. Quer, B. Cabot, J.I. Esteban, Hepatitis C viral quasispecies, Journal of Viral Hepatitis 6 (1999) 3-16.
48. M. Eigen, The Origin of Genetic Information - Viruses As Models, Gene 135 (1993) 37-47.
49. M. Eigen, P. Schuster The Hypercycle, Springer-Verlag, Berlin, 1979.
50. M.A. Nowak, What Is a Quasi-Species, Trends in Ecology & Evolution 7 (1992) 118-121.
51. B.D. Walker, B.T. Korber, Immune control of HIV: the obstacles of HLA and viral diversity, Nat Immunol 2 (2001) 473-475.
52. K. Yusim, C. Kesmir, B. Gaschen, M.M. Addo, M. Altfeld, S. Brunak, A. Chigaev, V. Detours, B.T. Korber, Clustering patterns of cytotoxic T-lymphocyte epitopes in human immunodeficiency virus type 1 (HIV-1) proteins reveal imprints of immune evasion on HIV-1 global variation, J. Virol. 76 (2002) 8757-8768.
53. B. Gaschen, J. Taylor, K. Yusim, B. Foley, F. Gao, D. Lang, V. Novitsky, B. Haynes, B.H. Hahn, T. Bhattacharya, B. Korber, Diversity considerations in HIV-1 vaccine selection, Science 296 (2002) 2354-2360.
54. B.J. Thomson, R.G. Finch, Hepatitis C virus infection, Clin Microbiol Infect 11 (2005) 86-94.

55. A.U. Neumann, N.P. Lam, H. Dahari, D.R. Gretch, T.E. Wiley, T.J. Layden, A.S. Perelson, Hepatitis C viral dynamics in vivo and the antiviral efficacy of interferon-alpha therapy, Science 282 (1998) 103-107.
56. N.M. Dixit, A.S. Perelson, HIV dynamics with multiple infections of target cells, Proc. Natl. Acad. Sci. U. S. A. 102 (2005) 8198-8203.
57. K.A. Powers, N.M. Dixit, R.M. Ribeiro, P. Golia, A.H. Talal, A.S. Perelson, Modeling viral and drug kinetics: hepatitis C virus treatment with pegylated interferon alfa-2b, Semin. Liver Dis. 23 Suppl 1 (2003) 13-18.
58. A.U. Neumann, N.P. Lam, H. Dahari, M. Davidian, T.E. Wiley, B.P. Mika, A.S. Perelson, T.J. Layden, Differences in viral dynamics between genotypes 1 and 2 of hepatitis C virus, J. Infect. Dis. 182 (2000) 28-35.
59. J.E. Layden-Almer, R.M. Ribeiro, T. Wiley, A.S. Perelson, T.J. Layden, Viral dynamics and response differences in HCV-infected African American and white patients treated with IFN and ribavirin, Hepatology 37 (2003) 1343-1350.
60. J.E. Layden, T.J. Layden, K.R. Reddy, R.S. Levy-Drummer, J. Poulakos, A.U. Neumann, First phase viral kinetic parameters as predictors of treatment response and their influence on the second phase viral decline, J. Viral Hepat. 9 (2002) 340-345.
61. M. Tsiang, J.F. Rooney, J.J. Toole, C.S. Gibbs, Biphasic clearance kinetics of hepatitis B virus from patients during adefovir dipivoxil therapy, Hepatology 29 (1999) 1863-1869.
62. S. Lewin, T. Walters, S. Locarnini, Hepatitis B treatment: rational combination chemotherapy based on viral kinetic and animal model studies, Antiviral Res. 55 (2002) 381-396.

SHORT AND LONG-TERM DYNAMICS OF CHILDHOOD DISEASES IN DYNAMIC SMALL-WORLD NETWORKS

JOSÉ VERDASCA

Centro de Matemática e Aplicações Fundamentais
Complexo Interdisciplinar da Universidade de Lisboa
*Av. Professor Gama Pinto 2, P-1649-003 Lisboa, Portugal**

We have performed individual-based lattice simulations of SIR and SEIR dynamics to investigate both the short and long-term dynamics of childhood epidemics. In our model, infection takes place through a combination of local and long-range contacts, in practice generating a dynamic small-world network. Sustained oscillations emerge with a period much larger than the duration of infection. We found that the network topology has a strong impact on the amplitude of oscillations and in the level of persistence. Diseases do not spread very effectively through local contacts. This can be seen by measuring an *effective* transmission rate $\beta_{\rm eff}$ as well as the basic reproductive rate R_0. These quantities are lower in the small-world network than in an homogeneously mixed population, whereas the average age at infection is higher.

1. Introduction

The recurrent outbreaks of measles and other childhood diseases is one of the most striking features of the pre-vaccination records. Despite continuing efforts over more than seven decades, the question of what is the mechanism behind these oscillations has not yet received a fully satisfactory answer[1].

Measles, mumps, varicella (chickenpox) and rubella are examples of diseases that confer lifelong immunity, and can be analyzed within the SIR (Susceptible, Infected and Recovered) or SEIR (comprising an additional class, the Exposed) general frameworks. If the population is constant, the mean-field implementation of SIR and SEIR scheme disregarding heterogeneity in contact structure leads to systems of respectively two and three coupled ODE's. Although based on this most unrealistic description of human contacts, mean-field deterministic models still capture most *static*

*Present address: Centro de Astrobiologia, Instituto nacional de Técnica Aeroespacial, Ctra. de Torrejón a Ajalvir, km 4, 28850 Torrejón de Ardoz, Madrid, Spain.

properties of epidemics of typical childhood diseases including the threshold values for the spread of an epidemic, its final size as well as the average age at which infection is acquired. However, these models fail in that they predict oscillations that are invariably damped. This, of course, contradicts the available records which evidence self-sustained oscillations of roughly constant period during the pre-vaccination era.

After a period during which complicated age-structured models were favored, seasonally forced models have become the framework of choice to explain the historic time-series. In fact, when subject to parametric forcing, deterministic SIR and SEIR ODE's display a rich dynamical behaviour including period-doubling cascades to chaos, quasiperiodicity, multistability between cycles of different periods, etc. [2,3]. When the drive is sinusoidal, the level of forcing at which complex behaviour is observed is deemed unreasonable. However, when a more realistic formulation is used, like an alternating sequence of periods of high and low transmissibility mimicking the opening and closing of schools, the levels of forcing required to obtain complex behaviour is considerably lowered[2,4]. If epidemics correspond to periodic orbits perturbed by noise, as opposed to chaotic solutions[5,6], then only periods which are integer multiples of the forcing period are allowed. More likely, the observation of both integer and non-integer periods in incidence time series of rubella and chickenpox is the signature of an autonomous system oscillating with a frequency that may or may not become locked with an external drive[7]. In this picture, seasonality is not ruled out completely but is reduced to a synchronizing role, *e.g.* by forcing the maxima to occur at precise times of the year.

By definition, there is one important feature that cannot be described by deterministic models. That is the pattern of disease persistence, *i.e.* the probability that a recurrent epidemic goes extinct after a given number of cycles. Persistence is an emergent property arising from the interaction of stochastic effects and dynamics. Its meaningfulness derives from being the key ingredient in the definition of the Critical Community Size (CCS), the population number below which a particular disease cannot be sustained. But persistence is also a tool to assess the relative merits of stochastic versus deterministic models. Indeed, by far the most serious shortcoming of deterministic forcing is that during the minima often the fraction of infective individuals falls below 10^{-10}, meaning that the global human population would not be enough to sustain recurrent epidemics.

It is well known that, like forcing, stochastic effects also tend to sustain the oscillations[8]. However, the implementation of stochastic SIR and SEIR dynamics disregarding heterogeneity in contact structure, generates only fluctuations which are much too small and irregular when compared to real epidemics unless immigration of infectives from outside is introduced.

Realizing that deterministic forced models fail short of explaining the patterns of persistence and that stochasticity alone could not provide for the necessary agreement with data, researchers began to explore the role of spatial factors in the persistence and dynamics of epidemics. The traditional way to account for an explicit spatial dependence in population ecology and epidemiology is through the use of metapopulation, or patch, models. These are based on a coarse-grained distribution of the global population over a number of interacting subpopulations – the patches. Within-patch dynamics is built into the model *a priory* and can be made as complex as one wishes to from the start, with heterogeneity effects restricted to the coupling between the patches. The coarse-graining procedure limits the ability of the model to assess the impact of the structure of individual contacts on the overall dynamics. It says very little about how emergent complex behaviour on a global scale can arise from simple interaction rules, either strictly local or not. To address this issue one must consider instead a network of interacting individuals.

Nevertheless, some interesting results have been achieved by adding metapopulation structure to stochastic models, either in conjunction with external forcing or not. Lloyd and May[9] simulated a two-patch stochastic SEIR model, each one having at least 10^6 nodes. The oscillations they obtained were too irregular and their amplitude too small when compared with data records. Moreover, strong coherence could not be obtained unless a considerable level of seasonal forcing was applied. However, in that case the number of infective individuals dropped to the unrealistic levels predicted by deterministic models. Bolker and Grenfell[10] considered a similar model but allowed for contacts inside each subpopulation and between individuals belonging to different subpopulations. By increasing the ratio of between- and within-patch contacts they could increase the levels of persistence. Although their study clearly indicated that adding structure to the network of contacts could indeed enhance persistence it greatly overestimated the size of the population needed to sustain recurrent epidemics.

It is interesting to note that the modelling of disease spread as a combination of local and long-range interactions in the context of patch models[10] actually precedes the introduction by Watts and Strogatz of a class of networks that interpolates between regular lattices and random networks[12] – small-world (SW) networks – and the acknowledgment of the importance of the small-world phenomenon on the spread of epidemics that soon followed[13,14,15,16,17]. Subsequently Boots & Sasaki[16] have used SW networks to analyze the selection of a particular strain of a pathogen and Keeling[18] considered a network of nodes which had many of the properties of a SW to calculate epidemic thresholds and determine a number of properties of the endemic state.

More recently, probabilistic cellular automata (PCA) models of infectious diseases evolving on SW networks were proposed by Johansen[19,20], Kuperman and Abramson[21] and He and Stone[22]. These particular models belong to the SIS (Susceptible-Infective-Susceptible) class, that is they apply to diseases which do not confer lifelong immunity. Therefore they are not suited to describe typical childhood diseases. A more problematic feature common to all these studies is that they predict oscillations with periods on the scale of infection and/or immune periods. In contrast, recurrent epidemics of childhood diseases have periods from a dozen up to a few hundred times the infection cycle. Notwithstanding, these models are certainly of great interest as toy models of generic SIS dynamics.

Whether recurrent epidemics are governed by an external drive or by the intrinsic nonlinear dynamics has been the focus of intensive debate practically since the dawn of theoretical epidemiology. Here we make a further contribution to this still unfinished debate by presenting a stochastic model that discards external factors and considers instead the heterogeneity of the contact network. As reported in a previous publication[23] the stochastic implementation of SIR dynamics in a small-world network can describe the onset of epidemic cycles in a population, without considering any exogenous factors such as seasonal forcing or immigration of infectives from outside. In our model, infection takes place through a combination of local rules and long-range contacts, generating a dynamic small-world network. In sharp contrast to the previous studies in the same vein[19,20,21], we observe the emergence of a characteristic time scale which is not that of birth (replenishment of susceptibles) neither is it related in any trivial way to the period of infection.

The results in this paper are arranged in two main parts. The first one is devoted to the SIR implementation. A first set of simulations of long-term behaviour shows that the network topology has a strong impact both on the amplitude of oscillations and the level of persistence. Then, we consider the evolution of an epidemic in a closed population, with birth and death rates set to zero. We show that the basic reproductive rate, R_0, increases from its minimum value in a regular lattice with local contacts only up to the maximum, mean-field value, as the percentage of long-range infection is increased from zero to one. In the second part of the paper we present simulations of the more realistic SEIR version. We calculate the period and amplitude of oscillations, R_0, as well as the average age at infection for realistic demographic and etiological parameters corresponding to measles, rubella and chickenpox.

The detailed structure of the paper is the following: in Section 2 we describe the algorithm in detail; in Section 3 we present the results of the SIR model and in Section 4 those of the SEIR model. This is followed by a Discussion and, finally, the Conclusion.

2. PCA Model

2.1. *General description*

Here we describe the probabilistic cellular automata (PCA) implementation of the SIR model that has been briefly outlined in a preceding paper[23].

Individuals live on a square lattice of $N = L \times L$ sites. The bonds between sites are connections along which the infection may spread to other individuals. Infection proceeds either locally, within a prescribed neighbourhood, or through a link established at random between any two individuals. We introduce a small-world parameter p_{SW} defined as the fraction of attempts at infection carried out through a random link; $p_{\mathrm{SW}} = 0$ corresponds to a regular lattice where each individual contacts with his k nearest neighbours only, while $p_{\mathrm{SW}} = 1$ corresponds to a random network. For intermediate values of p_{SW} the network of contacts is neither fully ordered nor completely random.

2.2. *Algorithm*

We choose first between *birth*, *death* and *infection* events, the latter being either *local* or *long-range* with respective probabilities $p_{\mathrm{inf}}^{loc} = (1 - p_{\mathrm{SW}})\beta_0$ and $p_{\mathrm{inf}}^{lr} = p_{\mathrm{SW}}\beta_0$. (Note that we call long-range link any connection established at random to any node on the lattice, not only those that connect to sites lying outside the established neighbourhood). The total population number if fixed, therefore $p_{birth} \equiv p_{death}$. $\beta_0 = p_{\mathrm{inf}}^{loc} + p_{\mathrm{inf}}^{lr}$ is the total probability of an attempt at infection while $(1 - p_{birth} - p_{death} - p_{\mathrm{inf}}^{loc} - p_{\mathrm{inf}}^{lr})$ is the probability that nothing happens. There is no restriction associated to the fact that the sum of probabilities cannot exceed one because it is possible to attribute any weight to any one of the events - birth, death or infection - by a suitable choice of the time scale; the probability of not realizing any event will change accordingly and a sweep through the lattice will simply correspond to a different time unit. In one time unit, or PCA step, we perform N attempts to realize one given event, as follows: we generate a random number r uniformly distributed between 0 and 1. If $0 < r \leq p_{birth}$ we make an attempt to realize a birth event; if $p_{birth} < r \leq p_{birth} + p_{death}$ an individual picked at random will die; if $p_{birth} + p_{death} < r \leq p_{birth} + p_{death} + p_{\inf}^{loc}$ an attempt at local infection is made while if $p_{birth} + p_{death} + p_{\inf}^{loc} < r \leq p_{birth} + p_{death} + p_{\inf}^{loc} + p_{\inf}^{lr}$ the attempt will be long-ranged. Finally, if $p_{birth} + p_{death} + p_{\inf}^{loc} + p_{\inf}^{lr} < r \leq 1$, we just carry on and generate a new random number. The realization of the events is the following:

(1) Attempt at local infection or long-range infection: First choose a site i at random; Then,

  (a) If that site is occupied by a recovered (R) individual or an infectious (I) individual, do nothing.

  (b) If it is occupied by a susceptible (S) then choose another site j from a list of k possible neighbours, in the case of local infection, or from all of the N sites, with equal probability, in the case of long-range infection. In the simulations presented here, the local range comprises nearest neighbours, next nearest neighbour and third nearest neighbours ($k = 12$). If,

    i. Site j is occupied either by another susceptible or a recovered individual: do nothing.

    ii. The site is occupied by an infective: The first individual becomes infected.

(2) A death event is chosen. Then one picks an individual at random who dies irrespectively of his present state. The probabilities of death are then proportional to the density of S, I and R individuals. Susceptible and infective individuals who die become recovered; recovered individuals remain in that same class.

(3) Birth event: one looks at the lattice (at random) for a recovered individual. Once found, that individual becomes susceptible. The trial only ends when one actually finds a recovered individual, so that the birth rate, as it should, is independent of any density.

After a time during which he stays infectious to others, the individual recovers. He becomes immune for life and cannot be infected again. For childhood diseases these periods vary only by a small amount among individuals within a population. So, it is more realistic to assume a constant infectious period – deterministic recovery – than a constant probability of recovery leading to an exponential distribution of infectious periods[24]. We model deterministic recovery by associating a counter n_r to every individual. Upon infection, the counter for that individual is updated at each time step, $n_r \rightarrow n_r + 1$. At each PCA step we move to the recovered class those infectives for which the counter has reached the fixed infectious period τ, and reset their counter. Stochastic recovery, on the other hand, is modelled by a Poisson process: we add a new event to the list above – a recovery event – taking place with probability γ. When such event is chosen, one picks an individual at random; if infective, that individual is moved into the recovered class, otherwise nothing happens.

The SEIR model comprises one additional class, the exposed (E). After being infected, the individuals enter a latency period during which they cannot be re-infected yet they are unable to transmit the infection to others. In our simulations the individuals move deterministically from the exposed to the infectious class after τ_{lat} steps and then stay infectious for τ_{inf}.

One important difference between the model and the population dynamics in developed countries is the implicit assumption of the so-called Type II mortality, where individuals die with equal probability independently of their age, as opposed to Type I mortality where all individuals live up to the same age and then die[1].

3. Results: SIR model

3.1. *Persistence*

Recurrent epidemics can persist in finite populations because of the finite birth rate that allows for the renewal of susceptibles. The way in which persistence varies with the small-world parameter $p_{\rm SW}$ depends crucially on the rate at which fresh susceptibles are supplied by birth. As shown in Fig. 1, for intermediate values of μ the persistence coefficient is zero at low $p_{\rm SW}$, and approaches one over a narrow range of $p_{\rm SW}$. At values $\mu \lessapprox 0.001$ a maximum starts to develop. Barely noticeable at $\mu = 0.0006$, it is already quite pronounced at $\mu = 0.0004$. At $p_{\rm SW} = 1.0$, only about half the runs survive up to the maximum ascribed time, whereas within the range $p_{\rm SW} = 0.2 - 0.3$ more than 90 % of the runs reach $t_{\rm max}$. The relevant fact is that at low enough birth rates there is an optimal value of $p_{\rm SW}$ for the disease to persist in finite populations. Conversely, for $\mu \gtrapprox 0.001$ the persistence

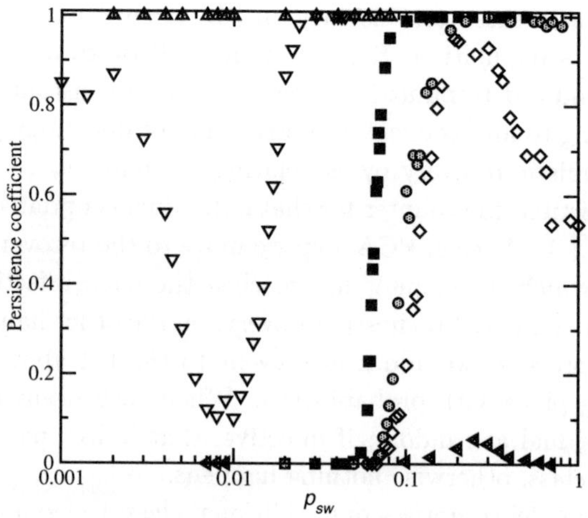

Figure 1. Persistence coefficient, defined as the fraction of runs that attain a prescribed time $t_{max} = 20000$ of a total of $n = 100$ runs started, calculated on a 400×400 lattice with deterministic recovery. The initial fraction of susceptibles was $s_0 = 0.165$. The probability of infection is $\beta_0 = 0.66$ and the infectious period $\tau = 16$. The different curves correspond to the following birth rates: $\mu = 0.0002$ (triangles left), $\mu = 0.0004$ (diamonds), $\mu = 0.0006$ (circles), $\mu = 0.001$ (squares), $\mu = 0.002$ (triangles down) and $\mu = 0.005$ (triangles up).

coefficient shows a minimum at finite p_{SW}. The logarithmic scale in Fig. 1 enables us to highlight this remarkable symmetry of behaviour but here we must note that what happens at high birth rates is irrelevant for childhood diseases. Indeed, for τ given in days (which is not unreasonable) we get a birth rate of $\mu = 0.002$ day^{-1} corresponding to an average lifespan of 1.4 years! Even for the lowest μ with which we can still observe a reasonable level of persistence, the lifespan would only be around 7 years. This feature is a distinguishing property of the PCA implementation of the SIR model, namely that to obtain realistic patterns of persistence one has to choose birth rates at least one order of magnitude higher than the real values. We shall see in Section 4, that the adoption of the more realistic SEIR dynamics permits to avoid this problem.

The fact that persistence depends on the fraction of long-range contacts implies that the Critical Community Size (CCS) also depends on the structure of the contact network. In order to demonstrate how the PCA can be used to estimate the Critical Community Size, and how the CCS depends on the fraction of long-range infection we choose two examples, at $p_{SW} = 0.07$ and $p_{SW} = 0.1$, with a birth rate $\mu = 0.001$ (squares in 7break Fig. 1). While at $p_{SW} = 0.1$ there is a sharp rise in persistence, at $p_{SW} = 0.07$ a much smoother curve is obtained (Fig. 2). The population size at which the persistence rises above 50 % can be taken as a estimate of the CCS but we might as well choose a different threshold.

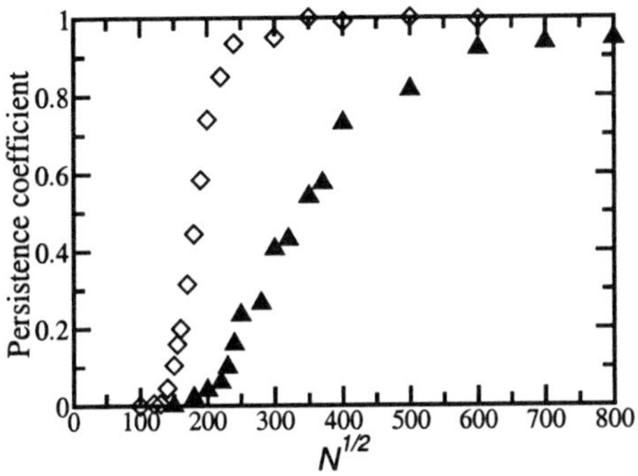

Figure 2. Persistence coefficient as a function of lattice size for $p_{SW} = 0.07$ (triangles) and $p_{SW} = 0.1$ (diamonds) at $\mu = 0.001$. The other parameters are the same as in Fig. 1 except $n = 200$.

3.2. *Amplitude of oscillations*

Fig. 3 shows a typical output of the SIR model implemented on a fairly large lattice. The time series shown are for the susceptible and infective fraction. It is impossible to run the three-state SIR implementation for a long time using realistic demographic and etiological parameters. Extinction is almost certain to occur after only a few cycles. Thus the values of the birth rates used here are much larger that real ones, again for a τ of a few days.

When p_{SW} is small, the oscillations have large amplitudes and are strongly synchronized. Keeping all other parameters unchanged but setting $p_{SW} = 1.0$ we observe that oscillations become much more irregular and their amplitude is considerably diminished. In Fig. 4 we plot the root mean square (RMS) amplitude of the oscillations in the fraction of susceptibles (a) and infectives (b) as a function of the birth rate for different values of p_{SW}. As μ is decreased the RMS amplitudes are amplified and this trend is the more pronounced the smaller the value of p_{SW}.

Looking at the data from a different perspective now, one detects quite clearly the enhancement of stochastic fluctuations growing into fully devel-

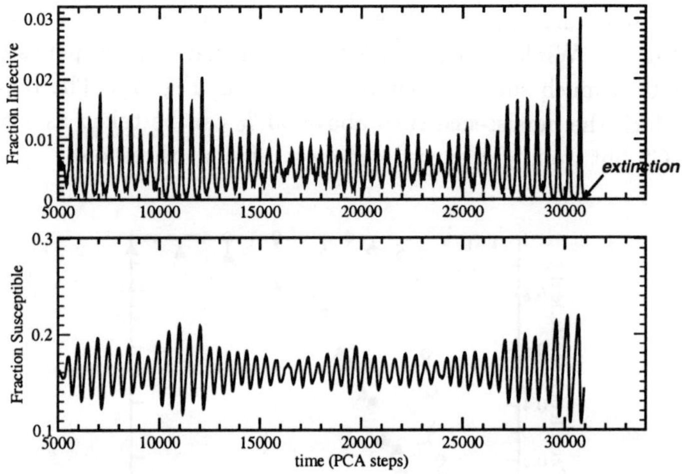

Figure 3. Typical evolution of the fraction of infectives (top) – number of infective individuals divided by the system size N – and the fraction of susceptibles (bottom), obtained with the SIR PCA model implemented on a small-world network of $N = 160000 \equiv 400 \times 400$ individuals. In the simulations, 90 % of the contacts are local and the remaining 10 % long-range ($p_{SW} = 0.1$). Recovery was stochastic with a probability $\gamma = 0.0625 \equiv 1/16$. The other parameters are $\mu = 0.0004$ and $\beta_0 = 0.66$. The system oscillates for more than 50 cycles before extinction occurs after about 30950 PCA steps.

oped oscillations as the relative weight of long-range infection is decreased. The smaller the birth rate the more accentuated the effect becomes: at $\mu = 0.001$ the amplitude of the susceptible oscillations at $p_{SW} = 0.08$ is approximatively four times that of the homogeneous mixed population while at $\mu = 0.0004$ the same ratio is almost five. The increase in the amplitude of oscillations in the fraction of infectives (Fig. 4 (b)) follows that of susceptibles but it is less marked.

3.3. *Effective transmission rate*

Diseases do not spread very effectively on lattices when only local contacts are allowed because infective individuals tend to interact mostly with other already infected individuals. To evaluate the impact of the aggregation of infectives and susceptibles into clusters on the spread of the disease brought about by the local contact rules we can estimate the transmission rate directly from the simulations. This is done by observing that once the transients have died out, the mean number of new infections that take place

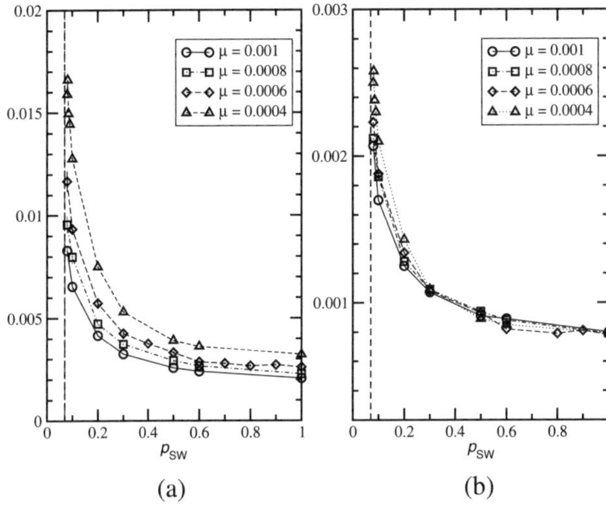

Figure 4. RMS amplitude of oscillations in the fraction of susceptibles (a) and infectives (b) as a function of the fraction of long-range contacts for decreasing values of the birth rate. Other parameters the same as in Fig. 3. Stochastic recovery. Each point represents an average over 10 runs started from different initial conditions.

in a short time interval Δt, $nni_{\Delta t}$, is approximatively proportional to the product of the mean number of infective individuals by the mean number of susceptibles during that same time interval:

$$nni_{\Delta t} \sim \overline{I}_{\Delta t} \times \overline{S}_{\Delta t}. \tag{1}$$

The sensitive issue here is to choose a suitable value for Δt, which must be much smaller than the average period of oscillations or otherwise we would not get an instantaneous measure of transmission, but at the same time considerably larger than the infection period so that stochastic effects get averaged out. There is no recipe to pick up the right value, but choosing a Δt of a few dozen time steps usually guarantees that the proportionality (1) is verified. In that case, we can define an *effective* transmission rate as:

$$\beta_{\text{eff}}(t) = \frac{nni_{\Delta t}(t)}{\overline{I}_{\Delta t}(t)\overline{S}_{\Delta t}(t)}. \tag{2}$$

This instantaneous transmission rate fluctuates wildly on the time scale of the birth rate but once the transients have died out we can calculate the temporal mean. This (averaged) effective transmission rate stays always below the mean-field transmissibility β_0. Indeed, by aggregating infectives and susceptibles into clusters, the structure of local infection on a *regular* network structure acts to keep the number of contacts that can actually result in transmission, namely those between a susceptible and an infective, well below the level that would result if the individuals were homogeneously distributed on the lattice irrespectively of their disease status. Although, as we shall see below, the dynamic small-world structure of contacts exhibits some features of the well-mixed situation, *locally* the structure of contacts remains highly clustered. As long as infected individuals remain in contact mostly with other already infected individuals the progression of the disease stalls but once an individual belonging to an infectious cluster establishes a shortcut that propagates the disease into a region where susceptible individuals predominate, the disease will perhaps get a new boost that will carry it through one more cycle.

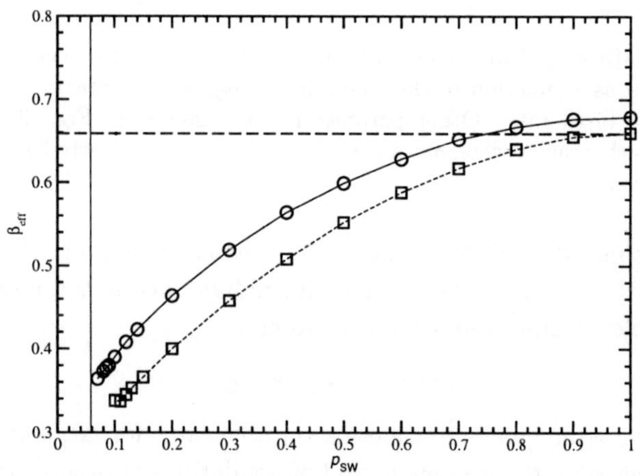

Figure 5. Effective transmission rate β_{eff} vs. the small-world parameter p_{SW}. The reduction of β_{eff} that arises both for deterministic (circles) and stochastic (squares) recovery is due to the clustering of infectives and susceptibles. The dashed line indicates the value of the transmissibility, or total probability of realizing an infection event used in the simulations: $\beta_0 = 0.66$. Note the almost perfect agreement with the mean-field result $\beta_{\text{eff}} \equiv \beta_0$ observed at $p_{\text{SW}} = 1.0$ for stochastic recovery. The birth rate is $\mu = 0.0006$, the infectious period for deterministic recovery $\tau = 16$ and $\gamma = 0.0625 \equiv 1/\tau$ for stochastic recovery. Each point represents an average over 10 runs.

As shown in Fig. 5, β_{eff} increases smoothly as p_{SW} increases. For values of p_{SW} to the left of the lower end of the curves only a very small fraction of the runs last for more than a few cycles. With such a small persistence it becomes impossible to follow the long term dynamical regime and compute repeated time averages of non-transient oscillations. The two curves shown in Fig. 5 correspond to stochastic and deterministic recovery. Their shape is identical, but the value of β_{eff} for stochastic recovery always stays below that obtained with deterministic recovery. The values of β_{eff} for stochastic recovery converge to the mean-field value for $p_{\text{SW}} \to 1$ with a small discrepancy due to finite-size effects, whereas for deterministic recovery β_{eff} is still about 3 % above β_0 in the same limit. Moreover we found that the amplitude of oscillations is also larger for deterministic recovery than for stochastic recovery, the differences being small at large p_{SW} but increasing significantly as p_{SW} is lowered. This is in agreement with a previous study of a homogeneously mixed stochastic model, showing that sharp distributions of the infection period result in larger fluctuations than those with a smooth distribution[24]. In our model this effect actually intensifies when contact structure is introduced.

3.4. *Basic reproductive rate*

It is incontrovertibly accepted that in order to figure out if a given pathogen will be able to succeed in a host population the crucial quantity to compute is the basic reproductive rate, R_0. This is the number of secondary infections produced by an infected individual in a population entirely composed of susceptibles.

The analysis of steady state solution of the SIR deterministic ODE's considering only weak homogeneous mixing – meaning that the rate of new infectives is proportional to the total number of susceptibles – and Type II survival, gives[1]:

$$R_0 = \frac{1}{\mu A}, \tag{3}$$

where A is the average age at infection, in the case of constant population size. Under the more stringent condition imposed by the mean-field approximation, one has also[1]

$$R_0 = \frac{\beta}{\gamma + \mu}. \tag{4}$$

Note that relations (3) and (4) suppose stochastic recovery.

The definition of R_0 rests on the existence of a pristine susceptible population, *i.e.* it is valid only at vanishing infective fraction. However, it is a consequence of mean-field models and a fact confirmed by the analysis of many real epidemics, that following the introduction of a pathogen in a population consisting entirely of susceptible individuals the number of infective individuals grows exponentially in the early stages:

$$I(t) \sim e^{\Lambda t}, \tag{5}$$

where $\Lambda = (R_0 - 1)/\tau$. So, with the exception of incipient epidemics ($R_0 \to 1$), very soon after an epidemic has started from a single infective, the number of infectives is already too large for the definition of R_0 to apply. What can be measured directly from the PCA in long-time simulations, and in most field studies is rather R_t, the effective reproductive rate at a given time t during the evolution of the disease when there are also infective and recovered individuals present. Upon the further assumption of weak homogeneous mixing, we can write

$$R_{t\to\infty} = R_0 s^* \tag{6}$$

where s^* is the fraction of susceptibles in the endemic steady-state. If the rate at which susceptible individuals are infected is exactly balanced by the rate at which new susceptibles are born then $\bar{R}_{t\to\infty} = 1$, where the bar denotes the temporal mean. That is, each primary infection will produce on average exactly one secondary infection. Under these precise conditions, the basic reproductive rate is simply $R_0 = 1/s^*$.

The depletion of susceptibles during the epidemic implies that $R_t < R_0$ always. If R_t is maintained below 1 for sufficiently long, then the pathogen will become extinct. However, as we can see in Fig. 6, R_t calculated directly from the PCA oscillates around an average value which, within statistical errors, is exactly one. R_t drops below the self-sustained threshold for half the period of oscillations only to rise again above it in the next half-cycle.

This result has drastic implications on the assessment of the risk of recurrent outbreaks of infectious diseases based on calculations of R_t. It is usually assumed that as soon as R_t drops below 1, an epidemic is on its way to be contained. Very recently this criterium was used to judge the outcome of the SARS epidemic[25,26] based on data spanning only a few weeks. In this vein, the results in Fig. 6 should act as a warning particularly when, as it often happens, one has to deal with short time series or otherwise incomplete data.

Now we must make the distinction between the long-time measurements, using time averages over (extended) time series of recurrent epidemics and short-time measurements of R_0. For the long-term measurements we have

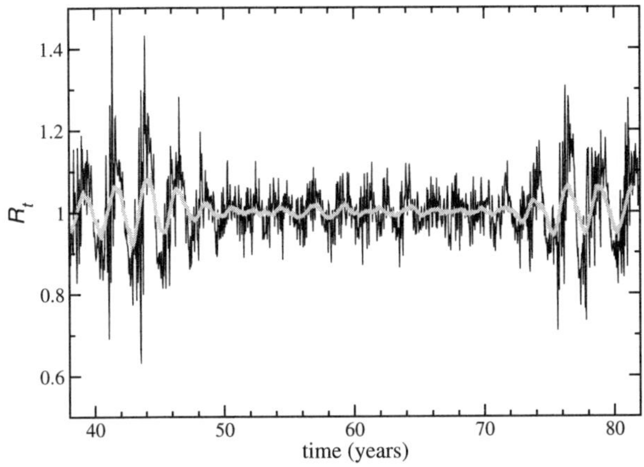

Figure 6. Evolution of the effective reproductive rate in a SEIR simulation. R_t is calculated by counting the number of secondary infections caused by every infective individuals and taking averages. The result is a very noisy time series which nevertheless shows an underlying structure composed of cycles with the same period as the oscillations in incidence. Smoothing the data by averaging it out with a moving window brings forward the pattern of oscillations (thick plain curve). In the SIR model we observe the same effect only the data contains a lot more noise.

different possibilities to estimate R_0: We can either i) compute A directly from the PCA by the method described in the next Section and then use Eq. 3 to evaluate R_0. ii) Estimate the effective transmissibility β_{eff} as described in Section 3.3 and then compute R_0 from (4) with $\beta \equiv \beta_{\text{eff}}$ or iii) use the fact that $R_t = 1$ once an asymptotic regime is attained and estimate $R_0 = 1/s^*$.

As an alternative to perform lengthy simulations of recurrent epidemics for the purpose of computing R_0, we can run the PCA in a closed population. Discarding births and deaths by setting $\mu = 0$, we can iv) compute the R_0 from the final size equation (FSE)[27]

$$\ln(s_\infty) = R_0(s_\infty - 1) + \ln(s_0), \tag{7}$$

where $s_\infty = S_\infty/N$, with S_∞ the number of susceptibles left once the epidemic has died out. The initial fraction of susceptibles is $s_0 = (N-1)/N$ since every new run begins with a single infective. The last form of computing R_0 that we considered was to v) fit the exponential growth law, Eq. (5), to the early stages of evolution of an epidemic ravaging a wholly susceptible population. The results are presented in Fig. 7. They show how R_0 can be used to assess the departure from mean-field behaviour as the fraction of long-range infection is decreased, and also that in a structured population with the characteristics of a dynamic small-world, the magnitude of R_0 is always below the value that would be obtained in an homogeneously mixed population. Different procedures of estimating R_0 that would give the same result under the mean-filed approximation will now yield markedly distinct values. This is because they still appeal to an approximation at some stage but not all at the same stage. Undoubtedly, this is a blow to the worth of R_0 evaluated in practical situations. Indeed, with the exception of contact tracing in the very early stages of an epidemic, which is a very difficult task to perform, R_0 can only be determined through indirect methods. Our simulations show that evaluating R_0 from different sources, for instance from A obtained from serological studies, or from the equilibrium number of susceptibles obtained from historic time series can lead to *intrinsically* different results as a consequence of the heterogeneity of contacts.

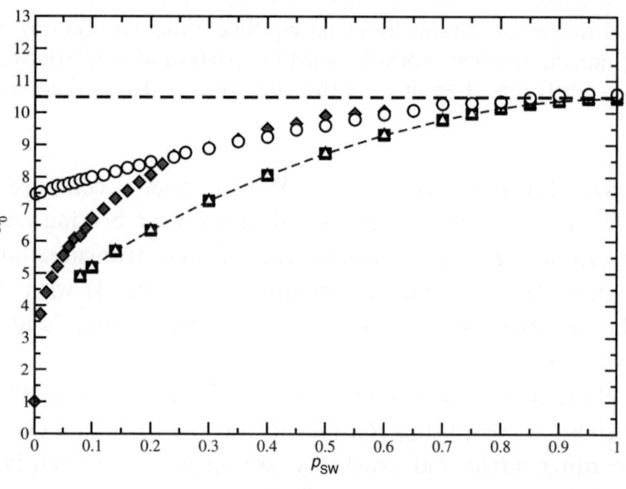

Figure 7. R_0 as a function of the small-world parameter p_{SW}. The two upper curves were calculated from simulations in a closed population ($\mu = 0$), from the final size equation (7) (circles) and by fitting exponential growth using Eq. (5) (diamonds). At p_{SW} the number of infectives grows in time as a power law and therefore $\Lambda \equiv 0$ implying $R_0 \equiv 1$. The curves obtained from the analysis of long-term dynamics using $R_0 = 1/s^*$ (black squares) and from A and Eq. (3) (white triangles) are almost indistinguishable. In the limit of a random network the four curves converge to $\beta_0/(\gamma + \mu)$, indicated by the dashed line, with a small discrepancy attributed to final size effects. SIR model in a 400×400 lattice with stochastic recovery; parameters are the same as in Fig. 5.

3.5. *Average Age at Infection*

The moment of their lives when susceptible individuals acquire infection is a very important epidemiological quantity. The average age at infection, A is inferred from serological surveys and can be compared to the output of epidemiological models.

We evaluate the impact of network structure on A by calculating it directly from the simulations as follows: a counter is associated to every susceptible and the moment when this individual is infected is recorded. Averaging over every single individual who has become infected so far one obtains a quantity that shows a slow but unremitting trend towards asymptotic behaviour characterized by small, rapid fluctuations around a steady-state value. Once the long term evolution appears to stabilize we compute the time average.

The average age at infection was found to be linear in $1/\mu$ down to only 8 % of long-range contacts. Below that value of $p_{\rm SW}$ persistence was too low to obtain meaningful averages. The slope of the lines in Fig. 8 gives $1/R_0$. Moreover, the ratio $r = R_t A\mu/s^*$ varies from 0.9999 ± 0.0002 at $p_{\rm SW} = 1.0$ to $r = 0.9994 \pm 0.0001$ at $p_{\rm SW} = 0.08$ showing that mean-field relations hold to an excellent approximation down to surprisingly low values of $p_{\rm SW}$. Deviations to mean-field behaviour do occur but are only slight even in relatively small populations. From the data in Fig. 8 we can extrapolate the average age at infection at low birth rates. Setting the time scale by taking $L \equiv 1/\mu = 61$ years we obtain $A = 6$ years for $p_{\rm SW} = 1.0$, $A = 7.7$ years for $p_{\rm SW} = 0.3$ and $A = 10.2$ years for $p_{\rm SW} = 0.08$ showing that the average age at infection increases significantly when we consider a local, clustered network of contacts.

4. Results: SEIR model

4.1. *Epidemic cycles*

We now present numerical simulations of the more sophisticated SEIR model, with etiological and demographic parameters corresponding to measles, rubella and chickenpox, in developed countries, in the pre-vaccination era. Just as in the SIR case we observe sustained oscillations in incidence. The oscillations obtained from the SEIR model are less coherent than those obtained with the SIR version. Their aspect is actually closer to the observed time series. But the most consequential finding of the SEIR simulations is the realization that one can only obtain sustained oscillations with amplitudes compatible with the existing data records, and

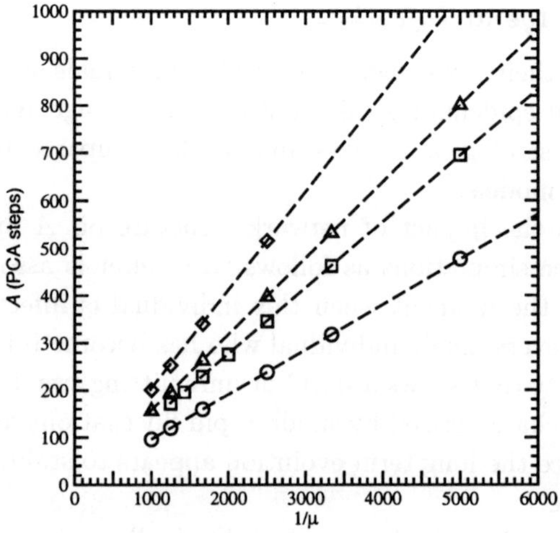

Figure 8. The average age at infection measured in PCA steps as a function of life expectancy $1/\mu$ in the same units, for $p_{SW} = 1.0$ (circles), 0.3 (squares), 0.2 (triangles) and 0.08 (diamonds). The slope of the lines gives $1/R_0$. The parameters are $\beta_0 = 0.66$, $\gamma = 0.0625$; the initial infective fraction was $s_0 = 0.165$ and 50 infectives were present on the lattice. The values of A were obtained from a temporal average of data sampled from $t = 10000$ to $t = 50000$ at 10 steps interval and further averaging over 10 realizations starting from different initial conditions.

for realistic values of the model parameters – (*e.g.* life expectancy, latency and infectious periods) – if one chooses values of p_{SW} in the small-world region[13]. To illustrate this feature, we show in Fig. 9 two time series for measles, one obtained for $p_{SW} = 0.2$ and the other for $p_{SW} = 1.0$. The amplitude of the oscillations is almost double in the small-world network.

For $p_{SW} = 1.0$ the frequency distribution is peaked around 1.5 years while for $p_{SW} = 0.2$ the peak is at about 2.5 years. For measles, almost every data record in developed countries points to cycles of almost exactly 2 years, in between those two values. Agreement with the observed periods can be improved, but only to some extent, by fine tuning the transmissibility. Indeed, increasing β_0 has the effect of decreasing the period making it more in line with the observations. The resulting time series are shown in Fig. 10. Nevertheless, we must stress that β_0 is nothing like a free parameter. First of all it cannot be changed at will in order to tune the period because the amplitude of oscillations also changes and the average size of outbursts must remain comparable to those observed in cities with the same population size[23]. Secondly, since the transmissibility of typical childhood

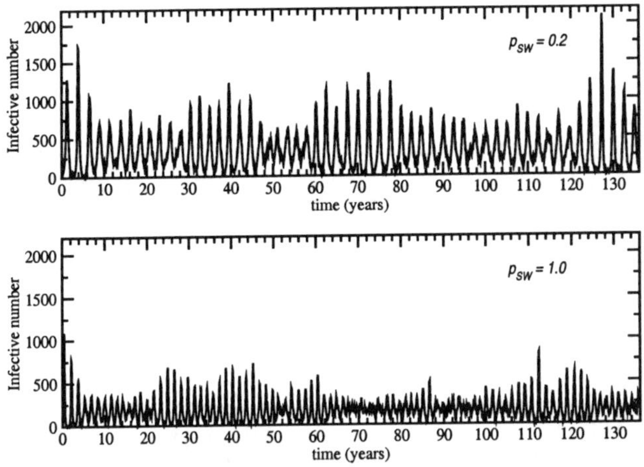

Figure 9. Time series of the number of infectives obtained from simulations of SEIR dynamics in a 1000 × 1000 lattice with $\mu = 1/61$ yr^{-1}, $\tau_{lat} = 6$ days, $\tau_{inf} = 8$ days and $\beta_0 = 3.92$ day^{-1}.

diseases must be similar (see Section 4.3 below), differences in period, average age at infection, etc. between them must be achievable with values of β_0 of the same order of magnitude and respecting the expected infectious rank of those diseases. Finally, and most importantly, although β_0 is a parameter that can only be very loosely inferred from the data, there are indirect estimates of β_{eff} from R_0 and A, using mean-field relations, that set an order of magnitude for β_0.

4.2. *Estimates of A and R_0*

We have computed the average age at infection obtaining $A = 1.6$ years for $p_{\text{SW}} = 1.0$ and 3.6 years for $p_{\text{SW}} = 0.2$ for the simulations shown in Fig. 10, and $A = 1.9$ years for $p_{\text{SW}} = 1.0$, and 4.1 years for $p_{\text{SW}} = 0.2$ for the simulations in Fig. 9. While the latter is just barely above the lower bound of the interval commonly accepted to correspond to measles data – between 4 to 6 years – the values obtained with the homogeneous mixed population lie notably outside the realistic range. A further refinement consisting in the inclusion of protection by maternal antibodies in newborns could increase both values by 3 to 9 months putting the result for $p_{\text{SW}} = 0.2$ well inside the interval of realistic values. However, this is expected to have a deleterious repercussion on the period of oscillations, increasing it above what is consistent with data records for measles.

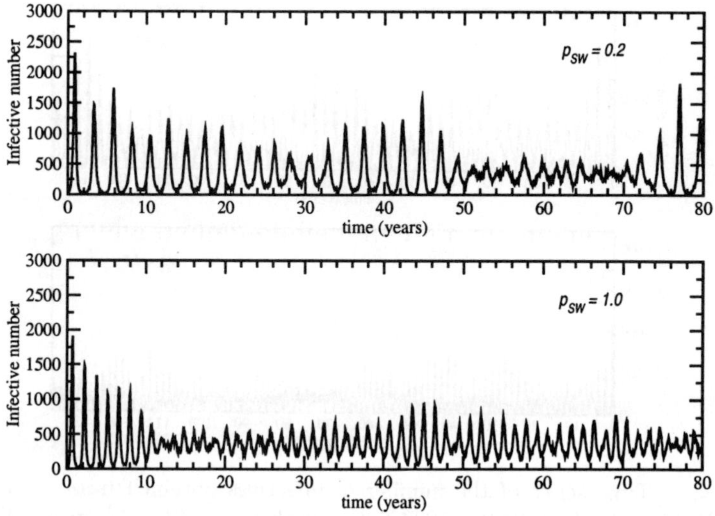

Figure 10. Same parameters as in Fig. 9 except that now $\beta_0 = 4.75$ day^{-1}. The period of oscillations shrinks when β_0 increases. Note the change of scale in both axes with respect to Fig. 9.

For the simulation at $p_{SW} = 0.2$ in Fig. 9, the average fraction of susceptibles was $s^* = 0.0657$ giving a value for $R_0 = 1/s^*$ of 15.2 while for $p_{SW} = 1.0$ we get $s^* = 0.0312$ yielding $R_0 = 32.1$. Whereas the former lies inside the reported range of R_0 for measles, $14 - 18$[1], the latter is way off range. For the SEIR model, the mean-field expression for R_0 is slightly more complicated than in the SIR case[1]:

$$R_0 = \frac{\beta}{\tau_{inf}^{-1} + \mu} \left(\frac{\tau_{lat}^{-1}}{\tau_{lat}^{-1} + \mu} \right) . \tag{8}$$

Solving for β and using the values of R_0 calculated above we obtain $\beta = 4.0$ and $\beta = 1.9$ for $p_{SW} = 1.0$ and 0.2 respectively. While the former is close to the β_0 used in the simulations, the latter is practically one half. The reason for this is that the transmission rate computed from Eq. (8) is rather the β_{eff} introduced in Section 3.3, and only in the mean-field case do we have $\beta_{\text{eff}} \equiv \beta_0$. The effective transmission rates obtained directly from the simulations by the method of Section 3.3 were $\beta_{\text{eff}} = 1.86$ for $p_{SW} = 0.2$ and $\beta_{\text{eff}} = 3.93$ in the homogeneous mixed case, in good agreement with the respective β's calculated from Eq. (8) with $R_0 = 1/s^*$ as input.

4.3. *Comparison between childhood diseases*

In this section we show that the ability of the PCA to describe sustained oscillations is not restricted to measles but extends to other childhood diseases conferring life-long immunity. In Figs. 11 and 12 we show long-term runs for etiological parameters corresponding to rubella and chickenpox, respectively. We kept the small-world parameter fixed at $p_{SW} = 0.2$, since childhood diseases must share a common contact structure. The PCA is nevertheless capable of discriminating between these different diseases, in terms of period, average age at infection and basic reproductive rate.

The periods estimated from the Fourier transform of the time series in Figs. 11 and 12 were $T \approx 4.4$ years for rubella and $T \approx 3.4$ for chickenpox; the average age at infection was $A \approx 6.2$ for rubella and $A \approx 4.7$ for chickenpox. Like for measles, A is considerably underestimated by the PCA, most studies in developed countries giving a lower bound for A of about 9 years for rubella and 6 years for chickenpox. Still, the relative values of A for the three diseases agree with the data[1]. The basic reproductive rate computed from the inverse fraction of susceptibles is about 10 for rubella and 13 for chickenpox. The simulations for measles gave $R_0 \approx 15$. Again, the values for measles and chickenpox are very good while the reproductive rate of rubella is above that reported.

Based on a purely qualitative evaluation of the mode of transmission we can rank common childhood diseases in roughly two classes of infectiousness. Measles and chickenpox are transmitted through aerosol droplets and therefore the most infectious. In the second group we have mumps and rubella which require direct contact with droplets generated by sneezing and coughing. In our simulations, this hierarchy of infectiousness was strictly respected, with $\beta_0^{\text{measles}} > \beta_0^{\text{chickenpox}} > \beta_0^{\text{rubella}}$. The numerical values used in the simulations give a quantitative estimate of infectiousness:

$$\frac{\beta_0^{\text{measles}}}{\beta_0^{\text{rubella}}} = 3.7, \quad \frac{\beta_0^{\text{measles}}}{\beta_0^{\text{chickenpox}}} = 1.9,$$

and the respective effective transmission rates obtained for a fraction of 20 % of long-range contacts give:

$$\frac{\beta_{\text{eff}}^{\text{measles}}}{\beta_{\text{eff}}^{\text{rubella}}} = 2.5, \quad \frac{\beta_{\text{eff}}^{\text{measles}}}{\beta_{\text{eff}}^{\text{chickenpox}}} = 1.7.$$

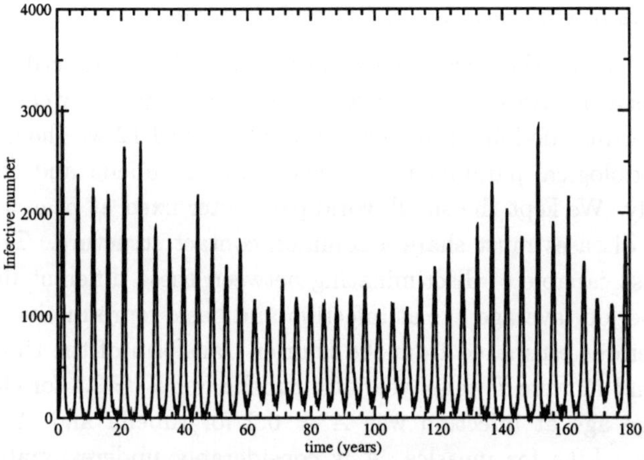

Figure 11. Time series of the number of infectives for rubella, which has longer latency and infectious periods than measles and also lower infectiousness. Simulation in a 1000×1000 lattice with $p_{\mathrm{SW}} = 0.2$. The birth rate $\mu = 1/61$ yr^{-1} is the same as in the measles simulations, but now $\tau_{lat} = 12$ days, $\tau_{inf} = 12$ days and β_0

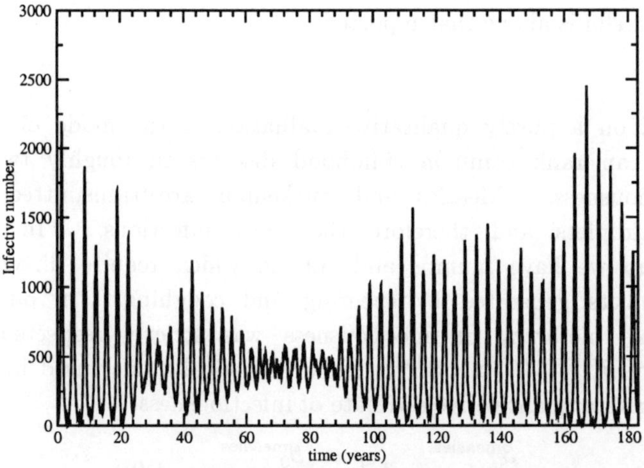

Figure 12. Time series of the number of infectives for etiological parameters corresponding to chickenpox. Demographic parameters and small-world parameter are the same as for rubella and measles; $\tau_{lat} = 10$ days, $\tau_{inf} = 10$ days and $\beta_0 = 2.4$ day^{-1}.

Our results suggest that measles is about three to four times more infectious then rubella and about twice as infectious as chickenpox. Given the crudeness of these estimations we can consider that the ratios agree quite well with those reported by Keeling & Grenfell[28]: 3.4 and 2.4 respectively for the transmissibility ratios and 2.5 and 1.4 for the effective transmission rate.[a] In another study[4] a value of 3.8 for the ratio of β_0's between measles and rubella was obtained by fitting the output of a seasonally forced SIR model to the data.

5. Discussion

The results in this paper show that both the short and long-term dynamics of childhood diseases conferring life-long immunity can be described by taking into account a dynamic small-world network of contacts. Moreover, the simulations make clear that the ability of recurrent epidemics to survive for a large number of cycles depends strongly on the level of heterogeneity in contact structure. The fraction of runs that survive up to a maximum ascribed time has a non-trivial dependence on the fraction of long-range contacts displaying, at low enough birth rates, a maximum at finite fraction of shortcuts. On the other hand, the persistence varies with the population size, and this gives the rationale on which to base a network-dependent Critical Community Size. The oscillations get more synchronized and their amplitude is strongly enhanced when p_{SW} is decreased below $p_{SW} = 1$, *i.e.* as the weight of local contacts is increased. The amplitude cannot keep growing because the troughs between epidemic surges will get so deep, the number of surviving infectives so close to zero, that a stochastic fluctuation will eventually drive the epidemic extinct. This explains why an increase in amplitude is correlated with a decrease in persistence.

The small-world network allows us to go step by step from a dynamical system with a large number of degrees of freedom – which for $p_{SW} = 0$ are the $3N$ possible states of the lattice – to the case where the evolution can be captured by only a couple of ordinary differential equations. How this this contraction of phase space happens is an important but difficult question that deserves further investigation. We do know, however, the properties of the flow of the low dimensional deterministic system, namely that for the SIR ODE's the only attractor is a fixed point (s_0, i_0) and that there is a saddle point at $(1, 0)$ with its stable manifold along the $i_0 = 0$ axis.

[a] Although their study focus on mumps instead of rubella, almost surely the two diseases have very similar values of infectiousness.

The fraction of infectives in the endemic steady-state is proportional to the birth rate, $i_0 = \mu(R_0 - 1)/\beta$, therefore as $\mu \to 0$, i_0 gets asymptotically close to the stable manifold. Since the real part of the pair of complex conjugate eigenvalues vanishes at zero birth rates one observes the critical slowing down of the dynamics at $\mu \to 0$. Small disturbances such as those provided by the stochastic drive will enable the system to make a large excursion in phase space before returning again to the vicinity of (s_0, i_0) where it is never allowed to settle.

It is clear that the progressive introduction of spatial degrees of freedom brings spatio-temporal coherence into the stochastic dynamics. The oscillations at small $p_{\rm SW}$ display a strong contribution from higher-order harmonics indicating that the system visits orbits located further away from the fixed point. Also, both the amplitude and period are larger than at larger values of $p_{\rm SW}$, the oscillations evolving on a slower time scale compared to the less structured ones observed in the limit of random mixing. Clustering lowers the effective transmission rate and the average infective number but has the collateral effect of deepening the troughs and rising the peaks of oscillations. In the epidemic lows, infectives and susceptibles keep close together and the epidemic almost dies out. But one shortcut that randomly links an infective to the middle of a susceptible cluster will be enough to cause a large outbreak that may even consume every susceptible in that cluster. However some susceptibles will still remain in small clusters scattered all over the lattice and, screened from infection, their numbers will steadily grow at the (slow) pace dictated by the birth rate. Eventually one or more of these clusters will reach a size large enough that there will be a high probability of a long-range infection event linking to an individual inside them. When this happens the conditions are set for the cycle to repeat itself. When the fraction of shortcuts is high, susceptibles are steadily consumed at an intermediate rate and do not have time to aggregate into medium-size clusters. On the average the infection does spread more effectively – the average infective number is higher – but the large outbreaks are suppressed and what we observe instead are small, rapid fluctuations around an endemic state.

The simple fact that the PCA implementation of SIR and SEIR dynamics leads to sustained, fully developed oscillations as a consequence of heterogeneity in the network of contacts is by far the most impressive difference between our results and the output of deterministic mean-field models. However there are quantitative differences arising in quantities that are very relevant for epidemiologists. The effective transmission rate

β_{eff} is always smaller than the mean-field value β_0. In this respect our long-term simulations corroborate the results obtained from the simulation of a single epidemic outbreak[17]. R_0 is also lower than in the limit of random mixing. Moreover, in structured populations it is very important to distinguish between measurements of R_0 obtained from long time series and those focusing on a single outbreak because these two methods were shown to present the greater differences.

The SIR implementation can be used to access the qualitative impact of the network structure but to make the predictions of the PCA quantitative SEIR dynamics is required. Long time series, more than twice the length of the best available records can be easily obtained for realistic values of the model parameters and show a good agreement with observed values of the period and basic reproductive rate for measles and chickenpox. For rubella the agreement was only reasonable. In all the three cases the average age at infection is systematically underestimated, the discrepancy reaching 30 % in the case of rubella. The values obtained could be improved by accounting for immunity by maternal antibodies but still with this respect the PCA behaves no better than the mean-field models.

Different childhood diseases evolve on social networks that are similarly organized, transmission occurring predominantly in schools and households with a few exceptions corresponding to the long-range contacts. The structure of the contact patterns is thus identical and so are the demographic parameters. A realistic network model of childhood diseases must be able to discriminate between different diseases, in terms of T, A and R_0, purely as a result of the different latency and infectious periods plus the transmissibility β_0. The PCA satisfies this requirement and yields a reasonable estimate for the infectiousness of each disease.

6. Conclusion

The individual-based network model presented in this paper combines local structure with casual, long-range links. The latter are shortcuts through which the disease can propagate into regions where susceptibles predominate. Correlations that build up in the system due to network structure cause deviations from mean-field behaviour, but in the relevant limits the mean-field results are restored. Surprisingly, we found that, when the long-term evolution is considered, even with a small fraction of shortcuts the mean-field relations between the average age at infection, the basic reproductive rate and the average number of susceptibles still hold. This consistency may explain why, although based on a unrealistic description of human contacts, deterministic models featuring homogeneous mixing remained for so long the basic conceptual tool of theoretical epidemiology. Their flaws, particularly the inability to describe recurrent epidemics, were exposed in the present work, in the context of childhood diseases. These are highly infectious diseases for which it is not unreasonable to assume that anyone who engages in any basic form of social interaction is equally at risk. Thus, we have disclosed the weaknesses of mean-field models in the case where they are certainly the less severe. Simulation of individual-based stochastic models becomes imperative in order to capture the dynamical complexity of infections like HIV of Hepatitis that spread on networks characterized by an extreme heterogeneity, like the network of sexual partnerships or needle sharing by intravenous drug users.

Acknowledgments

The author received a fellowship from Fundação para a Ciência e Tecnologia (FCT), ref. SFRH/BPD/5715/2001. Financial support by the FCT under project POCTI/ESP/44511/2002 is also gratefully acknowledged.

References

1. Anderson and May (1991), Infectious Diseases of Humans, Oxford University Press, Oxford.
2. Earn, David J. D. *et al.* (2000), A Simple Model for Complex Dynamical Transitions in Epidemics, *Science* **287**, 667-670.
3. Greenman, Jon, Kamo, Masashi, and Boots, Mike (2004), External Forcing of Ecological and Epidemiological Systems: a Resonance Approach, *Physica* D **190**, 136-151.
4. Keeling, Matt J., Rohani, Pejman and Grenfell, Bryan T. (2001), Seasonally Forced Disease Dynamics Explored as Switching Between Attractors, *Physica* D, **148**, 317-335.
5. Olsen, L. F. and Schaffer, W. M. (1990), Chaos versus Noisy Periodicity: Alternative Hypothesis for Childhood Epidemics, *Science* **249**, 499-504.
6. Rohani, Pejman, Keeling, Matthew J. and Grenfell, Bryan T. (2002), The Interplay between Determinism and Stochasticity in Childhood Diseases, *The American Naturalist*, **159**, 469-481.
7. Blasius, Bernd, Huppert, Amit and Stone, Lewi (1999), Complex Dynamics and Phase Synchronization in Spatially Extended Ecological Systems, *Nature*, **399**, 354-359.
8. Bailey, Norman T. J. (1975), The Mathematical Theory of Infectious Diseases, Charles Griffin, London.
9. Lloyd, Alen L. and May, Robert M. (1996), Spatial Heterogeneity in Epidemic Models, *J. Theor. Biol.*, **179**, 1-11.

10. Bolker, Benjamin and Grenfell, Bryan (1995), Space, Persistence and Dynamics of Measles Epidemics, *Phil. Trans. R. Soc. Lond.* B, **348**, 309-320.
11. Keeling, Matt J., Rohani, Pejman and Grenfell, Bryan T. (2000), Seasonally Forced Disease Dynamics Explored as Switching between Attractors, *Physica D*, **148**, 317-335.
12. Watts, D. J. and Strogatz, S. H. (1998), Collective Dynamics of 'Small-world' Networks, *Nature*, **393**, 440-442
13. Watts, D. J., Small Worlds (1999), Princeton University Press, Princeton, NJ.
14. Moore, Christopher and Newman, M. E. J. (2000), Epidemics and Percolation in Small-world Networks, *Physical Review* E, **61**, 5678-5682.
15. Newman, M. J., Jensen, I. and Ziff, R. M. (2002), Percolation and Epidemics in a Two-dimensional Small World, *Physical Review* E, **65**, 021904.
16. Boots, Michael and Sasaki, Akira (1999), "Small worlds" and the evolution of virulence: infection occurs locally and at a distance, *Proc. R. Soc. Lond.* B, **266**, 1933-1938.
17. Kleczkowski, A. and Grenfell, Bryan T., (1999), Mean-field-type equations for spread of epidemics: the "small world" model, *Physica A* **274**, 355-360.
18. Keeling, M. J. (1999), The effect of local spatial structure on epidemiological invasions, *Proc. R. Soc. Lond.* B, **266**, 859-867.
19. Johansen, Anders (1996), A Simple Model of Recurrent Epidemics, *J. Theor. Biol.*, **178**, 45-51.
20. Johansen, Anders (1994), Spatio-temporal Self-organization in a Model of Disease Spreading, *Physica* D **78**, 186-193.
21. Kuperman, Marcelo and Abramson, Guillermo, Small World Effect in an Epidemiological Model, *Phys. Rev. Lett.*, **86**, 2909-2911.
22. He, David and Stone, Lewi (2003), Spatio-temporal synchronization of recurrent epidemics, *Proc. R. Soc. Lond.* B, **270**, 1519-1526.
23. Verdasca, J. *et al.* (2005), Recurrent epidemics in small world networks, *J. Theor. Biol.*, **233**, 553-561.
24. Lloyd, Alun L. (2001), Realistic Distribution of Infectious Periods in Epidemic Models: Changing Patterns of Persistence and Dynamics, *Theor. Pop. Biol.*, **60**, 59-71.
25. Riley, S. R. et al., Transmission Dynamics of the Etiological Agent of SARS in Hong Kong: Impact of Public Health Interventions, *Science* **300**, 1961-1966.
26. Lipsitch, M. et al., Transmission Dynamics and Control of Severe Acute Respiratory Syndrome, *Science* **300**, 1966-1970
27. Diekmann O. and Heesterbeek, J. A. P. (2000), Mathematical Epidemiology of Infectious Diseases, Wiley, New York.
28. Keeling, Matt J. and Grenfell, Bryan T. (2000), Individual-based Perspectives on R_0, *J. theor. Biol.*, **203**, 51-61.

CLONAL EXPANSION OF CYTOTOXIC T CELL CLONES: THE ROLE OF THE IMMUNOPROTEASOME

MICHAL OR-GUIL

Institute for Theoretical Biology
Systems Immunology Group
Invalidenstr. 43, 10115 Berlin, Germany
E-mail: m.orguil@biologie.hu-berlin.de

FABIO LUCIANI

Institute for Theoretical Biology
Systems Immunology Group
Invalidenstr. 43, 10115 Berlin, Germany
E-mail: f.luciani@biologie.hu-berlin.de

JORGE CARNEIRO

Gulbenkian Institute of Science
Theoretical Immunology Group
Apartado 14, P-2781-901 Oeiras, Portugal
E-mail: jcarneir@igc.gulbenkian.pt

In the present model, we investigate the homeostasis and the clonal expansion of cytotoxic T cell clones taking into account peptide processing and the effect of changes in the cellular proteasome composition which happen during an infection. The model is based on a classical theoretical description of T cells competing for resources. It shows that the shaping of different peptide distributions by different proteasome compositions can strongly influence the dynamics of the T cell repertoire. We found that the immunoproteasome, which is upregulated during an immune response, may represent a major contribution to the selection of appropriate T cell clones, enhancing the effectiveness of the immune response by several orders of magnitude.

1. Introduction

Intracellular pathogens are fought by the immune system via cytotoxic T cells. These have the function of recognizing cells that are infected, e.g. by viruses, and kill them by inducing apoptosis. T cells possess on their membrane T cell receptors than can recognize what other cells are presenting: All cells degrade their cytosolic protein content by breaking the proteins into fragments. A fraction of these fragments is loaded on MHC I receptors, forming MHC-peptide complexes (MHCp), which are presented on the surface of the cell. With that, the cell shows a sample of its cytosolic content to the "outside world".

T cell receptors are able to bind to the MHCp's and signal the T cell that something was recognized on the cell membrane. The recognition strength is a function of the binding affinity and of the density of specific MHCp - the higher these two quantities, the higher the probability that a T cell will be activated by the contact with a cell presenting peptides[1]. When infection is absent, the presented peptidic repertoire consists of solely self-peptides, and no immune response are normally elicited.

The repertoire of T cells is very large, consisting of 10^8 cells divided into about 10^7 different clones in mice. Each clone carries a specific T cell receptor, which has its own specific MHCp-recognition properties. T cells mature in the thymus where they are selected such that T cell receptors with high-affinity toward peptides presented in the thymus are eliminated from the repertoire. Absence of immune responses in the absence of foreign antigens is, at least in part, due to the fact that T cells produced by the thymus have low affinity for self-antigens. The thymic output thus influences the clonal repertoire of naive cells, i.e. T cells that have not yet been activated by recognition of foreign peptides.

It has been shown that the repertoire of T cells is maintained in the periphery by contact with the self-peptides presented by antigen presenting cells (APC). Thus, recognition of self-peptides with low-affinity is paramount for a T cell to survive and proliferate, and thus for the persistence of the clone (see sketch in Figure 1).

The presented peptide distribution consists hence of a resource for the T cells. Since multiple T cell clones might recognize the same peptides, and the number of MHCp and space on antigen presenting cells is restricted, there must be competition for these common resources. Indeed, competition for resources, such as MHCp, biological space, and cytokines are known to play a role in the dynamics of T cell clones at the healthy state, leading to a homeostatic maintenance of total T cell numbers and diversity[2,3].

Consider now that the healthy steady state of a host is perturbed by an infection. Each body cell infected by the pathogen will process its proteins and present foreign peptides on its surface. APCs, which take up infected cells, home into secondary lymphoid tissues, activating naive T cells which recognize the foreign peptides with high affinity, together with other self-peptides presented. If these T cells become activated by such a contact, they undergo fast proliferation and differentiation into effector cells. These cells are able to move toward the infection sites and kill infected cells which present the foreign peptides.

The machinery processing intracellular proteins is very complex. An important player hereby is the proteasome, an ubiquitous, barrel shaped protease which degrades the proteins, producing fragments which can be transported into the endoplasmic reticulum to be loaded onto MHC I molecules. Hence, the fragments produced by the proteasome will help to define what peptides will be finally presented. However, there is not only one proteasome, but a whole "zoo", provided with different caps influencing the entrance of proteins into the barrel, where they are cleaved by sequence specific cleavage units. Two different types of proteasome "barrels" with different cleavage units exist: The constitutive and the immunoproteasome. While they are both present in thymic cells, usually the constitutive proteasome prevails. Only during an infection will the immunoproteasome be upregulated. At the same time, also the number of APCs and the number of MHCp's mounted on each APC increases[4,5].

Hence, the antigen processing and presenting machinery changes at the presence of an ongoing immune response. We will assume that these changes elicit a change in the peptide distributions presented, leading to a "peptide redistribution" (see e.g. Refs. 6, 7, 8, 9, 10 for discussions on the change in immunoproteasome products).

Leon et al.[11] have argued that peptide redistribution on MHC class II during infections facilitate immune responses by CD4 T cells. Here, we hypothesize that the mere change in the presented peptide distribution on MHC class I might also render the immune response by CD8 T cells more effective. Based on a previous mathematical description of the peripheral T cell repertoire dynamics we show that low precursor frequency clones recognizing pathogenic peptides increase their proliferation rates by outcompeting larger clones with whom they compete for the common self-peptidic resources in case of a peptide redistribution at the onset of an infection.

2. The model

2.1. *T cell dynamics*

The model presented in the following describes the dynamics of naive N and the effector E cytotoxic T cells, and is based on earlier published models[12]. It assumes that both naive and effector T cell pools require MHCp contact

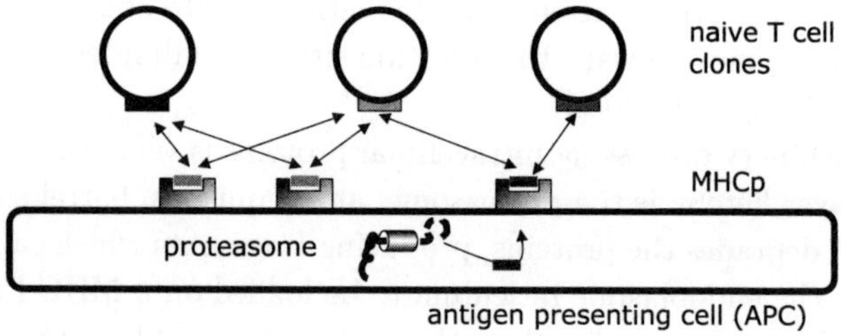

Figure 1. T cells competing for peptides presented on the surface of antigen presenting cells (APCs). T cells recognize, via their T cell receptors, peptides bound on MHC I molecules (MHCp). A T cell receptor might bind different peptides, with possibly different binding affinities. The binding affinity is a crucial parameter determining recognition. The repertoire of peptides presented by APCs originates from the degradation of cytosolic proteins, which can be either pathogenic or belong to "self". The constantly presented self-peptides trigger the proliferation of recognizing T cell clones, thus shaping their repertoire. Proteasomes are part of the protein degradation and presentation machinery and cleave proteins into fragments. Therefore, we assume that the repertoire of peptides presented is influenced by the proteasomes present. A partial replacement of constitutive proteasomes with immunoproteasomes is supposed to induce a new peptide distribution on the surface of APCs. This is the basis for a reshaping of the T cell repertoire.

to maintain their homeostatic size, their proliferation rate being dependent on the strength of recognition K_{ij} of a peptide j by a clone i, as well as of the amount of the peptide presented. A fraction $1 - \phi$ of the proliferating naive cells differentiate into effector cells:

$$\frac{dN_i}{dt} = H_i + (2\phi - 1)\,\rho_N \sum_{j=1}^{N_{pep}} K_{ij}\, F_j N_i - \delta_N N_i \tag{1}$$

$$\frac{dE_i}{dt} = 2(1-\phi)\,\rho_N \sum_{j=1}^{N_{pep}} K_{ij}\, F_j N_i + \sum_{j=1}^{N_{pep}} \rho_E\, K_{ij}\, F_j E_i - \delta_E E_i \tag{2}$$

The indexes i, j denote T cell clones i, $i = 1 \cdots N_{clone}$ and MHCp of specificity j, $j = 1 \cdots N_{pep}$, respectively. The thymic output of naive cells is constant and given by H_i. Moreover, a death rate constant for naive, δ_N, and for effector cells, δ_E, is considered. Naive cells N_i are activated with rate constant ρ_N, while effector cells E_i are characterized by a proliferation rate constant ρ_E, where $\rho_E > \rho_N$. F_j is the amount of free MHCp sites on the APCs with specificity j[13]:

$$F_j = \frac{M_T P_j}{1 + \sum_{i=1}^{N_{clone}} K_{ij}(N_i + E_i)} \tag{3}$$

P_j denotes the fraction of MHCp's presenting peptide j, hence $\sum_{j=1}^{N_{pep}} P_j = 1$. The peptide concentration is limited by the MHC availability, which is defined by the maximum abundance M_T. Equation (3) describes the competition for MHCp: When the numbers of T cells recognizing peptide j increases, more MHCp sites will be occupied by T cells, and the number of free complexes diminishes.

Note that the T cell response is simply proportional to the sum of the free MHCp weighted by the pairwise affinity. This is different from the model of Leon et al.[11], who assumed that the T cell response was proportional to the strength of the APC-T cell contact, which was on its turn a non-linear function of the MHCp density weighted by the affinity.

2.2. *Pathogen dynamics and foreign peptides*

A foreign pathogenic peptide P_P will be presented only when infected cells are present. The number of infected cells grows as the pathogen spreads. We assume that the amount of pathogenic peptide presented grows linearly with the number of infected cells I for small cell numbers, but saturates at $P_0 \leq 1$ for high values of I. This assumption is based on experimental data showing a saturation threshold in the amount of a single peptide presented on the surface of APCs[14].

$$P_P(t) = \frac{P_0 I(t)}{I(t) + \theta}$$

The peptide concentration increases with the increase of the number of infected cells, and saturates when P_0 is reached. The parameter θ gives the amount of infected cells needed to mount 50% of the maximum MHCp load.

The dynamics of the number of infected cells I is described by a standard model[13,15].

$$\frac{dI}{dt} = r(1 - cI)I - D(\sum_{i=1}^{N_{clone}} K_{iP} E_i)I$$

Infected cells replicate with maximum rate constant r. The replication is controlled by a maximum capacity of the system to contain infected cells.

This is accounted for by the parameter c, where $1/c$ is the maximum number of infected cells in the organism. Moreover, the death rate of infected cells is proportional to the amount of specific effector cells and to their recognition strength K_{iP}. We neglect here the contribution of the self-antigens to the killing function, which is a necessary condition to avoid autoimmunity in the healthy steady state.

The parameter values used in this work are summarized in Table 1.

Table 1. Parameters

Parameter	Description	Dimension	Default value
δ_N	Death rate naive	day^{-1}	0.005
δ_E	Death rate effector	day^{-1}	0.4
ϕ	Naive renewal fraction		0.99
k_h	Recognition strength self-peptide	day^{-1} $site^{-}1$	$5 \cdot 10^{-2}$
k_s	Recognition strength foreign peptide	day^{-1} $site^{-}1$	$5 \cdot 10^{-5}$
ρ_E	Proliferation rate effector	day^{-1}	100
ρ_N	Proliferation rate naive	day^{-1}	10
H_i	Thymic output	$day^{-1}clone^{-1}$	1
M_T	Number of MHCp		10^5
θ	Sensitivity of pathogen presentation		1
P_0	Max. foreign peptide fraction		0.07
r	Pathogen replication	day^{-1}	10
c	Pathogen capacity	$cell^{-1}$	10^{-7}
D	Clearance of infected cells	day^{-1} $cell^{-1}$	$1.2 \cdot 10^{-4}$

3. A caricature of the cytotoxic T cell and peptide repertoire

Describing the dynamics of an immune response poses a challenge due to the high diversity of the cytotoxic T cell and peptide repertoire. In order to capture the main features of these dynamics, we have chosen a simplified representation of these repertoires. We therefore regard only two constitutively presented self-peptides, A and B, and only one peptide P originating from an intracellular pathogen. These three peptides are presented in MHCp-complexes with the relative amounts P_A, P_B, and P_P.

Furthermore, we abstain from modeling single clones, but divide the cytotoxic T cell repertoire in classes representing subgroups of clones of T cells possessing certain recognition properties. T cells which do not recognize any of the regarded three peptides are not considered. We define four classes $i \in \{A, B, PA, PB\}$. These classes are able to recognize the peptides A, B, P and A, and P and B, respectively. That is, all classes recognize one self-peptide, while two recognize the pathogenic peptide as well (see Figure 2). The self-peptides are recognized with strength k_s, while we assume the recognition of the pathogenic peptide to be much higher ($k_h \gg k_s$).

The recognition strength matrix K will then read:

$$K = \begin{pmatrix} k_s & 0 & 0 \\ 0 & k_s & 0 \\ k_s & 0 & k_h \\ 0 & k_s & k_h \end{pmatrix}$$

Finally, we set the thymic input as $H_A = H_B = 1000$ and $H_{PA} = H_{PB} = 1$. These values represent a situation in which the majority of cells does not recognize a specific foreign peptide.

We assume that the expression of immunoproteasome during an infection elicits a change in the peptide frequencies. To let this change correspond to a symmetry operation in which the indices A and B in the equations Eqs. (1, 2) are interchanged, we suppose that the redistribution of peptides during an infection corresponds to interchanging the relative amounts P_A and P_B. The ratio between these amounts was chosen to be 1:10 (Ref. 16).

Figure 3 illustrates the two "phases" considered here. They are determined by the distribution of the three peptides. In the health case (Phase I) no pathogenic peptide is present. In Phase II, the pathogen is present and the amount of pathogenic peptide P is given by Eq. (2.2). In Phase IIa, the relative amount of self-peptides does not change. In Phase IIb, the amounts interchange (see Table 2). In order to mimic the increase in total MHCp-numbers, M_T is increased by a factor of two during Phase II.

Table 2. Peptide amounts considered

Phase	Definition	P_A/P_B	P_P
I	Healthy state	10	0
IIa	Immune response, no immunoproteasome	10	$0 < P_P \leq P_0$
IIb	Immune response, with immunoproteasome	0.1	$0 < P_P \leq P_0$

4. Results

With the proposed mathematical model we want to test the hypothesis that a change in peptide composition at the onset of an infection enhances the

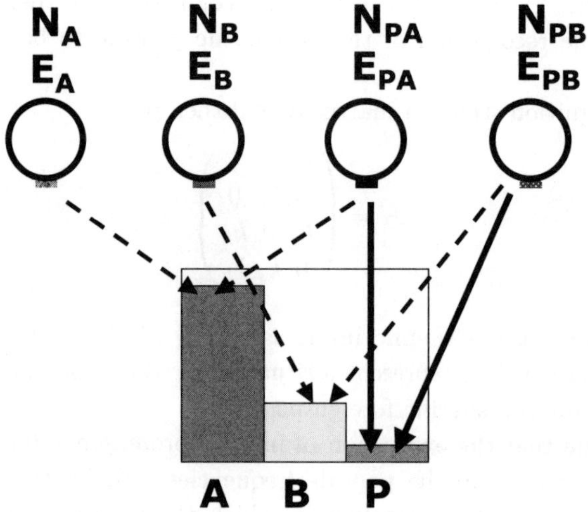

Figure 2. Schematic representation of the classes considered in the model. We consider naive (N_i) and effector (E_i) cytotoxic T cells. Each of the four classes $i \in \{A, B, PA, PB\}$ represent all T cell clones with certain recognition properties, symbolized by arrows pointing to the peptides. Dashed arrow: Weak recognition of self-peptides A resp. B. Solid arrow: Strong recognition of the pathogenic peptide P. The amount of cells produced by the thymus is considered to be a thousand times higher for N_A and N_B than for N_{PA} and N_{PB}.

effectivity of the immune response. To determine a possible change in effectivity, we compare the immune responses with and without redistribution. We start a simulation with a T cell repertoire given by the steady state of Eqs. (1, 2) at a peptide distribution at Phase I and $I = 0$. This represents the cytotoxic T cell distribution of a healthy individual. Starting from this state, we simulate the entering and growth of a pathogen, and observe the mounting of an immune response. This is done both in case of a change in self-peptide amounts (Phase IIb) and without (Phase Ia) for otherwise identical parameters. The effectiveness with which the immune response is mounted and the pathogen is depleted are then compared.

The simulations were performed with a fourth-order Runge-Kutta algorithm. For $t < 0$, the system is considered to be at the steady state of Phase I. At $t = 0$, the system is perturbed by the introduction of infected cells and hence foreign peptide P_P, and the increase of M_T.

In the simulation, the number of infected cells I can assume values

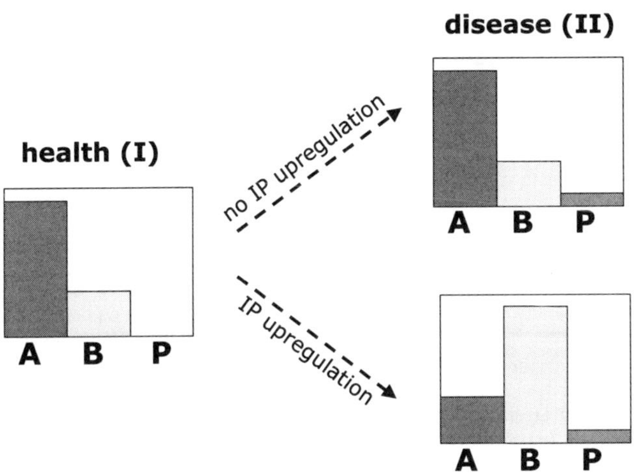

Figure 3. Schematic representation of the quantity of peptide copies presented by APCs. We take only three peptides into account. A and B represent self-peptides, while P represents a peptide originating from an intracellular pathogen. The relative amounts of A and B are determined by the amounts of constitutive proteasome and immunoproteasome present during protein degradation. In the case of health, we define that more A than B is presented (Phase I). During infection, also P is presented (Phase II). We assume that the relative amounts of A and B stayed constant would immunoproteasome not be upregulated, but are quickly interchanged when it is upregulated. Simulations start always at the equilibrium state obtained with the peptide distribution at Phase I. To investigate the role of the immunoproteasome on the immune response, we compare the evolution of the immune response in the cases with and without immunoproteasome upregulation.

smaller than 1. Therefore, a cutoff was introduced and I was set to zero for $I < 1$. When $I = 0$, if the self-peptide amount has been interchanged at the onset of the infection, it is changed back to its original value. As soon as the infection vanishes, M_T is set to its original value as well.

Infection is combated by the highly specific T cell effector classes E_{PA} and E_{PB}. Thus, the immune response shall be depicted by the time course of the sum of these effector cells and of the infected cells. Figure 4 shows these time courses for two simulations which differ only by the peptide redistribution at $t = 0$. In case the redistribution does not take place (panel a), the infection does not vanish. In a time scale of years, the system settles into a steady state with $I > 0$ (not shown). This case is thus characterized

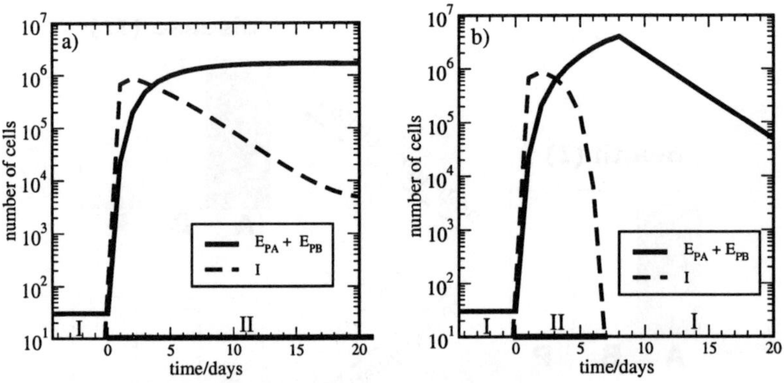

Figure 4. Computer simulation of the time course of an immune response, comparing the cases without (a) and with (b) immunoproteasome upregulation during infection. Solid line: Sum of all effector T cells able to recognize the pathogenic peptide, E_{PA} and E_{PB}. Dashed line: Number of infected cells I. At $t < 0$, no pathogenic peptide is present, and the system is at equilibrium (Phase I). At $t = 0$, infected cells are introduced, causing the pathogenic peptide to appear and the immune response to be initiated (Phase II). The heavy line on the x-axis shows the duration of Phase II. Panel a): No change in relative self-peptide amount is considered. The immune response is not sufficient to clear the infected cells. Panel b): At $t = 0$, the relative amounts of self-peptides A and B change. The immune response succeeds in clearing the infected cells within 7 days. Parameters as in Table 1. During Phase II, M_T was increased by a factor of two.

by a chronic presence of infected cells. In case of redistribution of self-peptides (panel b), the infection vanishes around $t = 7$ days. The peak of the immune response is at day 8, as suggested from experimental data of in vivo mice infections[17,18]. After that, the system returns slowly to the original steady state. Comparison of these two cases shows impressively how a peptide redistribution can affect the outcome of an immune response.

Figure 5 now depicts the time course of each T cell class individually in the same two simulations. In both cases, a fast proliferation of the foreign peptide specific effector cells is observed soon after the onset of the immune response at $t = 0$. This high proliferation rate is due to the high recognition strength of foreign peptides. With that, a previously small class can rapidly overtake large fractions of the T cell repertoire. Note also the change in immunodominance: In panel b), the class E_{PA} is the largest effector class, while in panel d) it is E_{PB}. This suggests that the self-peptide distribution can have a strong effect in shaping the clonal distribution of responding

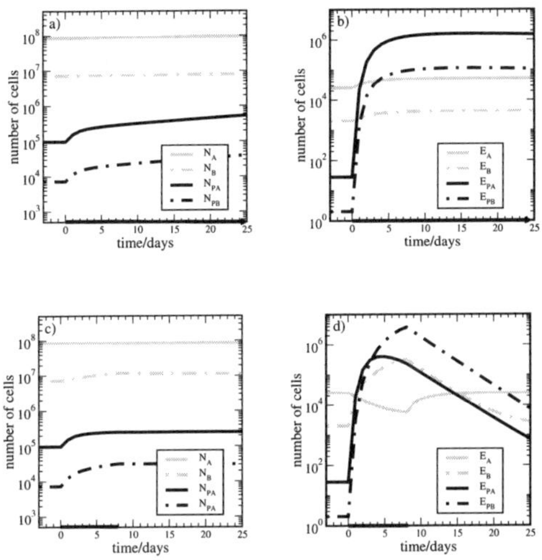

Figure 5. Computer simulation of the time course of an immune response, comparing the cases with and without immunoproteasome upregulation during infection. The time course of the cytotoxic T cells is depicted for the two cases shown in Figure 4. In panels a) and b) (naive and effector T cells), upregulation of immunoproteasome is not considered. In panels c) and d), the relative amount of self-peptides changes during Phase II (heavy line on the x-axis).

cells. Note that the effector classes weakly activated by self-peptides alone do not change much in size. The pool of naive cells is not strongly affected by the infection dynamics due to their slower proliferation rates.

It should be remarked that the higher effectivity of the immune response in case of peptide redistribution is not a consequence of this special choice of parameters. Simulations have been performed extensively, and this effect was observed in all of them[a]. Thus, this behavior seems to be robust.

5. Discussion

We have proposed a possible role for upregulation of immunoproteasome, showing how the capability to fight a pathogen can be improved by reshaping the self-peptide distribution at the onset of an infection. The proliferation rate, the clone abundance at the peak of the immune response, and the size of clones in the model shown are in accordance with experimental values[18,19,17].

[a]The absence of peptide redistribution did not always lead to "chronic infection" though

Intuition might help to understand the proposed higher effectivity of the immune response when peptides are redistributed: The redistribution leads to an abundance of the peptide B at the onset of infection. Suddenly, there is plenty of peptide B, and all the clones recognizing B will be incited to expand. The clones with the highest proliferation rate, namely those recognizing the foreign peptide, will rapidly be able to take advantage of the increased resource. The other cells recognizing B possess a much slower proliferation rate. They will compete with the former for this resource, but only in that long time scale will the rapidly growing clones PA and PB be suppressed by this competition. On the other hand, all clones recognizing A will be suppressed (notwithstanding activation by P) due to the reduction of this resource. This suppression will take place only at the slow time scales given by the death rates. But during an immune response, time scales are paramount. The growing pathogen has to be caught as soon as possible, less it grows even more. These mechanisms lead to the impressive difference in infection outcomes depicted in Figure 4.

This effect was found to be robust in a system where competition for self-peptides influences the expansion rate of cells activated by pathogenic peptides. While it has been shown experimentally that contact with self-peptides is a necessary condition for the establishment of the homeostatic equilibrium of the naive T cell repertoire[3,20], it remains to be proved that this is also the case for an expanding effector cell population. Note, however, that our results suggest that a small amount of competition for self might result in a significant increase in the numbers of effector cells, since a small change in proliferation rate makes up for a large difference in absolute numbers after a few generations.

The shown effect is based on the *change* in peptide distribution and the *short* time scale reaction to that change. Thus, we can conclude that a phenomenon as the one shown here would not be seen in long time scales: Once a chronic infection is established, it does not matter if the peptides presented are being produced by the constitutive or the immunoproteasome[21]. Summarizing, we presented a mechanism which suggests a new role of immunoprotesome induction: A clever way to redistribute resources and thus improve the effectiveness of acute immune responses.

Acknowledgments

Michal Or-Guil and Fabio Luciani thank the Volkswagen Foundation for financial support.

References

1. C. A. Janeway, P. Travers, M. Walport, and M. Shlomchik. *Immunobiology. The Immune System in Health and Disease.* Garland Publications, New York, London, 5 edition, 2001.
2. A. R. Almeida, B. Rocha, A. A. Freitas, and C. Tanchot. Homeostasis of T cell numbers: from thymus production to peripheral compartmentalization and the indexation of regulatory T cells. *Semin. Immunol.*, 17:239–249, 2005.
3. S. C. Jameson. T cell homeostasis: Keeping useful T cells alive and live T cells useful. *Semin. Immunol.*, 17:231–237, 2005.
4. P. M. Kloetzel and F. Ossendorp. Proteasome and peptidase function in MHC-class-I-mediated antigen presentation. *Curr. Opin. Immunol.*, 16:76–81, 2004.
5. B. Strehl, U. Seifert, E. Kruger, S. Heink, U. Kuckelkorn, and P. M. Kloetzel. Interferon-gamma, the functional plasticity of the ubiquitin-proteasome system, and MHC class I antigen processing. *Immunol. Rev.*, 207:19–30, 2005.
6. C. Kesmir, V. Van Noort, R. J. De Boer, and P. Hogeweg. Bioinformatic analysis of functional differences between the immunoproteasome and the constitutive proteasome. *Immunogenetics.*, 55:437–449, 2003.
7. M. Basler, N. Youhnovski, M. Van Den Broek, M. Przybylski, and M. Groettrup. Immunoproteasomes down-regulate presentation of a subdominant T cell epitope from lymphocytic choriomeningitis virus. *J. Immunol.*, 173:3925–3934, 2004.
8. S. Bulik, B. Peters, C. Ebeling, and H. Holzhutter. Cytosolic processing of proteasomal cleavage products can enhance the presentation efficiency of MHC-1 epitopes. *Genome. Inform. Ser. Workshop. Genome. Inform.*, 15:24–34, 2004.
9. F. Ossendorp, N. Fu, M. Camps, F. Granucci, S. J. Gobin, P. J. Van den Elsen, D. Schuurhuis, G. J. Adema, G. B. Lipford, T. Chiba, A. Sijts, P. M. Kloetzel, P. Ricciardi-Castagnoli, and C. J. Melief. Differential expression regulation of the α and β subunits of the PA28 proteasome activator in mature dendritic cells. *J. Immunol.*, 174:7815–7822, 2005.
10. F. Luciani, C. Kesmir, M. Mishto, M. Or-Guil, and R. J. De Boer. A mathematical model of protein degradation by the proteasome. *Biophys. J.*, 88:2422–2432, 2005.
11. K. Leon, A. Lage, and Carneiro J. Tolerance and immunity in a mathematical model of t-cell mediated suppression. *J. Theor. Biol.*, 225:107–126, 2003.
12. R. J. De Boer and A. S. Perelson. T cell repertoires and competitive exclusion. *J. theor. Biol.*, 169:375–390, 1994.
13. R.J. De Boer and A.S. Perelson. Towards a general function describing T cell proliferation. *J. Theor. Biol.*, 175:567–576, 1995.
14. E. J. Wherry, M. J. McElhaugh, and L. C. Eisenlohr. Generation of CD8($^+$) T cell memory in response to low, high, and excessive levels of epitope. *J. Immunol.*, 168:4455–4461, 2002.
15. A. Scherer and S. Bonhoeffer. Epitope down-modulation as a mechanism for the coexistence of competing T-cells. *J. theor. Biol.*, 233:379–390, 2005.

16. M. Skoberne and G. Geginat. Efficient in vivo presentation of Listeria monocytogenes- derived CD4 and CD8 T cell epitopes in the absence of IFN-gamma. *J. Immunol.*, 168:1854–1860, 2002.
17. V. P. Badovinac, A. R. Tvinnereim, and J. T. Harty. Regulation of antigen-specific $CD8^+$ T cell homeostasis by perforin and interferon-gamma. *Science*, 290:1354–1358, 2000.
18. R. J. De Boer, D. Homann, and A. S. Perelson. Different dynamics of $CD4^+$ and $CD8^+$ T cell responses during and after acute lymphocytic choriomeningitis virus infection. *J. Immunol.*, 171:3928–3935, 2003.
19. K. Murali-Krishna, J. D. Altman, M. Suresh, D. J. Sourdive, A. J. Zajac, J. D. Miller, J. Slansky, and R. Ahmed. Counting antigen-specific CD8 T cells: a reevaluation of bystander activation during viral infection. *Immunity*, 8:177–187, 1998.
20. W. Dummer, B. Ernst, E. LeRoy, D. Lee, and C. Surh. Autologous regulation of naive T cell homeostasis within the T cell compartment. *J. Immunol.*, 166:2460–2468, 2001.
21. U. Kuckelkorn, T. Ruppert, B. Strehl, P. R. Jungblut, U. Zimny-Arndt, S. Lamer, I. Prinz, I. Drung, P. M. Kloetzel, S. H. Kaufmann, and U. Steinhoff. Link between organ-specific antigen processing by 20S proteasomes and $CD8(^+)$ T cell-mediated autoimmunity. *J. Exp. Med.*, 195:983–990, 2002.

MODELING PLAGUE DYNAMICS: ENDEMIC STATES, OUTBREAKS AND EPIDEMIC WAVES*

FRANCISCO A. B. COUTINHO*, EDUARDO MASSAD*+, LUIZ F. LOPEZ*
AND MARCELO N. BURATTINI*

**School of Medicine, The University of São Paulo*
Av. Dr. Arnaldo 455, São Paulo, 01246-903, SP, Brazil
+Department of Infectious and Tropical Diseases
Diseases, London School of Hygiene & Tropical Medicine, London, UK.
E-mail: edmassad@usp.br

This paper is intended to contribute to the understanding of plague dynamics through a mathematical modelling of its natural history. Plague is a very complex system, comprising up to three type of hosts and two vectors. This implies in a multicompartment system with several (73 in our model) parameters, few of each have been measured. Therefore, we used reasonable values for those less known parameters and studied the behaviour of the system. The model was applied to describe two distinct epidemiological scenarios; one epidemic outbreak of short duration (Florence, 1348), and one endemic state (Viet Nam, present day).

1. Introduction

Plague is a severe infection in humans caused by a gram-negative bacteria called *Yersinia pestis*[1]. Its most common form, the bubonic plague can be successfully treated with antibiotics; but if untreated plague can cause a letality rate of above 50%. Its physiopathology is relatively simple but its epidemic dynamics is very complicated, involving other mammal reservoirs and arthropod vectors. Indeed, plague epidemics is probably the most paradigmatic among human scourges, causing outbreaks of biblical proportions (as a matter of fact the book of Samuel noticed the *enlargement of inguinal glands* indicating the probability of plague occurrence[2]).

Since the first reported plague outbreak of universal proportion, the Justinian pandemic, which caused an estimated number of 40 million deaths

*This work was partially supported by grants LIM 01/HC-FMUSP, CNPq, Fapesp and PRONEX.

worldwide in the sixth or seventh century, two other major *visitations* (to use a term applied by Daniel Defoe in his *Journal of the Plague Year*[3]) occurred: the Black Death, which swept away a quarter or more of the Medieval Europe population and the nineteen century pandemic caused by the spread of rat-infested steamships from China and which caused an estimated number of more than 12 million deaths, most of which occurred in India[4]. It was during this last pandemic that the causative organism was discovered by Alexandre Yersin[5] in 1894 and 4 years later Simond suggested that fleas transmitted the disease[6]. The pandemic then spread by rat on board ships to the rest of the world where it established in sylvatic rodents[7]. In between the major plague pandemics, minor outbreaks occurred in several places of the old world, causing comprehensible anxiety in the affected populations.

After the introduction of strict regulations of rat-control in ports around 1900, plague spreading was greatly reduced. However, by 1910 plague has become established among rodents populations around the world, causing zoonotic foci which keeps endemic states and minor outbreaks throughout the world.

This paper is intended to contribute to the understanding of plague dynamics through a mathematical modelling of its natural history. Our model reproduces the current epidemiological beliefs, according to which, plague is either an unstable epidemic with short duration outbreaks and long intervals without any case or an endemy with very low prevalence. As we shall see, the first behavior is related to the fact that plague is usually very close to or below the threshold for the establishment of the infection. Our model has the advantage of reproducing also endemic states with very low prevalence of human cases, simply by including wild reservoirs of plague into it. These are exemplified by two different situations. The epidemic simulated is that of Black Death in Florence, when 50% of that city population was wiped out by the disease. The endemic situation analised in this paper is that of Viet Nam, which has a stable endemic state with very low prevalence of human cases and which is maintained by wild rodents reservoirs of the plague bacillus.

After this introduction, we briefly describe the natural history of the spread of *Yersinia pestis*, considering all the possible transmission pathways from hosts-to-vectors-to-hosts. In section 3 we present the model which is intended to mimic the plague dynamics. Unfortunately, the complexity of this vectors-hosts system produces a model with 14 ordinary differential equations, containing 73 parameters, some of which has not yet been measured (see discussion in subsection 5.1). However, as we shall see, this limitation does not invalidate the qualitative results of our model. This kind of complexity occurs in many other vector-borne infections of interest in public health and veterinary, like yellow fever, Lyme disease, among others, as discussed in Lopez *et al.*[22]. Section 4 describes the analysis of the threshold condition for the establishment of the infection in plague-free populations, using the methods described in Lopez *et al.*[22]. Again, the resulting expression for the basic reproduction number, R_0, is too complicated to be of any use and therefore we considered only particular cases. In section 5 we present numerical simulations of the dynamic model, in an attempt to describe two different epidemiological situations, namely, an epidemic (Florence, 1348) and an endemic state (Viet Nam, 1989). We did not intend to fit any real-world data but rather to present a qualitative analysis of the dynamic of two distinct epidemiological scenarios. Our results suggest that the presence of wild animal reservoir and wild vectors are necessary to have an endemic state, like in Viet Nam. In contrast, when the infection is transmitted only from domestic rodents through domestic fleas, the case of Florence in 1348, only one epidemic outbreak occurs due to the sudden arrival of infected domestic fleas and domestic rats. Finally, in the discussion section we summarize our results and discuss their implications.

2. Natural History of the Ecology and Spread of *Y. pestis*

Considering the complexity of *Y. pestis* spreading , it is important to describe its ecology and life cycle to understand plague dynamics and the design of control strategies.

Y. pestis is a zoonotic infection that affects both wild and domestic rodents and their fleas. Foci of infection are maintained worldwide in animal reservoirs and transmission to humans occurs by bites of infected fleas, ingestion of infected animals and by direct transmission among humans[7]. Plague spreads from rodents to humans when humans intrude into the natural (wild) cycle, or when domestic rodents become infected and transmission is established in domestic settings[8]. Human-to-human transmission typically results from close exposure to persons with pulmonary plague[8]. The most important animal reservoirs are the domestic black rat *Rattus rattus* and the brown sewer rat *Rattus norvegicus*, from which plague is transmitted to man[1]. In the wild the infection is maintained by other rodents, like *Rattus exulans* in Asia, *Mastomus* ssp. in Africa[7] and *Tatera indica* in India[8]. In 1903 the principal rat flea vector of plague, *Xenopsylla cheopis,* was described by Rothschild[6], and later it was demonstrated that X. *astia,* X. *brasiliensis* and other species maintain the disease in various species of rats. Some 220 species of rodent fleas can harbor the plague bacilli and approximately 30 species can transmit the disease[9].

Rats contaminated with the infection develop an acute and often fatal disease, with high levels of circulating bacilli (the septicaemic form). Its fleas then abandon the corpses as they cool down and can eventually infest humans hosts. Infected fleas have their proventriculus organs blocked by the rapidly reproducing bacteria. These mass of bacilli are regurgitated and then infects the new hosts[6]. This first infection normally generates the bubonic form of the disease in man.

In Figure 1, we summarize the natural history of the plague cycle of transmission.

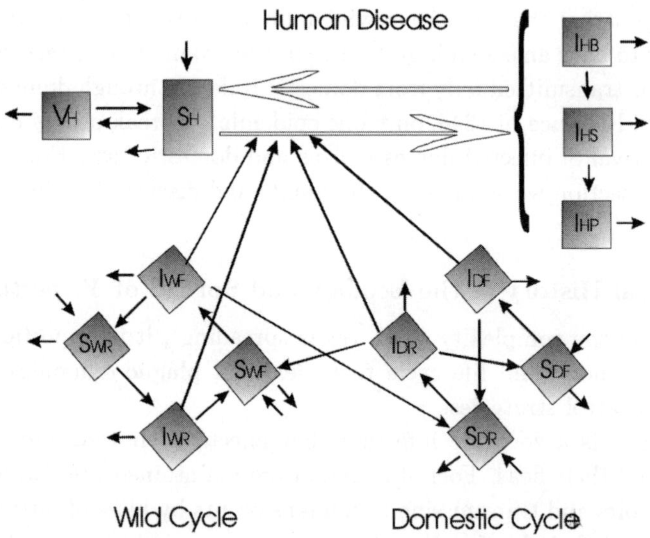

Figure 1. Block diagram summarizing the natural history of the plague cycle of transmission.

Note that we consider two cycle of transmission, one in the wild and one domestic. In the former, wild rodents can be infected both through the bites of a wild flea vector and by direct contact with other infected rodents. The infected rodents can then infect men through the bites of their wild fleas or by the rarer form of direct contact. Also, the domestic rodents can be contaminated through the bites of wild fleas and initiate the domestic cycle of plague transmission. In this, domestic rodents infect other rodents both by the bites of domestic fleas and by direct contact. Human hosts, in turn, can be infected both by the bites of domestic fleas and by direct contact with domestic rodents. The bubonic form of the disease can eventually evolve to the septicaemic form of the infection or progress to the pulmonary form when he/she is infective to other human hosts by direct contact.

Fatality rate of untreated plague may reach 50% of the bubonic form and almost 100% of septicaemic and pulmonary forms. Streptomycin has been the drug of choice for plague treatment since its introduction in 1948[10,11]. No other drug has been demonstrated to be more efficacious or less toxic[10]. Patients successfully treated recover to become susceptible again.

Control and prevention include vaccination of humans, rodent and flea elimination, and antibiotic therapy.

3. The Model

We begin by defining the variables and parameters which determine the model dynamics.

We divide the human hosts into five possible states, namely, susceptible, denoted S_H; vaccinated individuals, denoted V_H; infected individuals who develop the bubonic form, denoted I_{HB}; infected individuals who develop the pulmonary form, denoted I_{HP}; and infected individuals who develop the septicemic form, denoted I_{HS}. People recoverd from the infeection are denoted as R_H. Other rarer clinical forms will not be considered in our model.

The remaining species involved in plague transmission are divided in two classes, namely, susceptible and infected. So we have, S_{DR} and S_{WR} denoting susceptible domestic and wild rodents, respectively; I_{DR} and I_{WR} denoting infected domestic and wild rodents, respectively; S_{DF} and S_{WF} denoting susceptible domestic and wild fleas, and finally, I_{DF} and I_{WF} denoting infected domestic and wild fleas, respectively. The model's dynamics is described by the following set of differential equations:

$$
\begin{aligned}
\frac{dS_H}{dt} &= -a_{WFS_H} b_{WFS_H} I_{WF} \frac{S_H}{N_H} - a_{DFS_H} b_{DFS_H} I_{DF} \frac{S_H}{N_H} - \\
&\quad (\beta_{WRHB} + \beta_{WRHS}) I_{WR} \frac{S_H}{N_H} - (\beta_{DRHB} + \beta_{DRHS}) I_{DR} \frac{S_H}{N_H} - \\
&\quad \beta_{I_{HP}S_H} I_{HP} \frac{S_H}{N_H} - \nu S_H - \mu_H S_H + \varphi V_H + r_H \left(1 - \frac{N_H}{K_H}\right) (S_H + V_H + R_H) \\
\frac{dV_H}{dt} &= \nu S_H - (\varphi + \mu_H) V_H \\
\frac{dI_{HB}}{dt} &= a_{WFS_H} b_{WFS_H} I_{WF} \frac{S_H}{N_H} + a_{DFS_H} b_{DFS_H} I_{DF} \frac{S_H}{N_H} + \beta_{WRHB} I_{WR} \frac{S_H}{N_H} + \\
&\quad \beta_{DRHB} I_{DR} \frac{S_H}{N_H} - \sigma_B I_{HB} - (\mu_H + \alpha_B) I_{HB} - (\gamma_{BS} + \gamma_{BP}) I_{HB}
\end{aligned}
$$

$$
\begin{aligned}
\frac{dI_{HS}}{dt} &= \beta_{WRHS} I_{WR} \frac{S_H}{N_H} + \beta_{DRHS} I_{DR} \frac{S_H}{N_H} - \sigma_S I_{HS} - (\mu_H + \alpha_S) I_{HS} + \\
&\quad \gamma_{BS} I_{HB} - \gamma_{SP} I_{HS} \\
\frac{dI_{HP}}{dt} &= \beta_{I_{HP}S_H} I_{HP} \frac{S_H}{N_H} - \sigma_P I_{HP} - (\mu_H + \alpha_P) I_{HP} + \\
&\quad \gamma_{BP} I_{HB} + \gamma_{SP} I_{HS} \\
\frac{dR_H}{dt} &= \sigma_B I_{HB} + \sigma_S I_{HS} + \sigma_P I_{HP} - \mu_H R_H \\
\frac{dS_{DR}}{dt} &= -a_{WFDR} b_{WFDR} I_{WF} \frac{S_{DR}}{N_{DR}} - a_{DFDR} b_{DFDR} I_{DF} \frac{S_{DR}}{N_{DR}} - \beta_{DRDR} I_{DR} \frac{S_{DR}}{N_{DR}} - \\
&\quad \mu_{DR} S_{DR} + \frac{r_{DR}}{1 + b_{DF} N_{DF}} \left(1 - \frac{N_{DR}}{k_{DR}}\right) S_{DR} \\
\frac{dI_{DR}}{dt} &= a_{WFDR} b_{WFDR} I_{WF} \frac{S_{DR}}{N_{DR}} + a_{DFDR} b_{DFDR} I_{DF} \frac{S_{DR}}{N_{DR}} + \beta_{DRDR} I_{DR} \frac{S_{DR}}{N_{DR}} - \\
&\quad (\mu_{DR} + \alpha_{DR}) I_{DR} \\
\frac{dS_{DF}}{dt} &= -a'_{DFHB} c_{HBDF} S_{DF} \frac{I_{HB}}{N_H} - a'_{DFHS} c_{HSDF} S_{DF} \frac{I_{HS}}{N_H} - \\
&\quad a'_{DFHP} c_{HPDF} S_{DF} \frac{I_{HP}}{N_H} - a'_{DFWR} c_{WRDF} S_{DF} \frac{I_{WR}}{N_{WR}} - \\
&\quad a'_{DFDR} c_{DRDF} S_{DF} \frac{I_{DR}}{N_{DR}} - \mu_{DF} S_{DF} + r_{DF} \left(1 - \frac{N_{DF}}{k_{DF} + b_{DR} S_{DR}}\right) S_{DF} \\
\frac{dI_{DF}}{dt} &= a'_{DFHB} c_{HBDF} S_{DF} \frac{I_{HB}}{N_H} + a'_{DFHS} c_{HSDF} S_{DF} \frac{I_{HS}}{N_H} + \\
&\quad a'_{DFHP} c_{HPDF} S_{DF} \frac{I_{HP}}{N_H} + a'_{DFWR} c_{WRDF} S_{DF} \frac{I_{WR}}{N_{WR}} + \\
&\quad a'_{DFDR} c_{DRDF} S_{DF} \frac{I_{DR}}{N_{DR}} - (\mu_{DF} + \alpha_{DF}) I_{DF} \\
\frac{dS_{WF}}{dt} &= -a'_{WFHB} c_{HBWF} S_{WF} \frac{I_{HB}}{N_H} - a'_{WFHS} c_{HSWF} S_{WF} \frac{I_{HS}}{N_H} - \\
&\quad a'_{WFHP} c_{HPWF} S_{WF} \frac{I_{HP}}{N_H} - a'_{WFWR} c_{WRWF} S_{WF} \frac{I_{WR}}{N_{WR}} - \\
&\quad a'_{WFDR} c_{DRWF} S_{WF} \frac{I_{DR}}{N_{DR}} - \mu_{WF} S_{WF} + r_{WF} S_{WF} \left(1 - \frac{N_{WF}}{k_{WF} + b_{WR} S_{WR}}\right) \\
\frac{dI_{WF}}{dt} &= a'_{WFHB} c_{HBWF} S_{WF} \frac{I_{HB}}{N_H} + a'_{WFHS} c_{HSWF} S_{WF} \frac{I_{HS}}{N_H} + \\
&\quad a'_{WFHP} c_{HPWF} S_{WF} \frac{I_{HP}}{N_H} + a'_{WFWR} c_{WRWF} S_{WF} \frac{I_{WR}}{N_{WR}} + \\
&\quad a'_{WFDR} c_{DRWF} S_{WF} \frac{I_{DR}}{N_{DR}} - (\mu_{WF} + \alpha_{WF}) I_{WF} \\
\frac{dS_{WR}}{dt} &= -a_{WFWR} b_{WFWR} I_{WF} \frac{S_{WR}}{N_{WR}} - a_{DFWR} b_{DFWR} I_{DF} \frac{S_{WR}}{N_{WR}} - \beta_{WRWR} I_{WR} \frac{S_{WR}}{N_{WR}} - \\
&\quad \mu_{WR} S_{WR} + \frac{r_{WR} S_{WR}}{1 + b_{WF} N_{WF}} \left(1 - \frac{N_{WR}}{k_{WR}}\right) \\
\frac{dI_{WR}}{dt} &= a_{WFWR} b_{WFWR} I_{WF} \frac{S_{WR}}{N_{WR}} + a_{DFWR} b_{DFWR} I_{DF} \frac{S_{WR}}{N_{WR}} + \\
&\quad \beta_{WRWR} I_{WR} \frac{S_{WR}}{N_{WR}} - (\mu_{WR} + \alpha_{WR}) I_{WR}
\end{aligned}
\tag{1}
$$

where the definition and the biological meaning of each of the parameters used are described in Table 1.

In the table the symbols are defined as follows: a_{ij} stands for the average daily biting rate of i-type infected fleas in j-type susceptible rodents/humans, for example, a_{WFS_H} means the biting rate of infective wild fleas on susceptible humans; a'_{ij} stands for the average daily biting rate of i-type susceptible fleas in j-type infected rodents/humans (this difference was included in order to differentiate the biting rates of infective and susceptible fleas – infective fleas bite with rates much higher than non-infected ones), for example, a'_{WFHB} means the biting rate of susceptible wild fleas on infected humans with the bubonic form of the disease; b_{ij} is the probability of j getting the infection when bitten by i; c_{ij} is the probability of j getting the infection when biting i; γ_{ij} stands for the clinical evolution from form i to form j; and β_{ij} is the number of infective contacts between i and j. The meaning of subscripts and the remaining parameters are described in the Table 1. The way those terms appear in the equations is described below.

Table 1. Parameters

Parameter	Biological Meaning	Parameter	Biological Meaning
a_{WFS_H}	wild fleas in susc. hum.	γ_{BS}	from bubo. to scept.
a_{DFS_H}	dom. fleas in susc. hum.	γ_{BP}	from bubo. to pulmo.
a_{WFDR}	wild fleas in dom. rod.	γ_{SP}	from scept. to pulmo.
a_{DFDR}	dom. fleas in dom. rod.	β_{WRHB}	wild rod. and bub. hum.
a_{WFWR}	wild fleas in wild rod.	β_{WRHS}	wild rod. and scep. hum.
a_{DFWR}	dom. fleas in wild rod.	β_{DRHB}	dom. rod. and bub. hum.
a'_{DFHB}	dom. fleas in bub. hum.	β_{DRHS}	dom. rod. and scep. hum.
a'_{DFHS}	dom. fleas in scep.hum.	$\beta_{I_{HP}S_H}$	pul. hum and susc.hum.
a'_{DFHP}	dom. fleas in pul. hum.	β_{DRDR}	dom. rod. and dom. rod.
a'_{DFWR}	dom. fleas in wild rod.	β_{WRWR}	wild rod. and wild rod.
a'_{DFDR}	dom. fleas in dom. rod.	σ_B	recovery from bubonic
a'_{WFHB}	wild fleas in bub. hum.	σ_S	recovery from scept.
a'_{WFHS}	wild fleas in scep.hum.	σ_P	recovery from pulmonar
a'_{WFHP}	wild fleas in pul. hum.	μ_H	nat. mort. of humans
a'_{WFWR}	wild fleas in wild rod.	μ_{DF}	nat. mort. of dom. fleas
a'_{WFDR}	wild fleas in dom. rod.	μ_{DR}	nat. mort. of dom. rod.
b_{DFWR}	dom.rod. from wild fleas	μ_{WF}	nat. mort. of wild fleas
b_{WFWR}	wild rod. from wild fleas	μ_{WR}	nat. mort. of wild rod.
b_{DFDR}	dom.rod. from dom.fleas	α_B	plague mort. bubonics
b_{WFDR}	dom.rod. from wild fleas	α_P	plague mort. pulmonar
b_{WFS_H}	susc.hum. from wild fleas	α_S	plague mort. sceptic.
b_{DFS_H}	susc.hum. from dom. fleas	α_{DF}	plague mort. dom. fleas
c_{HBDF}	dom.fleas from bub.hum.	α_{DR}	plague mort. dom. rod.
c_{HSDF}	dom.fleas from scep.hum.	α_{WF}	plague mort. wild fleas
c_{HPDF}	dom.fleas from pul.hum	α_{WR}	plague mort. wild rod.
c_{WRDF}	dom.fleas from wild rod.	r_H	birth rate of humans
c_{DRDF}	dom.fleas from dom. rod.	r_{DR}	birth rate of dom. rod.
c_{HBWF}	wild fleas from bub.hum.	r_{WR}	birth rate of wild rod.
c_{HSWF}	wild fleas from scep.hum.	r_{DF}	birth rate of dom. fleas
c_{HPWF}	wild fleas from pul.hum.	r_{WF}	birth rate of wild fleas
c_{WRWF}	wild fleas from wild rod.	φ	loss of immunity
c_{DRWF}	wild fleas from dom. rod.	ν	vaccination rate
b_{DR}	rats density-dependence	k_H	carr.capac.of humans
b_{DF}	fleas density-dependence	k_{DR}	carr.capac.of dom.rod.
b_{WR}	wild rats density-dependence	k_{WR}	carr.capac.of wild.rod.
b_{WF}	wild fleas density-dependence	k_{DF}	carr.capac.of dom.fleas
		k_{WF}	carr.capac.of wild fleas

Let us briefly describe a few features of the model.

First, consider the first term on the right hand side of the first equation of system (1):

$$-a_{WFS_H} b_{WFS_H} I_{WF} \frac{S_H}{N_H}$$

The product $a_{WFS_H}I_{WF}$ is the total number of bites per unit time infected wild fleas inflict on humans. Of this a fraction $\frac{S_H}{N_H}$ is on susceptible humans, of which only a fraction b_{WFS_H} result in new infections (bubonic form). By the same token, consider the first term on the right hand side of the nineth equation of system (1):

$$-a'_{DFHB}c_{HBDF}S_{DF}\frac{I_{HB}}{N_H}$$

The product $a'_{DFHB}S_{DF}$ is the total number of bites per unit time susceptible domestic fleas inflict on humans. Of this a fraction $\frac{I_{HB}}{N_H}$ is on infected humans with the bubonic form, of which only a fraction c_{HBDF} result in new infections in the fleas.

Humans are subject to a density dependent birth rate and a linear mortality rate. The population dynamics in the absence of disease is

$$\frac{dN_H}{dt} = r_H N_H \left(1 - \frac{N_H}{K_H}\right) - \mu_H N_H, \tag{2}$$

where r_H is the birth rate of humans, N_H is the total human population, K_H is a constant and the human carrying capacity is $\frac{r_H - \mu_H}{r_H}K_H$.

Of course, equation (2) could be written as

$$\frac{dN_H}{dt} = r_H N_H - \left(\mu_H - \frac{r_H N_H}{K_H}\right) N_H, \tag{3}$$

which can be interpreted as a density dependence in the mortality rate.

Note that, in the full model, as we consider that nobody is born infected, vaccinated or recovered, the full density dependence term, for instance in the vaccinated class, in the form (3) should be

$$\frac{dV_H}{dt} = \nu S_H - \varphi V_H + r_H V_H - \left(\mu_H - \frac{r_H V_H}{K_H}\right) V_H \tag{4}$$

However, under the assumption of no birth in these classes, $r_H = 0$ and equation (4) reduces to

$$\frac{dV_H}{dt} = \nu S_H - \varphi V_H - \mu_H V_H \tag{5}$$

Note that, although nobody is born vaccinated or recovered, these classes contributes to the births in the susceptible state, with the same birth rate of susceptible humans. In contrast, infected individuals do not reproduce in this state.

For the animal populations of rodents and fleas, we choose also to assign a density dependence in the birth rates, now including a term $(1+b_{DF}N_{DF})$, where b_{DF} is a constant, reflecting the negative influence of fleas on domestic rodents, such that the birth rate r_{DR} is multiplied by the term

$$\frac{1}{1 + b_{DF}N_{DF}}\left(1 - \frac{S_{DR}}{k_{DR}}\right) \tag{6}$$

Analogously, we included a term $(k_{DF} + b_{DR}S_{DR})$, where b_{DR} is a contant, reflecting the positive influence of domestic rodents on fleas, such that the birth rate r_{DF} is multiplied by a new term

$$\left(1 - \frac{S_{DF}}{k_{DF} + b_{DR}S_{DR}}\right) \tag{7}$$

So, the term (6) reflects the reduction in the birth rate of rodents when parasitized by fleas and the term (7) reflects the increase in the carrying capacity of fleas due to the presence of rodents[12,13,14]. The same applies to the wild populations.

We shall describe how we simulated plague as a characteristically epidemic disease, with sharp outbreaks, followed by long periods without new cases. This is done by choosing values of the parameters that sets the subsystem (humans-domestic rats-domestic fleas and no wild rodents and wild fleas) slightly below the threshold. Thus, an initial amount of disease in the vector and in the mammal reservoir (for instance, the arrival to a port of a ship pested with infected rats and fleas) rapidly triggers an outbreak. After the outbreak the infection disappears because the disease is bellow the threshold.

If in contrast, we include the wild cycle of transmission (rodents and fleas), that is, the whole system, all the populations (including humans) reach an equilibrium characterizing an endemic state, with prevalence levels depending on how above the threshold the system is.

4. Threshold conditions for the establishment of the infection

The central parameter related to the intensity of transmission of infections is the so called basic reproduction number (R_0), defined by Macdonald[15] as the number of secondary infections produced by a single infective in an entirely susceptible population. Originally applied in the context of malaria, a vector-borne infection, R_0 is a function of the vector population density (mosquitoes) as related to the host population (humans), m, the average daily biting rate of the vector, a, the host susceptibility, b, the vector mortality rate, μ, the parasite extrinsic incubation period in days,

n, and the parasitemia recovery rate, r, according with the (now) historical equation:

$$R_0 = \frac{ma^2 b \exp\left[-\mu n\right]}{r\mu} \tag{8}$$

(actually, Macdonald denoted R_0 as z_0 in his original paper). From the definition of the basic reproduction number it can be demonstrated that if R_0 is not greater than one, that is, when an index case (the first infective individual) is not able to generate at least one new infection, the disease dies out. Hence, in the original Macdonald analysis, R_0 coincides with the threshold for the infection persistence.

To obtain R_0 for a complex system such as plague is exceedingly complicated (see Lopez *et al.*[16]). However, it is possible to obtain a thershold condition as outlined below.

The complete expression for the threshold condition of this model contains around 140 terms and, therefore, it is not very useful. In what follows, we outline the method to get a correspondent threshold and calculate it for a simplified particular case.

We begin by linearizing the equations of system 1 for the infected individuals around the solution in which there is no disease, that is, $I_{ij} = 0 + I^*_{ij}$. As we are considering a vaccination rate, ν, the proportion of susceptible individuals in the population is

$$\frac{S_H}{N_H} = \frac{\varphi + \mu_H}{\nu + \varphi + \mu_H} = 1 - p \tag{9}$$

where p is the proportion of vaccinated individuals of the population.

The linearized system for the equations of the infected individuals is then:

$$\begin{aligned}
\frac{dI^*_{HB}}{dt} &= \left(a_{WFS_H} b_{WFS_H} I^*_{WF} + a_{DFS_H} b_{DFS_H} I^*_{DF} + \beta_{WRHB} I^*_{WR} + \beta_{DRHB} I^*_{DR}\right)(1-p) \\
&\quad - \sigma_B I^*_{HB} - (\mu_H + \alpha_B) I^*_{HB} - (\gamma_{BS} + \gamma_{BP}) I^*_{HB} \\
\frac{dI^*_{HS}}{dt} &= \left(\beta_{WRHS} I^*_{WR} + \beta_{DRHS} I^*_{DR}\right)(1-p) - \sigma_S I^*_{HS} - (\mu_H + \alpha_S) I^*_{HS} + \\
&\quad \gamma_{BS} I^*_{HB} - \gamma_{SP} I^*_{HS} \\
\frac{dI^*_{HP}}{dt} &= \beta_{I_{HP} S_H} I^*_{HP} (1-p) - \sigma_P I^*_{HP} - (\mu_H + \alpha_P) I^*_{HP} + \gamma_{BP} I^*_{HB} + \gamma_{SP} I^*_{HS} \\
\frac{dI^*_{DR}}{dt} &= a_{WFDR} b_{WFDR} I^*_{WF} + a_{DFDR} b_{DFDR} I^*_{DF} + \beta_{DRDR} I^*_{DR} - \\
&\quad (\mu_{DR} + \alpha_{DR}) I^*_{DR} \\
\frac{dI^*_{WR}}{dt} &= a_{WFWR} b_{WFWR} I^*_{WF} + a_{DFWR} b_{DFWR} I^*_{DF} + \beta_{WRWR} I^*_{WR} - \\
&\quad (\mu_{WR} + \alpha_{WR}) I^*_{WR} \\
\frac{dI^*_{DF}}{dt} &= a'_{DFHB} c_{HBDF} I^*_{HB} + a'_{DFHS} c_{HSDF} I^*_{HS} + a'_{DFHP} c_{HPDF} I^*_{HP} + \\
&\quad a'_{DFWR} c_{WRDF} I^*_{WR} + a'_{DFDR} c_{DRDF} I^*_{DR} - (\mu_{DF} + \alpha_{DF}) I^*_{DF} \\
\frac{dI^*_{WF}}{dt} &= a'_{WFHB} c_{HBWF} I^*_{HB} + a'_{WFHS} c_{HSWF} I^*_{HS} + a'_{WFHP} c_{HPWF} I^*_{HP} + \\
&\quad a'_{WFWR} c_{WRWF} I^*_{WR} + a'_{WFDR} c_{DRWF} I^*_{DR} - (\mu_{WF} + \alpha_{WF}) I^*_{WF}
\end{aligned} \tag{10}$$

Next we assume, as usual, a solution of the form $I^*_{ij} = C_{ij} \exp\left[\lambda t\right]$, and substitute it in equation (10) to get the so called characteristic equation for λ[17,18]. In our case, the characteristic equation is an equation of 7th order. If all the roots of this equation have negative real parts, then the equilibrium solution is stable and the disease cannot invade the population. The threshold occurs when one or more of the roots crosses the imaginary axis so that their real parts become positive. In this case the introduction of a single infective case triggers an epidemic.

A condition to have all the roots with negative real parts can be obtained by applying the Routh-Hurwitz criteria[19]. Unfortunately, the resulting expression is too complicated to be reproduced here. However, we can send it if requested.

Let us, however, consider a reduced toy model with fewer host populations. Consider only the bubonic form in a domestic cycle of transmission with vaccination. In this case we have a situation in which the system has only one human host, one vector and one animal reservoir. This scenario can be obtained by making equal to zero all the parameters in the relevant equations, for instance, in the first equation of the system (10) we make $a_{WFS_H} b_{WFS_H} = 0$ and $\beta_{WRHB} = 0$. The resulting system is

$$\begin{aligned}
\frac{dI^*_{HB}}{dt} &= \left(a_{DFS_H} b_{DFS_H} I^*_{DF} + \beta_{DRHB} I^*_{DR}\right)(1-p) - \sigma_B I^*_{HB} - \left(\mu_H + \alpha_B\right) I^*_{HB} \\
\frac{dI^*_{DR}}{dt} &= a_{DFDR} b_{DFDR} I^*_{DF} + \beta_{DRDR} I^*_{DR} - \left(\mu_{DR} + \alpha_{DR}\right) I^*_{DR} \qquad (11) \\
\frac{dI^*_{DF}}{dt} &= a'_{DFHB} c_{HBDF} I^*_{HB} + a'_{DFDR} c_{DRDF} I^*_{DR} - \left(\mu_{DF} + \alpha_{DF}\right) I^*_{DF}
\end{aligned}$$

The threshold condition, Th, for the establishment of the infection is the set of parameters that breaks the stability of the trivial solution (that is no disease). This in then given by $Th > 1$, where

$$\begin{aligned}
Th = {} & \frac{\beta_{DRDR}}{(\mu_{DR}+\alpha_{DR})} + \frac{a_{DFDR} a'_{DFDR} b_{DFDR} c_{DRDF}}{(\mu_{DR}+\alpha_{DR})(\mu_{DF}+\alpha_{DF})} + \\
& \frac{a'_{DFHB} a_{DFS_H} c_{HBDF} b_{DFS_H}}{(\sigma_B+\mu_H+\alpha_B)(\mu_{DF}+\alpha_{DF})}(1-p) + \left(\frac{\beta_{DRHB} a'_{DFHB} c_{HBDF} a'_{DFDR} b_{DFDR}}{(\sigma_B+\mu_H+\alpha_B)(\mu_{DF}+\alpha_{DF})(\mu_{DR}+\alpha_{DR})} - \right. \\
& \left. \frac{\beta_{DRDR} a'_{DFHB} a_{DFS_H} c_{HBDF} b_{DFS_H}}{(\sigma_B+\mu_H+\alpha_B)(\mu_{DF}+\alpha_{DF})(\mu_{DR}+\alpha_{DR})}\right)(1-p) \qquad (12)
\end{aligned}$$

Assuming that the direct contagious between and among the mammals hosts are usually negligible, we note that the expression of Th reduces to a sum of two Macdonald like equations, that is

$$Th = \frac{a_{DFDR} a'_{DFDR} b_{DFDR} c_{DRDF}}{\left(\mu_{DR} + \alpha_{DR}\right)\left(\mu_{DF} + \alpha_{DF}\right)} + \frac{a'_{DFHB} a_{DFS_H} c_{HBDF} b_{DFS_H}}{\left(\sigma_B + \mu_H + \alpha_B\right)\left(\mu_{DF} + \alpha_{DF}\right)}(1-p) \qquad (13)$$

5. Simulating the model

5.1. *Model's dynamics*

In order to describe the qualitative behavior of the model, we numerically simulated the complete system of differential equations (1) to describe two distinct epidemiological scenarios; one epidemic of short duration and one endemic state. The values used for the parameters to simulate the model can be seen in Table 2, some of which were derived from the established literature on the natural history of plague[1,5,8]. The values of the other parameters are discussed below.

Table 2. Parameter values used in the simulations

Parameter	Epidemic *(Florence, 1348)*	Endemic *(Viet Nam, 1989)*	Parameter	Epidemic *(Florence, 1348)*	Endemic *(Viet Nam, 1989)*
a_{WFS_H}	10^{-5}	1.5×10^{-3}	γ_{BS}	1.4×10^{-3}	1.4×10^{-3}
a_{DFS_H}	5	5	γ_{BP}	1.4×10^{-3}	1.4×10^{-3}
a_{WFDR}	10^{-5}	10^{-4}	γ_{SP}	5×10^{-2}	5×10^{-2}
a_{DFDR}	10	10	β_{WRHB}	0	0
a_{WFWR}	5	10	β_{WRHS}	0	0
a_{DFWR}	0	0	β_{DRHB}	0	0
a'_{DFHB}	10^{-2}	10^{-2}	β_{DRHS}	0	0
a'_{DFHS}	10^{-2}	10^{-2}	$\beta_{I_HPS_H}$	10^{-4}	10^{-4}
a'_{DFHP}	10^{-2}	10^{-2}	β_{DRDR}	10^{-3}	10^{-3}
a'_{DFWR}	0	0	β_{WRWR}	10^{-3}	10^{-3}
a'_{DFDR}	0.1119	0.1119	σ_B	6.7×10^{-2}	2×10^{-1}
a'_{WFHB}	10^{-5}	10^{-5}	σ_S	10^{-2}	3.3×10^{-2}
a'_{WFHS}	10^{-5}	10^{-5}	σ_P	10^{-2}	3.3×10^{-2}
a'_{WFHP}	10^{-5}	10^{-5}	μ_H	6×10^{-5}	5×10^{-5}
a'_{WFWR}	5	5	μ_{DF}	6.7×10^{-2}	6.7×10^{-2}
a'_{WFDR}	10^{-3}	10^{-3}	μ_{DR}	1.8×10^{-3}	1.8×10^{-3}
b_{DFWR}	10^{-2}	10^{-2}	μ_{WF}	4×10^{-2}	4×10^{-2}
b_{WFWR}	10^{-2}	10^{-2}	μ_{WR}	2.7×10^{-3}	2.7×10^{-3}
b_{DFDR}	10^{-2}	10^{-2}	α_B	10^{-1}	2.5×10^{-2}
b_{WFDR}	10^{-2}	10^{-2}	α_P	2×10^{-1}	10^{-1}
b_{WFS_H}	10^{-2}	10^{-2}	α_S	2×10^{-1}	10^{-1}
b_{DFS_H}	10^{-1}	10^{-1}	α_{DF}	6.7×10^{-2}	6.7×10^{-2}
c_{HBDF}	10^{-1}	10^{-1}	α_{DR}	10^{-1}	10^{-1}
c_{HSDF}	10^{-1}	10^{-1}	α_{WF}	6.7×10^{-2}	6.7×10^{-2}
c_{HPDF}	10^{-2}	10^{-2}	α_{WR}	4×10^{-2}	4×10^{-2}
c_{WRDF}	1	1	r_H	2×10^{-3}	2×10^{-3}
c_{DRDF}	1	1	r_{DR}	1	1
c_{HBWF}	10^{-2}	10^{-2}	r_{WR}	1	1.7
c_{HSWF}	10^{-2}	10^{-2}	r_{DF}	1	1
c_{HPWF}	10^{-2}	10^{-2}	r_{WF}	1	1.7
c_{WRWF}	10^{-2}	10^{-2}	φ	5.5×10^{-3}	5.5×10^{-3}
c_{DRWF}	10^{-2}	10^{-2}	ν	variable	variable
b_{DR}	1	1	k_H	10^5	1.5×10^7
b_{DF}	2×10^{-4}	2×10^{-4}	k_{DR}	5×10^5	5×10^5
b_{WR}	1	1	k_{WR}	5×10^4	5×10^4
b_{WF}	2×10^{-4}	2×10^{-4}	k_{DF}	10^6	10^6
			k_{WF}	10^5	10^5

The parameters shown in Table 2 can be classified as those related to demography (birth and death rates and carrying capacities of humans, rodents and fleas), which are reasonably well known; those related to the clinical stages of the infection (specific mortality rates, recovery rates, transition rates between different stages and forms of the disease) which are also reasonably well known; and finally, those parameters related to the potentially infective contacts between the different actors involved in the biological cycle of plague (biting rates, direct transmission rates, probabilities of getting the infection after a contact), which to the best of our knowledge have not been established yet, but to which reasonable values can be assigned from other similarly transmitted infections.

The initial conditions, which distinguish the two epidemiological scenarios, are presented in Table 3.

Table 3. Initial conditions used in the simulations

Variable	Florence (1348)	Viet Nam (1989)
$S_H(0)$	90,000	1,000,000
$I_{HB}(0)$	0	0
$I_{HS}(0)$	0	0
$I_{HP}(0)$	0	0
$R_H(0)$	0	0
$S_{DR}(0)$	490,000	490,000
$I_{DR}(0)$	1,000	0
$S_{DF}(0)$	990,000	990,000
$I_{DF}(0)$	10,000	0
$S_{WR}(0)$	49,000	490,000
$I_{WR}(0)$	0	10
$S_{WF}(0)$	99,000	990,000
$I_{WF}(0)$	0	100

5.2. *Simulating an Epidemic (Black Death in Florence, 1348)*

The Black Death arrived in Italy through the port of Messina in October 1347, probably coming from Crimea[20]. Among the Italian cities, Florence, where the infection arrived in March, 1348, was particularly striken by the pestilence with about half its estimated 90,000 population dying from the disease in less than six months.

We simulated this epidemic of short duration by using the set initial conditions and carrying capacities shown in the first column of Table 3.

Results of those simulations are shown in Figures 2a and 2b.

In Figure 2a, we show the dynamics of infected humans. It can be noted that the epidemic reachs a peak of around 7,200 cases in the first weeks, dropping thereafter to undetectable levels one year after the beginning.

In Figure 2b, we show the accumulated number of deaths and the total population. The accumulated deaths totalized 45,000 after six months, and the total population drops to 53,000 in the first six months of the epidemics and started to recover thereafter. Both numbers are in agreement with the record for the period[20].

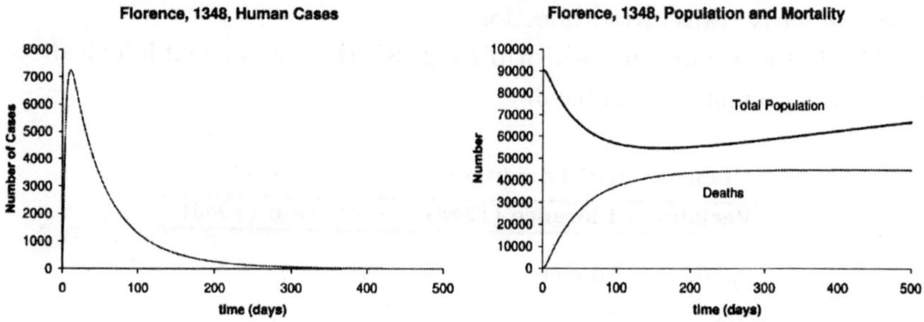

Figure 2. (a) Dynamics of infected humans and (b) accumulated number of deaths.

5.3. *Simulating an Endemy (Viet Nam, 1989)*

Current levels of plague endemicity in the world is restricted to some areas, in which the main factor for maintaining low levels of human disease is the presence of wild reservoirs[21]. In the first half of the twentieth century, India was burdened with the largest share of reported plague in the world, with an estimated total of 10 million deaths[21]. While plague ceased to occur in India in the 1950s, Viet Nam experienced a ressurgence in the plague in the 1960s and 1970s, with as many as 10,000 deaths a year. Since then, sporadic cases continue to be reported, with around 400 reported cases in Viet Nam in 1989[21].

Viet Nam was then chosen to exemplify endemic plague for being an area with persistent and stable transmission of yersinia infection, with around 40 deaths per year.

We simulated the Viet Nam case by setting the initial condition and carying capacities as in table 3, in which it can be noted that the main difference from the Florence epidemic is the absence of infected domestic fleas and rats in the beginning of the epidemic and a substantial number of infected wild fleas and rats.

Results of those simulations are shown in Figures 3a to 3c.

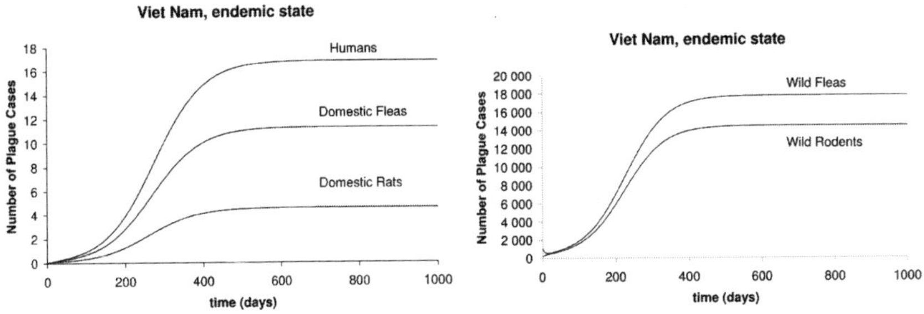

Figure 3.

In Figure 3a, we show the dynamics of infected humans, infected domestic fleas and infected domestic rodents. It can be noted that the human cases stabilizes at around the average 17 cases, the cases in domestic rats stabilizes at around 5 cases and the cases in domestic fleas stabilizes at around 11 cases.

In Figure 3b, we show the infection dynamics in the wild populations of infected rodents and infected fleas. The rodents stabilizes at around 14,500 cases (31% of the wild rodent population) and the fleas at around 18,000 cases (13% of the wild fleas population).

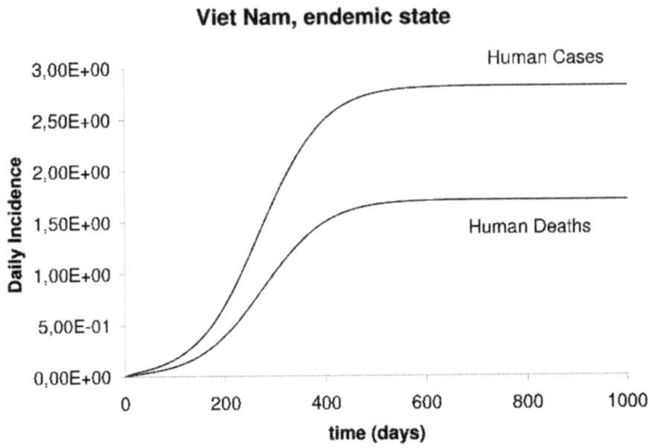

Figure 4.

Finally, in Figure 4, we show the daily incidence of human cases and of deaths. It can be noted that at equilibrium there are 2.8 new cases/day and 1.7 deaths/day. This corresponds, at equilibrium, to a coefficient of incidence of 1.7 cases per 100,000 inhabitants. The reported attack rate for Viet Nam, described in Dennis[8], in the period between 1980 and 1994 was about 0.6 reported cases per 100,000 inhabitants. It is interesting that the reported rate for endemic areas of the United States[5] among Native-Americans living in states like Arizona, New Mexico and Utah was 1.4 cases per 100,000, which contrasts with the figures among non-natives in the same states of 0.1 cases per 100,000 inhabitants. This tenfold difference is probably due to a possible direct contact with wild rodents or carnivores in the immediate vicinity of the Native-American's homes.

6. Discussion

In this paper we present a model to describe the dynamics of plague, one of the most complex vector-borne infections[22]. In fact, it involves three mammal hosts and two flea vectors, comprising a domestic and a wild cycle of transmission.

The model was applied to describe two distinct epidemiological scenarios; one epidemic outbreak of short duration (Florence, 1348), and one endemic state (Viet Nam, present day).

In order to simulate the epidemic outbreak in Florence we choose initial conditions so that the wild cycle is absent and a set of parameter that the makes the domestic subsystem (including humans) is below threshold ($R_0 < 1$). Then we asked and answered what would happened if a substantial number of infected fleas and/or rats would be introduced (by the arrival of a caravan or ship) in the community. It is found that an epidemic of great proportion occurs that vanishes after a period of time. So it is possible to understand the progress of an epidemic wave as a transport of infected fleas and/or rats from one location to another.

In contrast, in Viet Nam the wild cycle is the most important and with the same parameters as in Firenze, but different initial conditions and carrying capacities, the disease is sliglty above the threshold due to the wild reservoir/wild vector cycle of transmission. As a consequence, an endemic state is produced in the human population that is very difficult to eradicate unless effective control on the wild cycle of the disease is achieved.

In spite of the complex dynamics of plague and the consequent complexity of our model, we managed to reproduce distinct epidemiological scenarios by changing the initial conditions and carrying capacities. We believe that our model may be a useful tool for helping health authorities to better understand such an old and complex disease.

References

1. Butler, T. 1983. Plague and other Yersinia infections. Plenun Press. New York.
2. Slack, P. 1989. The black death past and present. Some historical problems. Transactions of the Royal Society of Tropical Medicine and Hygiene. 83:461-463.
3. Defoe, D. 1969. A Journal of the Plague Year. World's Oxford Classics. Oxford.
4. Politzer, R. 1954. Plague. WHO Monograph series 22:1. Geneva, World Health Organization.
5. Butler, T. 2000. Yersinia species, including plague. In: Principles and Practice of Infection Diseases (edited by G.L.Mandell, J.E.Bennett and R.Dolin) chap. 218, pp. 2406-2414. Churchill Livingstone. Philadelphia.
6. Beesley, W. N. 1998. Infestation by Fleas. In: Zoonosis (edited by S.R. Palmer, L. Soulsby and D.I.H. Simpson) chap. 66, pp. 873-879. Oxford University Press. Oxford.
7. Smith, M. D. and Thanh, N. D. 1996. Plague. In: Manson's Tropical Diseases (edited by G.C.Cook) chap. 50, pp. 918-924. Saunders. London.
8. Dennis, D. T. 1999. Plague. In: Tropical Infectious Diseases: Principles, Pathogens and Practices (edited by R.L. Guerrant, D.H.Walker and P.F.Weller) chap. 45, pp. 506-516. Churchill Livingstone. New York.
9. Service, N.W. 1986. Lecture Notes on Medical Entomology. Blackwell Scientific. Oxford.
10. Frean, J.A.; Arntzen, L.; Kapper, T. 1996. In vitro activity of fourteen antibiotics against 100 human isolates of Yersinia pestis from a South Africa plague focus. Antimicrobial Agents and Chemotherapy 40: 2646-2647.
11. Campbell, G. L. and Dennis, D. T. 1997. Plague and other yersinia infections. In Harrison's Principles of Internal Medicine 14th edition (edited by K.J.Isselbacher, E.Brownwald, J.D.Wilson). pp. 975. McGraw-Hill. New York.
12. Moore, J. 2002. Parasites and the Behavior of Animals. Oxford University Press, Oxford.
13. Barnett, S.A. 2001. The Story of Rats. Allen & Unwin, Adelaide.
14. Massad, E.; Coutinho, F. A.; Burattini, M. N.; Sallum, P. C.; Lopez, L.F.: A Mixed Ectoparasite-Microparasite Model for Bat-Transmitted Rabies. 2001. Theoretical Population Biology 60(4): 261-276
15. Macdonald, G. 1952. The analysis of equilibrium in malaria. *Trop.Dis.Bull.* **49**: 813-828.
16. Lopez, L. F.; Coutinho, F. A. B.; Burattini, M.N. and Massad, E. 2002. Threshold Conditions for Infection Persistence in Complex Host-Vectors Interactions. *Comptes Rendus Biologies Académie des Sciences Paris* 325: 1073-1084.
17. Massad, E.; Coutinho, F. A. B.; Burattini, M. N.; and Lopez, L. F. The Risk of Yellow Fever in a Dengue Infested Area. 2001. Transactions of the Royal Society of Tropical Medicine, 95(3): 370-374.

18. Lopez, L. F.; Coutinho, F. A. B.; Burattini, M. N. & Massad, E. Modeling the spread of infections when the contact rate among individuals is short ranged: propagation of epidemic waves. 1999. Mathematical and Computer Modelling 29: 55-69.
19. Murray, J. D. Mathematical Biology (2nd ed.), Springer Verlag, Berlin, 1993.
20. Ziegler, P. The Black Death, Penguin Books, 1984.
21. Gratz, N. G. Rodents as Carriers of Disease. In: Rodent Pests and their Control (ed. A.P. Buckle and R.H. Smith). CAB International, 1994.
22. Lopez, L. F.; Coutinho, F.A. B.; Burattini, M.N. and Massad, E. Threshold Conditions for Infection Persistence in Complex Host-Vectors Interactions. 2002. Comptes Rendus Biologies Académie des Sciences Paris 325: 1073-1084.

THE BASIC REPRODUCTION RATIO FOR A MALARIA MODEL

ANA PAULA P. WYSE AND LUIZ BEVILACQUA
LNCC – Laboratório Nacional de Computação Científica,
Av. Getúlio Vargas, 333
25651 - 075, Petrópolis – RJ, BR
E-mail: anawyse@lncc.br, bevi@lncc.br

MARAT RAFIKOV
UNIJUÍ – Universidade Regional do Noroeste do RS,
Rua São Francisco, 501
98700 - 000, Ijuí – RS, BR
E-mail: rafikov@unijui.tche.br

A mathematical model describing the dynamics of the relationship human – vector in malaria transmission is represented by a system of ordinary differential equations, which divides humans and mosquitoes population in three categories: susceptible, exposed and infectious. Different recovery rates are considered according to the treatment employed. For this model, the expression for the basic reproduction ratio is obtained by analyzing the infectious humans and mosquitoes equations isoclines; herewith the parameters that evaluate the epidemic occurrence are identified.

1. Introduction

Malaria is a contagious tropical disease that may spread throughout a human population through the biting of the *Anopheles* female mosquitoes. The protozoan *Plasmodium* inoculated during the biting is the infection agent. The malaria parasite completes a first complex maturation cycle in the mosquito's organism and starts a new cycle in humans after inoculation.

According to the World Health Organization[1], about 300 to 500 million of new cases are reported every year while 1 million persons die directly or indirectly from malaria in the same period.

Top economists and the United Nations have identified malaria as one of the top four causes of poverty. People faced with a high threat of malaria spend as much as a quarter of their incomes on medical visits, mosquito nets, medicines, laboratory tests and funerals for victims. They are less productive and lose income because of absences from work or inability to plant and harvest crops. Children lose out on educational opportunities, too. Governments in Africa south of the Sahara spend up to 40% of their health budgets on medical care for malaria victims and malaria control. In Africa alone, the total economic burden of malaria is estimated at US$12 billion annually.[1]

In Brazil, 99% of all cases (about 500 thousand cases per year) occurs in Amazonian region where socioeconomic and environmental conditions favor the proliferation of the *Anopheles* mosquito[2]. *Anopheles darlingi* is the most efficient vector of malaria in the Amazonian region, where the malaria transmission follows the precipitation seasonal fluctuation. The mosquito population density increases in the dry season that presents the climate conditions more favorable for mosquitoes breeding.

To analyze the disease evolution tendency affecting a susceptible population it is convenient to evaluate the basic reproductive number, R_0, that indicates how the disease progress in time, namely, either it will stabilize at an endemic level, or can take one of two opposite paths, one leading to complete eradication and the other where the entire population will eventually become contaminated.

According to Anderson and May[3], dealing with a disease transmited by microparasite (as the malaria case), R_0 is more precisely defined as the average number of secondary infections, when one infected individual is introduced into a host population where everyone is susceptible; R_0 is computed at the end of the infectiousness period of the individual. Thus, if $R_0 < 1$ the disease will disappear, if $R_0 = 1$ an equilibrium state may be attained and if $R_0 > 1$ the disease will spread out.

For control purposes only the information given by the inequality $R_0 > 1$ is not enough, the magnitude of R_0 is also important to estimate the control strategy. Moreover, the expression for R_0 contains the parameters involved in the disease transmission, providing important information to select the control variables that will be used to maximize efficient and reduce costs.

The basic reproduction ratio is usually derived algebraically by the analysis of the stability properties of the system, being mathematically defined as the dominat eigenvalue of a positive linear operator. A more direct derivation can, however, be obtained by a geometrical phase – plane analysis of the dynamical behaviour of the model[3].

2. The Mathematical Model

Mathematical models help to understand the dynamic relating mosquitoes density and malaria transmission. Using models, it is possible to simulate numerically the system behavior, helping to understand the interaction among the intervening variables. The relative weights of the main parameters involved in the dynamic process can also be estimated which provides a critical information in connection with model validation, that is how it approaches the related real phenomenon. To build up the mathematical model proposed, some assumptions should made:

- The proposed model considers two populations: H – Humans, with constant population size and V – Mosquitoes, with variable population size. The hypothesis that the human population size is constant while the mosquitoe's is variable lies on the fact that one human generation corresponds to many mosquitoes generations. Mosquitoes life span is much shorter than humans life span;
- In the human population, people of all ages and both of sex are considered;
- The host population is homogeneously mixed, in the sense that, on average, all hosts have intrinsically similar epidemiological characteristics independent of age, genetic formation, social habitats, geographical location, etc.
- In the mosquito population only the adult females are considered, for only those take blood meals to feed on protein needed to maturate their eggs;
- Human (H) and mosquito (V) populations, both are separated into three categories that represent the state variables: Susceptible (H_s and V_s), Exposed (H_e and V_e) and Infectious (H_i and V_i), where $H = H_s + H_e + H_i$ and $V = V_s + V_e + V_i$;
- Both populations are homogeneously distributed in space. This hypothesis is admissible when the region considered is small enough so that all characteristics are approximately invariant for the entire region;
- Healing does not lead to immunity. Humans cured return immediately to the susceptible category and can acquire the disease again as soon as he or she falls within the radius of action of infected mosquitoes;
- The disease does not reduce the fecundity of humans or mosquitoes;
- The mortality caused by the disease is negligible;

- All new-borns are susceptible in both populations (no vertical transmission or heredity). The infection of a susceptible mosquito occurs when it bites an infectious human and the infection for the susceptible human occurs when he or she is bite by an infectious mosquito;
- The number of bites of susceptible, exposed and infectious mosquitoes is assumed to be the same;
- The recovery span for humans depends on the effectiveness of the treatment. Several cases can occur: people which are effectively treated, people that are partially or inadequately treated in several scales, as many as necessary or desirable. The infectious period corresponding to humans that were never treated ends with its death;

Thus, the mathematical model can be put in the following way:

$$\begin{cases} \dfrac{dH_s}{dt} = \mu H - abV_i\dfrac{H_s}{H} - \mu H_s + \displaystyle\sum_{j=1}^{n}(\phi_j p_j)H_i \\ \dfrac{dH_e}{dt} = abV_i\dfrac{H_s}{H} - \mu H_e - \eta H_e \\ \dfrac{dH_i}{dt} = \eta H_e - \mu H_i - \displaystyle\sum_{j=1}^{n}(\phi_j p_j)H_i \\ \dfrac{dV_s}{dt} = \varepsilon V - f(V)V_s - acV_s\dfrac{H_i}{H} \\ \dfrac{dV_e}{dt} = acV_s\dfrac{H_i}{H} - f(V)V_e - \alpha V_e \\ \dfrac{dV_i}{dt} = \alpha V_e - f(V)V_i \end{cases} \tag{2.1}$$

where:

- μ – birth and death rate for humans per time unit, which is considered to be identic by the hypothesis that the human population size is assumed to be constant and the migration rates were not considered;
- a – rate of biting on humans by a single mosquito per time unit;
- b – relative number of infectious bites on susceptible human that produces an effective infection in him;
- c – relative number of bites of susceptible mosquito on infected people that produce an effective infection in the mosquito;
- η, α – rates in which exposed humans and mosquitoes become infectious per time unit, respectively;

$\frac{1}{\eta}, \frac{1}{\alpha}$ – length of intrinsic and extrinsic latent period, respectively;

p_j – relative number of infected humans under treatment j, where $\sum_{j=1}^{n} p_j = 1$;

ϕ_j – recovery times rate for persons that receive treatment j;

$\frac{1}{\phi_j}$ – length of infectious period for persons that receive treatment j;

ε – number of adult female mosquitoes generated by one adult female mosquito per time unit;

$f(V)$ – continuously differentiable function which represents the output of mosquitoes of the system through intracompartment and inter-compartment competition and death;

Adding up the relevant equations in (2.1) we find that the total population in both cases satisfy,

$$\begin{cases} \dfrac{dH}{dt} = 0 \\ \dfrac{dV}{dt} = \varepsilon V - f(V)V \end{cases} \tag{2.2}$$

where the human population size is constant according to the hypothesis of the model. The mosquitoes population is modeled by Verhulst – Pearl logistic growth model:

$$f(V) = (\delta + \delta_1 V), \tag{2.3}$$

where:

δ – death rate for mosquitoes per time unit;

δ_1 – intraespecific competition rate per time unit.

Then, the intraespecific competition relative to input rate is

$$\delta_1 = \left(\frac{\varepsilon - \delta}{\kappa}\right), \tag{2.4}$$

where κ is the carrying capacity.

Thus, the following equation is an example which describes the behavior of mosquitoes total population:

$$\frac{dV}{dt} = (\varepsilon - \delta)\, V - \left(\frac{\varepsilon - \delta}{\kappa}\right) V^2 \tag{2.5}$$

The non-zero steady state of system (2.5) for the mosquito equation is

$$V^*(t) = \kappa \tag{2.6}$$

3. The Basic Reproductive Rate

For the case of time dependent population density it is more convenient to normalize the intervening variables. That is, all variables relative to mosquitoes population will be divided by the total mosquito's population size and the variables relative to humans will be divided by the total human population. For this compute, the mosquito population will be considered constant because as t increase the mosquito's density approaches carrying capacity κ according to equation (2.6).

The new variables, therefore, represent the proportions or relative number of the different classes of mosquitoes and humans, susceptible, exposed and infectious referred to the respective total populations. Let make change of variables as follows:

$$h_s = \frac{H_s}{H}, \quad h_e = \frac{H_e}{H}, \quad h_i = \frac{H_i}{H}, \quad v_s = \frac{V_s}{V}, \quad v_e = \frac{V_e}{V}, \quad v_i = \frac{V_i}{V} \tag{3.1}$$

so that $h_s + h_e + h_i = 1$ and $v_s + v_e + v_i = 1$. Thus, substituting the new variables (3.1) in the system (2.1), we get

$$\begin{cases} \dfrac{dh_s}{dt} = \mu - abv_i h_s \dfrac{V}{H} - \mu h_s + \displaystyle\sum_{j=1}^{n} (\phi_j p_j) h_i \\ \dfrac{dh_e}{dt} = abv_i h_s \dfrac{V}{H} - \mu h_e - \eta h_e \\ \dfrac{dh_i}{dt} = \eta h_e - \mu h_i - \displaystyle\sum_{j=1}^{n} (\phi_j p_j) h_i \\ \dfrac{dv_s}{dt} = \varepsilon - f(V) v_s - acv_s h_i \\ \dfrac{dv_e}{dt} = acv_s h_i - f(V) v_e - \alpha v_e \\ \dfrac{dv_i}{dt} = \alpha v_e - f(V) v_i \end{cases} \tag{3.2}$$

where

$h_s(t)$ – Proportion of susceptible humans relative to the human's total population;

$h_e(t)$ – Proportion of exposed humans relative to the human's total population;

$h_i(t)$ – Proportion of infectious humans relative to the human's total population;

$v_s(t)$ – Proportion of susceptible mosquitoes relative to the mosquito's total population;

$v_e(t)$ – Proportion of exposed mosquitoes relative to the mosquito's total population;

$v_i(t)$ – Proportion of infectious mosquitoes relative to the mosquito's total population;

Requiring that $\frac{dh_s}{dt} = 0$, $\frac{dh_e}{dt} = 0$, $\frac{dh_i}{dt} = 0$ and solving the system for h_s, h_e, v_i, the h_i isocline is given by,

$$v_i^*(t) = \frac{-h_i^* \left(\mu^2 + (\mu + \eta) \sum_{j=1}^{n} (\phi_j p_j) + \mu\eta \right) H}{ab \left(-\eta + (\mu + \eta)\, h_i^*(t) + \sum_{j=1}^{n} (\phi_j p_j) \right) V}. \tag{3.3}$$

Requiring that $\frac{dv_s}{dt} = 0$, $\frac{dv_e}{dt} = 0$, $\frac{dv_i}{dt} = 0$ and solving the system for v_s, v_e, v_i, the v_i isocline is given by

$$v_i^*(t) = \frac{\alpha a c \varepsilon h_i^*}{f(V)\left(\alpha f(V) + \alpha a c h_i^* + f(V)^2 + f(V) a c h_i^*\right)} \tag{3.4}$$

Thus, two equilibrium points are identified: the disease free equilibrium $P_0 = (h_s^*, 0, 0, v_s^*, 0, 0)$ and the endemic equilibrium $P_1 = (h_s^*, h_e^*, h_i^*, v_s^*, v_e^*, v_i^*)$.

The intersection of the two isoclines, if it exists, represents the equilibrium state of the system.

Two isoclines will intersect at positive values of h_i and v_i, if the initial slope of the v_i isocline exceed that of the h_i isocline. Notice that the concavity of the h_i isocline is upwards, while that of the v_i isocline is downwards.

The slope of the h_i isocline evaluate at $h_i = 0$ is s_h and it is given by

$$s_h = \frac{H \left(\mu^2 + (\mu + \eta) \sum_{j=1}^{n} (\phi_j p_j) + \eta\mu \right)}{ab\eta V} \tag{3.5}$$

The slope of the v_i isocline evaluate at $h_i = 0$ is s_v and it is given by

$$s_v = \frac{\alpha a c \varepsilon}{f(V)^2(\alpha + f(V))} \tag{3.6}$$

If the initial slope of the v_i isocline lays below the slope of the h_i isocline, $(s_v < s_h)$, we necessarily have $R_0 < 1$ and the infection can not persist, remaining below the transmission threshold $R_0 = 1$. Otherwise, if the initial slope of the v_i isocline lays above the slope of the h_i isocline, $(s_v > s_h)$, we necessarily have $R_0 > 1$ and the infection persists.

Thus, if $s_v > s_h$, the inequality

$$\frac{\dfrac{\alpha a c \varepsilon}{f(V)^2(\alpha + f(V))}}{\dfrac{H}{V}\left(\dfrac{\mu^2 + (\mu+\eta)\displaystyle\sum_{j=1}^{n}(\phi_j p_j) + \eta\mu}{ab\eta}\right)} > 1 \tag{3.7}$$

is satisfied, and its left hand is the basic reproduction ratio

$$R_0 = \frac{\alpha a^2 c \varepsilon b \eta V}{f(V)^2\,(\alpha + f(V))\left(\mu^2 + (\mu+\eta)\displaystyle\sum_{j=1}^{n}(\phi_j p_j) + \eta\mu\right) H} \tag{3.8}$$

Note that the equation (3.8), the a term is squared indicating that the mosquito biting rate controls the transmission from humans to mosquitoes and mosquitoes to humans.

Now, it is important to note that the expressions (3.5) through (3.8) represent parametric families of curves depending on the parameters V and H. The parameter H is time invariant as assumed in the previous section, H is constant, and V is a function of t that approaches an equilibrium level κ according to equation (2.6).Therefore, the basic reproduction ratio is a constant in time because $\frac{V}{H}$ defines the number of female mosquitoes per human host at an equilibrium level, which is a constant.

The main interest here, however, is to find out what is the path of the disease evolution starting from the initial condition of a disease free population. Therefore the results are restricted to the case of H_s and V_s approximately equal to the total population.If a considerable proportion of individuals in a population is not susceptible, some authors suggest that the basic reproductive number can be given by the following expression:

$$R_1 = \frac{H_s}{H}\frac{V_s}{V}R_0, \tag{3.9}$$

where R_0 is given by (3.8).

4. Numerical Simulation

The phase –plane built here will help to understand the methodology used to calculate the basic reproductive number. The critical parameters to be estimated dealt with the *Anopheles darlingi* mosquito behavior and biological conditions.

4.1. *Parameters Estimation*

Data for parameter estimation are only enough to provide a first approximation of the complex reality. Nevertheless the approximations obtained help to draw reasonably realistic sceneries.

The value of $(\varepsilon - \delta)$ is estimated considering the fact that the mosquito's population increases exponencially in the absence of limiting factors. The evolution is given by:

$$\frac{dV}{dt} = (\varepsilon - \delta)\, V,$$

whose solution is $V = V(0)e^{(\varepsilon-\delta)t}$. Then,

$$\varepsilon = \frac{1}{t} ln \left(\frac{V}{V(0)} \right) - \delta.$$

Experimental research carried out by Santos et al.[4] on the biology of *Anopheles darlingi* has shown that the biological cycle encompassing the period starting at the egg stage till the adult stage takes, on average, 15.6 days (0.51 month) with a survival number of 57%. The average number of eggs in each posture reaches 110 and each female lays eggs two times in its life time[5]. Charlwood and Alecrim[6] estimated a survival number for adults of 80.4%. Considering equal probability that an egg will hatch into a male or a female mosquito , 50% of the eggs will turn out to breed female mosquitoes. Then:

- $\delta = 1 - 0,804^{30} = 0,9985/\text{month}$;
- $\varepsilon = \dfrac{1}{0,51} ln \left(\dfrac{220 * 0,5 * 0,57}{1} \right) - 0,9985 = 7,116/\text{month}.$

Since the mosquitoes total population is modeled by Verhulst - Pearl logistic growth model, as the time t increases, V approaches the carrying capacity κ. Thus, it is allowed to take $V = \kappa$. For the numerical simulation it is assumed $\kappa = 200$.

Taking the life mean expectancy of human population 60 years, then

$$\mu = \frac{1}{720} = 0,00139/\text{month}.$$

The *Anopheles darlingi* biting rate is estimated from the stability index $S = \dfrac{a}{\delta_d}$, where δ_d is the mosquito daily death rate. In the Amazonian region, it is estimated that S lies somewher between $5,01$ and 8^6. In the present simulation, this parameter is assumed $S = 7$. Then,

$$7 = \frac{a}{1 - 0,804} \Rightarrow a = 1,372/\text{day}.$$

Not all mosquitoes survive a month, thus it is normal the biting rate to be related to its daily survival rate. Then,

$$a = 1,372 * \sum_{t=1}^{30} 0,804^t = 5,628/\text{month}.$$

The coefficients b and c should vary within $[0, 1]$ according to the susceptibility degree of humans and mosquitoes, respectively. In this case, we have $b = c = 1$.

The malaria latent period is defined as the time from initial infection to the appearance of gametocytes in the blood (extrinsic latent period – for humans) or the appearance of sporozoites in the salivary glands (intrinsic latent period – for mosquitoes). Its duration vary depending on the environment temperature, but it is estimated to be in average 10 days (0.33 month) for *Plasmodium falciparum* and *Plasmodium vivax*[3]. Thus, the transition from exposed category to infectious one is given by:

$$\alpha = \eta = \frac{1}{0,33} = 3/\text{month}.$$

With regard to treatment, three cases are considered:

- Complete and efficient treatment: In this case, the individual remains infectious and a potential source of disease dissemination from the end of the latent period till the end of the treatment, when recovery is completed. The treatment begins when the patient detects the disease symptoms and takes approximately 15 days. To this time span we have to add the transmissible incubating period (asymptomatic) which is the difference between the total incubating period and the latent (non transmissible) period. Note that individual just look for treatment when the disease symptoms are manifested.

$$\phi_1 = \frac{1}{15 + 5} * 30 = 1,5/\text{month}$$

- Incomplet treatment: In this case the individual remains infeccious for a long time, because the recovery time is much longer as the treatment is conducted. For compute, the infeccious period is considered 2 years.

$$\phi_2 = \frac{1}{24} = 0,0416/\text{month}$$

- Absent treatment: When treatment is skipped, the individual remains infeccious until the end of its life. In this case,

$$\phi_3 = 0.$$

To compute it we have assumed that there is one female mosquito for each person. Then, $V/H = 1$.

The coefficients obtained with the previous calculations are displayed in the Table 1.

Table 1. Parameters of Model

Parameters	Values
ε	7.116
δ	0.9985
κ	200
μ	0.00139
a	5.628
b	1
c	1
α	3
η	3
ϕ_1	1.5
ϕ_2	0.0416
ϕ_3	0
V/H	1

4.2. *The Phase – Plane*

To draw the isoclines on the phase – plane, we used the parameters derived in Table 1. We will make comparisons between the different sceneries obtained by varying p_j. Each point in the plane corresponds to a particuar pair of values (h_i, v_i).

Here, the horizontal axis corresponds to the dynamical variable h_i, the proportion of infected people, and the vertical axis to the dynamical variable v_i, the proportion of infected mosquitoes. The intersection of the two isoclines represents the equilibrium state, a convergin point of at least a subset of trajectories.

The Figure 1 shows that the isoclines do not cross; all trajectories are directed towards the origin. This indicates that exists only the disease free equilibrium. Note that in this case, the slope of the v_i isocline is below than the slope of the h_i isocline, requiring $R_0 < 1$. Actually, $R_0 = 0.974$. In other words, each infectious person generate, on average, 0.974 new case of malaria and, consequently, the disease can not maintain itself and it tends to disappear.

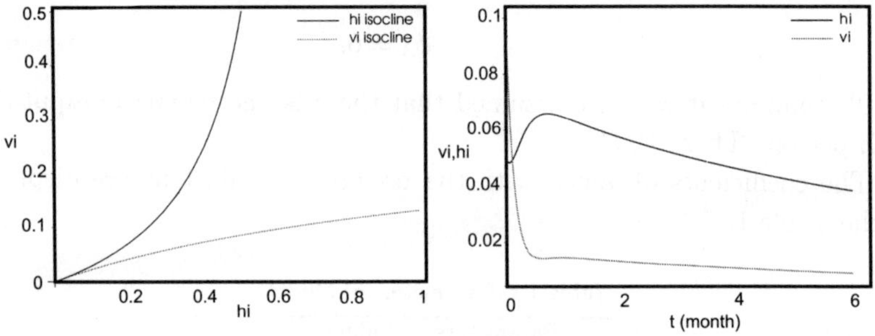

Figure 1. (left) Phase – plane of the dynamical variables h_i and v_i. (right) Time evolution of the infectious humans and mosquitoes population. The parameters used were that from Tab 1 and $p_1 = 0,9$, $p_2 = 0,08$ and $p_3 = 0,02$.

When the treatment is partially omited, the situation tends to be worst as it is shown in the Figure 2. Here the intersection of the two isoclines exists, representing the equilibrium point to which the trajectories converge. This indicate the existence of endemic equilibrium. Note that, in contrast to the previous case, the slope of the v_i isocline is greater than the slope of the h_i isocline, requiring $R_0 > 1$. Actually, $R_0 = 2.8$. It means that each infectious person generate, on average, 2.8 new cases of malaria and, because this, the disease is self sustained and it is maintained at a endemic level. Moreover, the initial slope of the v_i isocline is relatively small, but still exceeds that of the h_i isocline, and the equilibrium point exists, but in a relatively shallow canon, corresponding to Macdonald's 'unstable' malaria.[3]

If the initial slope of the v_i isocline significantly exceeds that of the h_i isocline, the equilibrium point rests in a relatively deep valley, corresponding to Macdonald's 'stable' malaria[3]. This situation is observed in Figure 3. As the Figure 2, the disease is maintained at a endemic level.

When there is no treatment at all to the infectious population, the entire population became infectious as can be seen in the Figure 4. Note that the human population became completely infectious but the mosquito's did not. The conclusion is that a few infectious mosquitoes are enough to produce an epidemic wave.

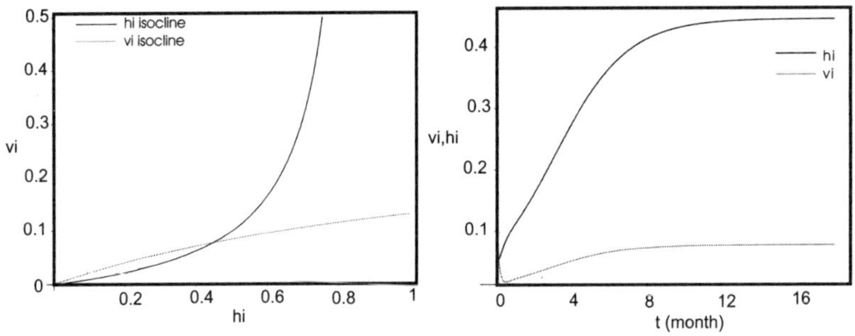

Figure 2. (left) Phase – plane of the dynamical variables h_i and v_i. (right) Time evolution of the infectious humans and mosquitoes population. The parameters used were that from Tab 1 and $p_1 = 0,3$, $p_2 = 0,5$ and $p_3 = 0,2$.

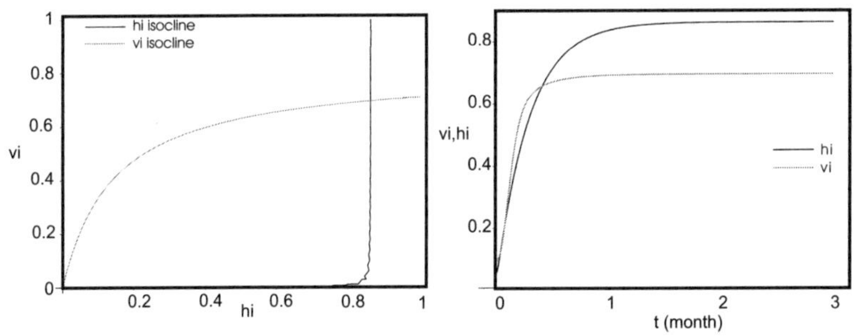

Figure 3. (left) Phase – plane of the dynamical variables h_i and v_i. (right) Time evolution of the infectious humans and mosquitoes population. The random parameters used were: $\varepsilon = 7.116$, $\delta = 0.09985$, $\kappa = 2000$, $\mu = 0.00139$, $a = 50.628$, $b = c = 1$, $\alpha = 30$, $\eta = 3$, $\phi_1 = 1.5$, $\phi_2 = 0.0416$, $\phi_3 = 0$, $p_1 = 0,3$, $p_2 = 0,5$ and $p_3 = 0,2$.

5. Discussion

Although the mathematical model described here do not contemplate all particularities of malaria transmission the dynamics of the whole system is suitable enough to give some necessary informations about the factors that influence the disease spread.

When control mechanisms are used to reduce disease spreading, it is necessary to know which parameters are more critical in the process. If the control drives the basic reproduction ratio to a value less than one, then the control is effective; otherwise it is not cost effective.

It is necessary a more detailed analysis of the parameters sensitivity to identify what kind of control is more efficient to combat malaria.

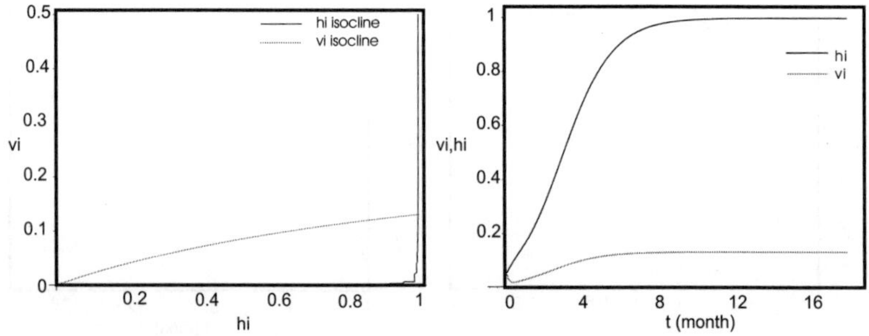

Figure 4. (left) Phase – plane of the dynamical variables h_i and v_i. (right) Time evolution of the infectious humans and mosquitoes population. The parameters used were that from Tab 1 and $p_1 = 0$, $p_2 = 0$ and $p_3 = 1$.

Acknowledges

We thank Dr. Manuel Cesário (UFAC – Brazil) for discussion about malaria treatment forms. Financial support for this research was provided by CNPq – National Consortium for Scientific and Technologycal Development and GEOMA – Thematic Grid for Research in Environmental Modelling for Amazonian.

References

1. Roll Back Malaria Department, Basic Facts on Malaria, World Health Organization, 2004.
2. Programa Nacional de Prevenção e Controle da Malária, Brasília: Ministério da Saúde. Fundação Nacional da Saúde – FUNASA, 2002.
3. R. M. Anderson e R. M. May, "Infectious Diseases of Humans", Oxford Science Publication, New York, 1991.
4. J. Santos et al, Biologia de anofelinos amazônicos 1- Ciclo biológico, postura e estádios larvais de *Anopheles darlingi* Root 1926 (Díptera: Culicidae) da Rodovia Manaus – Boa Vista, *Acta Amazônica*,11(4) (1981) 789-797.
5. W. Tadei et al, Entomologia da Malária em Áreas de Colonização da Amazônia, Programa de Pesquisa Dirigida – PPD, MCT, 1997.
6. J. D. Charlwood, W. A. Alecrim, Capture-recapture studies with the South America malaria vector *Anopheles darlingi*, *Ann. Trop. Med. Parasitol.*, 83 (1989) 569-576.
7. A. Kiszewski et al, A Global Index Representing the Stability of Malaria Transmission, *Am. J. Trop. Med. Hyg.*, 70(5) (2004) 486-498.

EPIDEMIOLOGICAL MODEL WITH FAST DISPERSION

MARIANO R. RICARD, CELIA T. GONZÁLEZ GONZÁLEZ
Department of Mathematics & Computer Science
Havana University, C.Habana 10400, Cuba.
E-mail: rricard@matcom.uh.cu; celia@matcom.uh.cu

RODNEY C. BASSANEZI
Department of Applied Mathematics, University of Campinas
CP 6065, São Paulo, Brasil.
E-mail: rodney@ime.unicamp.br

In this paper we study the effect of distance between infectives and susceptibles in SIS-type epidemic in which is considered heterogeneity respect to the viral load. We consider that the infected population diffuses randomly in a bounded spatial region with impenetrable boundary. Using multiple scales procedures we obtain an asymptotic expansion of the solution to the initial boundary value problem, which is treated with a fuzzy technique in order to determine a parameter showing the effect of the heterogeneity in the tendencies of the infected population and its spatial average.

1. Introduction

Our concern is a SIS (susceptible-infected-susceptible) -epidemiological model in which the rate of transmission and recuperation of the disease are considered fuzzy sets in order to reflect the *heterogeneity* respect to the *viral load* (**VL**) of the individuals in a population. This means that infected members with different VLs contribute differently to the propagation of the disease. The dynamics of the disease depends on a dissemination of a virus (or bacteria) transported by infected individuals in contact with susceptible ones. This is the so-called *direct transportation* dynamics. In other words, this means that the infection dynamics follows the *mass action law.*

Here we will consider an homogeneously distributed (total) population in which the class of infected, and hence the susceptible population, depend both on time and on the location. We want to explore the effect of distance between infectives and susceptibles under the assumption that both parts of the populations diffuse randomly, with equal diffusivities. We will obtain an *asymptotic expansion* (see [1]) of the solution to the *initial boundary value problem* (**IBVP**) modeling an infected population inside a bounded spatial region, assuming its boundary closed to the infected. In the model we are considering the growth speed of the infected slower than the speed of the dispersion process, and from this fact appears the small parameter with respect to which is considered the expansion of the solution. The reaction part in the model equation reflects the interaction between the susceptible and the infected population, as in the Kermack and Mckendric well-known paper [2] but the reader may see [3], [4] and references therein for a modern presentation. An important parameter in the study of epidemiological models is the so called *basic reproductive value* (**BRV**) of the disease R_0, defined as the number of secondary infections produced by one infected individual when it is introduced in a population entirely susceptible. The threshold $R_0 = 1$ is a bifurcation value determined by the appearance of a nontrivial stationary stable equilibrium point which reflects the establishment of the disease. Additionally, this threshold value serve as a condition to the appearance of a traveling wave solution to the resultant equation if we assume that the population is spatially distributed and the infected are randomly dispersed.

In this paper we consider a previously defined (see [5]) fuzzy integral number R_0^f which plays the role of the BRV when the population is heterogeneous respect to the VL. More precisely, reflecting the fact that individuals with greater VL should be more infective. Then, we will see that $R_0^f > 1$ is a sufficient condition to the appearance of a non-negative constant solution to the IBVP with Neumann conditions, which is the limit, as time tends to infinity, of any solution to the problem. We will determine a more appropriate fuzzy integral number which reflects the heterogeneity and the establishment of the desease along the spatial region.

2. Preliminaries

2.1. *SIS-model with heterogeneity*

The SIS-model system considered in [5]-[6], in which is considered the temporal evolution of the population with heterogeneity respect to the VL is

$$\begin{cases} \frac{dS}{dt} = -\beta(v)\ SI + \gamma(v)\ I \\ \frac{dI}{dt} = +\beta(v)\ SI - \gamma(v)\ I \end{cases} \tag{1}$$

where $S + I = 1$, S is the *fraction of the susceptible* population, I is the *fraction of infected*, v represents the VL, $\beta(v)$ is the *contact rate*, and $\gamma(v)$

is the *recovery rate*. Of course, the meaning of $\beta(v) > 0$ is that susceptible population can be infected. Further, $\gamma(v)^{-1}$ is the half-period of recovery and $\beta(v)/\gamma(v)$ is the fraction of the population that comes into contact with an infective individual during the period of infectiousness. Let us consider, as in [5], the rates as functions of the VL:

$$\beta(v) = \begin{cases} 0 \text{ if } v \leq v_{\min} \\ \frac{v - v_{\min}}{v_M - v_{\min}} \text{ if } v_{\min} < v \leq v_M \\ 1 \text{ if } v_M < v \leq v_{\max} \\ 0 \text{ if } v > v_{\max} \end{cases} \tag{2}$$

and, the decreasing function

$$\gamma(v) = \frac{(\gamma_0 - 1)}{v_{\max}} v + 1 \tag{3}$$

where γ_0 is the minimum value of the recovery rate and $0 < \gamma_0 < 1$. The value of $v_{\max}$ serve to focusing the analysis to a finite domain of value of the VL. So, $v_{\max}$ can be selected as any upper bound of all possible values of the VL. Furthermore, we will consider the (triangular) *distribution of the VL*, $\rho(v)$, centered in the average value $\overline{v}$ of the VL. More exactly, $\rho(v)$ is a function given by

$$\rho(v) = \begin{cases} 0 \text{ if } v \leq \overline{v} - \delta \\ \frac{1}{\delta}(v - \overline{v} + \delta) \text{ if } \overline{v} - \delta < v \leq \overline{v} \\ -\frac{1}{\delta}(v - \overline{v} - \delta) \text{ if } \overline{v} < v \leq \overline{v} + \delta \\ 0 \text{ if } v \geq \overline{v} + \delta \end{cases} . \tag{4}$$

There are two stationary points for system in Eq. (1) if

$$\sigma(v) > 0 \tag{5}$$

where

$$\sigma(v) \underset{\text{def}}{=} \beta(v) - \gamma(v) \tag{6}$$

is an increasing function. Assuming Eq. (5), the stationary point $(1,0)$ of system in Eq. (1) turns to be unstable while the second one, $\left(\frac{\gamma(v)}{\beta(v)}, \frac{\sigma(v)}{\beta(v)}\right)$ is asymptotically stable and also lies in the first quadrant of the SI-plane. This fact is a consequence of the inequality

$$R_0(v) = \frac{\beta(v)}{\gamma(v)} > 1 \tag{7}$$

representing the establishment of the disease for a given VL v. In [5] was introduced the *fuzzy BRV* as

$$R_0^f = \frac{1}{\gamma_0}\ FEV[\gamma_0\ R_0(v)] \tag{8}$$

where the *FEV* will be defined below. In that paper they proved the inequalities

$$R_0(\overline{v}) < R_0^f < R_0(\overline{v}+\delta) \tag{9}$$

and, from the Mean Value Theorem for increasing continuous function $R_0(v)$, they conclude the existence of a unique value $\widehat{v}$ of the VL, $\overline{v} < \widehat{v} < \overline{v}+\delta$, defined by

$$R_0^f = R_0(\widehat{v})\,. \tag{10}$$

Then, following Eqs. (9-10) was concluded in [5] that the BRV $R_0(\overline{v})$ for the average VL, results to be a subevaluation of the parameter R_0^f:

$$R_0(\widehat{v}) < R_0^f \tag{11}$$

The value R_0^f can be interpreted as the average number os secondary infections caused by the introduction of one infected individual in a fully susceptible population, but considering all VLs present in the population.

2.2. *Fuzzy expected value*

A proper subset $\mathcal{F}$ of the universe $\mathcal{U}$ can be represented through a real valued function, the so called *characteristic function* of the set:

$$S_{\mathcal{F}} : \mathcal{U} \to [0,1] \tag{12}$$

with the properties

$$S_{\mathcal{F}}(x) = 1 \text{ if } x \in \mathcal{F} \tag{13}$$

$$S_{\mathcal{F}}(x) = 0 \text{ if } x \notin \mathcal{F}\ . \tag{14}$$

We define a *fuzzy subset* of the universe $\mathcal{U}$, $\mathcal{U} \neq \emptyset$, or simply a *fuzzy set*, as a function

$$u : \mathcal{U} \to [0,1] \tag{15}$$

in which the number $u(x)$ represents the *membership grade* of the element x to the fuzzy subset u. Function u is often called the *membership function* of the fuzzy set. For example, functions β in Eq. (2), γ in Eq. (3), and ρ in Eq. (4) can be treated as fuzzy set membership functions.

We define a *fuzzy measure* μ on $\mathcal{U}$ as a non-negative real valued function defined on the corresponding power set $\mathcal{P}(\mathcal{U})$, i.e. the set of all parts of $\mathcal{U}$,

$$\mu : \mathcal{P}(\mathcal{U}) \to [0,1] \tag{16}$$

with the properties:

$$a)\ \mu(\emptyset) = 0 \tag{17}$$

$$b)\ \mu(\mathcal{U}) = 1 \tag{18}$$

$$c)\ \mu(A) \leq \mu(B) \quad \text{if} \quad A \subset B \subset \mathcal{U}\,. \tag{19}$$

As an example of such a fuzzy measure, we define the *measure of feasibility* with respect to the fuzzy set with membership function ρ, by the equality

$$\mu(A) = \sup_{s \in A} \rho(s)\,. \tag{20}$$

This measure is very convenient when we want to reflect the worst situation, because the measure of a given subset of the population takes the same value as those that more infected individual in the population have.

Now we can define the *fuzzy integral* or *fuzzy expected value* (**FEV**) of the fuzzy set u with respect to the fuzzy measure μ, as

$$FEV(u) = \sup_{\alpha \in [0,1]} \min\left[\alpha,\ \mu\{u \geq \alpha\}\right] \tag{21}$$

where

$$\{u \geq \alpha\} = \{x \in \mathcal{U}\ |u(x) \geq \alpha\}\ . \tag{22}$$

Fuzzy set theory (see [7], [8]) is used here to represent the uncertainties due to the heterogeneity in the model and for a linguistic evaluation of the results in terms of the intensity of the VL: weak, average or strong. We will introduce here the $FEV(u)$ as a *defuzzyfication method* for the fuzzy set u.

3. The model

We want to extend the model in [5] to the case in which the fraction of infected population I depends also in the spatial location and diffuses randomly, with a diffusion coefficient $k > 0$. Let us consider the case when $I = I(x,t)$ and x represents only one spatial coordinate. So, the equation for the infected population is

$$\frac{\partial I}{\partial t} = k\frac{\partial^2 I}{\partial x^2} + I\ \left[\sigma(v) - \beta(v)\ I\right] \tag{23}$$

where v is the VL, and $\beta(v)$, $\sigma(v)$ are fuzzy sets as in Eqs. (2-3-6) ([5]). To Eq. (23) we consider the IBVP: "Find the smooth solution to Eq. (23) that satisfies the initial condition

$$I(x,0;v) = J(x;v) \geq 0 \quad \text{for } 0 \leq x \leq l \tag{24}$$

and the boundary conditions:

$$\frac{\partial I}{\partial x}(0,t;v) = 0 \quad , \quad \frac{\partial I}{\partial x}(l,t;v) = 0 \quad \text{for } t \geq 0\text{''}. \tag{25}$$

Furthermore, we assume the existence of an heterogeneous distribution $\rho(v)$ of VL in the population which is represented by a fuzzy set. We expect that the establishment of the disease for a given VL will depend on the grade of membership of v to the fuzzy distribution $\rho(v)$. As we will see later, we restrict the analysis to the case in which the VL v varies in the range

$$\max\{v^*,\ \overline{v} - \delta\} \leq v \leq \min\{v_{\max},\ \overline{v} + \delta\} \tag{26}$$

where $\overline{v}$ is the average VL and δ is the dispersion, both defined in Eq. (4). From Eq. (2) and Eq. (3) follows that $\sigma(v)$ is an increasing function, hence the existence of a unique value, say v^*, $v_{\min} < v^* < v_M$ which is the root of the equation

$$\sigma(v^*) = 0\ . \tag{27}$$

So, $\beta(v^*) > 0$. Furthermore, the VL v^* can be easily calculated as

$$v^* = v_{\max}\ v_M\ \left[v_{\max} + (v_M - v_{\min})\ (1 - \gamma_0)\right]^{-1}. \tag{28}$$

In our presentation we will use a small parameter ε reflecting the fact that the evolution process is slower than diffusion as we shall see in the next subsection. So, we will consider a slightly different equation from Eq. (23).

3.1. *The nondimensional problem*

Let us consider new variables in Eq. (23) given by

$$x = l\xi \tag{29}$$

and

$$t = \frac{l^2}{k}\tau\ . \tag{30}$$

Let us now consider $v > v_{\min}$. Noting that I is a non-dimensional variable, and using the same letter to denote the new function $I(\xi,\tau) \underset{\text{def}}{=} I\left(l\cdot\xi, \frac{l^2}{k}\cdot\tau\right)$, we rewrite Eq.(23) in the form

$$\frac{\partial I}{\partial t} = \frac{\partial^2 I}{\partial \xi^2} + \epsilon\ g(I;v) \tag{31}$$

where, for any given v, we have the nondimensional small parameter

$$\epsilon \underset{\text{def}}{=} \frac{l^2}{k}\beta(v) \tag{32}$$

and

$$g(I;v) = I\ \left(\frac{\sigma(v)}{\beta(v)} - I\right)\ . \tag{33}$$

Due to the boundness of β it is easy to see that the assumption

$$\varepsilon \underset{\text{def}}{=} \frac{l^2}{k}\beta(v_{\max}) \ll 1 \tag{34}$$

implies, for any VL, the inequality

$$\epsilon \ll 1\ . \tag{35}$$

More exactly, ε represents a small parameter asymptotically equivalent to ϵ when both together tend to zero. So, in order to exclude the VL in the expression of the small parameter we will substitute Eq. (31) by the following

$$\frac{\partial I}{\partial t} = \frac{\partial^2 I}{\partial \xi^2} + \varepsilon\ g(I;v) \tag{36}$$

and we will consider the IBVP for this equation with the conditions

$$I(0,\xi;v) = J(\xi;v) \geq 0 \ \text{ for } 0 \leq \xi \leq 1 \tag{37}$$

corresponding to Eq. (24) and

$$\frac{\partial I}{\partial \xi}(\tau,0;v) = 0\ \ ,\ \ \frac{\partial I}{\partial \xi}(\tau,1;v) = 0 \ \text{ for } \tau \geq 0 \tag{38}$$

which correspond to Eq. (25). Note that the asymptotic expansion of the solution to Eqs. (31-37-38) will be the same as the corresponding to the solution of the problem Eqs. (36-37-38) according to the equivalence between the two small parameters if $\beta(v) > 0$.

We will find a uniform asymptotic expansion of the solution to the problem Eqs. (36-37-38) using perturbation techniques for parabolic PDEs.

3.2. *The multiple scales method*

Let us represent the solution to the Eq. (36) in the form of asymptotic expansion

$$I(\tau, \xi; v) = I_0(\tau, \xi; v) + \varepsilon\, I_1(\tau, \xi; v) + O(\varepsilon^2)\ . \tag{39}$$

in the small positive parameter ε [1]. One should note that the fact τ is an unbounded variable in the problem can cause the expansion in Eq. (39) turns to be nonuniform. In order to obtain a uniform expansion in a τ-interval with length $O\left(\varepsilon^{-1}\right)$ we will follow the procedure in [9]. Then, let introduce T_0 and T_1 as the scales involving the small parameter

$$T_0 = \tau \tag{40}$$

$$T_1 = \varepsilon \tau\ . \tag{41}$$

For simplicity we will only consider these two scales, but in general we can consider

$$T_n = \varepsilon^n\, \tau \tag{42}$$

and take the scales as independent variables. This is also the main idea in the multiple scales procedure for ODEs, as can be seen in [1], [10] or in the vast literature about perturbation methods in differential equations. We remark that the variable ξ do not introduce nonuniformities in the expansion of I due to the boundness of the ξ-range of values. We resume writing

$$I(\tau, \xi; v) = I(T_0, T_1, \xi; v) = I_0(T_0, T_1, \xi; v) + \varepsilon\, I_1(T_0, T_1, \xi; v) + O\left(\varepsilon^2\right) \tag{43}$$

and further the relations

$$\frac{\partial I}{\partial \tau} = \frac{\partial I}{\partial T_0} + \varepsilon\, \frac{\partial I}{\partial T_1} \tag{44}$$

$$\frac{\partial I}{\partial \tau} = \frac{\partial I_0}{\partial T_0} + \varepsilon \left(\frac{\partial I_0}{\partial T_1} + \frac{\partial I_1}{\partial T_0} \right) + O\left(\varepsilon^2\right)\ . \tag{45}$$

Then, subtituting Eq. (43) in Eq. (36) we obtain a hierarchy of IBVPs after equating the corresponding terms in powers of the small parameter.

More precisely, for the leading term $O(\varepsilon^0)$ we get:

$$\frac{\partial I_0}{\partial T_0} = \frac{\partial^2 I_0}{\partial \xi^2} \tag{46}$$

with the initial condition

$$I_0(0, 0, \xi; v) = J(\xi; v) \geq 0 \tag{47}$$

and the boundary condition

$$\frac{\partial I_0}{\partial \xi}(T_0, T_1, 0; v) = \frac{\partial I_0}{\partial \xi}(T_0, T_1, 1; v) = 0 \ . \tag{48}$$

We remark that $I_0(T_0, T_1, \xi; v)$ will depend parametrically on v, and this IBVP have a well known solution which can be obtained by separation of variables:

$$I_0(T_0, T_1, \xi; v) = \sum_{n=0}^{\infty} C_n(T_1; v) \exp(-\lambda_n^2 \ T_0) \cos n\pi\xi \tag{49}$$

where the coefficients $C_n(T_1; v)$ are determined by Eq. (47) considering T_1 and v as parameters. Here $\lambda_n = n\pi$ denote the sequence of eigenvalues of the Stürm-Liouville operator associated to the problem for I_0 after separation of variables. Let us take, following [9], the appropriate form

$$C_n(T_1; v) = U_0(T_1; v) \left\{ \frac{s_n\,(v)}{\int_0^1 J\,(\xi; v) \ d\xi} \right\} \tag{50}$$

where $s_n\,(v)$ are the Fourier coefficients of $J\,(\xi; v)$ respect to the orthonormal basis $\{X_n\}$ where $X_1 = 1$ and, for $n > 1$, $X_n = \sqrt{2} \cos n\pi\xi$. Furthermore, it is possible to rewrite Eq. (49) from Eq. (50) as follows

$$I_0(T_0, T_1, \xi; v) = U_0(T_1; v) \ G(T_0, \xi; v) \tag{51}$$

where

$$G(T_0, \xi; v) = \left(\int_0^1 J\,(\xi; v) \ d\xi \right)^{-1} \sum_{n=0}^{\infty} s_n\,(v) \exp(-\lambda_n^2 \ T_0) \cos n\pi\xi \tag{52}$$

is the density function corresponding to the spatial distribution of the infected population along the considered region. So,

$$\int_0^1 G(T_0, \xi; v) \ d\xi = 1 \tag{53}$$

and

$$\lim_{T_0 \to \infty} G(T_0, \xi; v) = 1 \ . \tag{54}$$

As ussual in the multiple scales method, we impose conditions in order that I_1 in Eq. (43) be bounded and so, the uniformity of the expansion can be guarranted. Of course, these conditions will be taked on $U_0\,(T_1)$ as we shall see later.

Let now consider the equation obtained from the terms having orden $O(\varepsilon)$ in both sides of Eq. (43). Then, we get

$$\frac{\partial I_1}{\partial T_0} = \frac{\partial^2 I_1}{\partial \xi^2} + \left[g(I_0, v) - \frac{\partial I_0}{\partial T_1} \right] \tag{55}$$

so, I_1 will depend explicitly on the VL v. Corresponding to Eq. (55) the initial and boundary conditions for $I_1$are taken equal to zero. Further, integrating Eq. (55) over the whole space and taking into account Eq. (53) and the null boundary conditions, we get for the unknown

$$U_1(T_0, T_1; v) = \int_0^1 I_1(T_0, T_1, \xi; v) \, d\xi \tag{56}$$

the following PDE in which no derivatives respect to the spatial coordinate are considered:

$$\frac{\partial U_1(T_0, T_1; v)}{\partial T_0} = -\frac{\partial U_0}{\partial T_1} + \int_0^1 g(I_0, v) \, d\xi \; . \tag{57}$$

In order to assure the uniformity of the expansion Eq. (43) in the form

$$I(T_0, T_1, \xi; v) = I_0(T_0, T_1, \xi; v) + O\left(\varepsilon\right) \tag{58}$$

we should first guarrantee the boundness of function I_1, and so the boundness of U_1 in the range of values of T_0 which length be of order $O\left(\varepsilon^{-1}\right)$. To assure so, we take

$$\frac{\partial U_0(T_1; v)}{\partial T_1} = \int_0^1 g\left(U_0(T_1; v) \; G(T_0, \xi); v\right) \, d\xi \; . \tag{59}$$

The above equation is not really well posed, due to the dependence on T_0 of the right side. To use this condition appropriately we will consider the asymptotic behavior of G as T_0 grows, and instead ot Eq. (59) we determine U_0 from

$$\frac{\partial U_0(T_1; v)}{\partial T_1} = \int_0^1 g\left(U_0(T_1; v); v\right) \, d\xi \tag{60}$$

where

$$g(U_0(T_1; v); v) = U_0(T_1; v) \; \left(\sigma\left(v\right) - \beta\left(v\right) \; U_0(T_1; v)\right) \tag{61}$$

according to Eq. (33). Further, we associate to Eq. (60) the initial condition

$$U_0(0;v) = \int_0^1 J(\xi;v)\ d\xi \tag{62}$$

which is compatible with Eq. (49) and Eq. (50).

From the right hand side in Eq. (60) follows

$$\frac{\partial U_0(T_1;v)}{\partial T_1} = U_0(T_1;v)\ \ (\sigma\,(v) - \beta\,(v)\ \ U_0(T_1;v)) \tag{63}$$

hence

$$U_0(T_1;v) = \frac{p_0\,(v)\ \ \sigma\,(v)\ \ \exp\left(\sigma\,(v)\ \ T_1\right)}{p_0\,(v)\ \ \beta\,(v)\ \ (\exp\left(\sigma\,(v)\ \ T_1\right) - 1) + \sigma\,(v)} \tag{64}$$

where

$$p_0\,(v) = \int_0^1 J(\xi;v)\ d\xi \tag{65}$$

is the right hand in Eq. (62). Finally, from Eq. (51), taking into account Eq. (52) and Eq. (64) we obtain

$$I_0(T_0,T_1,\xi;v) = \frac{\sigma\,(v)\ \ \exp\left(\sigma\,(v)\ \ T_1\right)}{p_0\,(v)\ \ \beta\,(v)\ \ (\exp\left(\sigma\,(v)\ \ T_1\right) - 1) + \sigma\,(v)} \times \tag{66}$$

$$\sum_{n=0}^{\infty} s_n\,(v) \exp(-n^2\pi^2\ T_0) \cos n\pi\xi\ .$$

If we return to the original variables in Eq. (43),

$$I(\tau,\xi;v) = \tag{67}$$

$$\frac{\sigma\,(v)}{p_0\,(v)\ \ \beta\,(v)\ \ (1 - \exp\left(-\sigma\,(v)\ \ \varepsilon\ \tau\right)) + \sigma\,(v)\ \ \exp\left(-\sigma\,(v)\ \ \varepsilon\ \tau\right)} \times$$

$$\sum_{n=0}^{\infty} s_n\,(v) \exp(-n^2\pi^2\ \tau) \cos n\pi\xi + O\,(\varepsilon)$$

which represents a uniform expansion of the solution as we mentioned before. From Eq. (67) can be expected that the limit

$$\lim_{\tau\to+\infty} I(\tau,\xi;v) = \frac{\sigma\,(v)}{\beta\,(v)} \tag{68}$$

takes place. In the bottom of the next subsection we will show the veracity of this assertion.

3.3. *Spatial average and spatially homogeneous stationary solution*

Here, we introduce the function

$$p(\tau; v) = \int_0^1 I(\tau, \xi; v)\, d\xi\ . \tag{69}$$

We will interpret this function as the *spatial average* **(SA)** *of the infected fraction of the population* in the considered bounded region. Note that if $I(\tau, \xi; v)$ is a solution to the heat equation Eq. (36) in which $\varepsilon = 0$ with the boundary conditions Eq. (38), then p is a constant function given by

$$p(\tau; v) = p_0(v) = \int_0^1 J(\xi; v)\, d\xi\ . \tag{70}$$

As we consider $\varepsilon > 0$, the average p have, in general, a non-constant with slow variation behavior. Nevertheless, even considering the nonlinearity in Eq. (36), we have a constant solution given by

$$\overline{I}(\tau, \xi; v) \underset{\text{def}}{=} \frac{\sigma(v)}{\beta(v)} \tag{71}$$

which represents a *spatially homogeneous stationary solution* to Eq. (36) with the Neumann conditions in Eq. (38). Note that this solution exists for a given VL v only under conditions in Eq. (26).

Now, let us estimate the SA in Eq. (69). To do so, we integrate in Eq. (36) and use the Eq. (38) to get:

$$p_\tau = \varepsilon \left(\sigma(v)\ p - \beta(v) \int_0^1 I^2(\tau, \xi; v)\, d\xi \right) \tag{72}$$

with the initial condition

$$p(0; v) = p_0(v)\ . \tag{73}$$

Further, from the bounds of $I(\tau, \xi; v)$ follow the inequalities

$$0 \leq \int_0^1 I^2(\tau, \xi; v)\, d\xi \leq \int_0^1 I(\tau, \xi; v)\, d\xi = p \tag{74}$$

hence, the solution to the equation

$$P_\tau = \varepsilon\left(\sigma(v) - \beta(v)\right) P = -\varepsilon\, \gamma(v)\ P \tag{75}$$

with initial condition $P(0;v) = p_0(v)$ is a subsolution to the problem in Eqs. (72-73). Then,

$$p_0(v)\ \exp\left(-\varepsilon\ \gamma(v)\ \tau\right) \leq p(\tau;v)\ . \tag{76}$$

On the other hand, as $\sigma(v) \leq 1-\gamma_0$ and $\gamma(v) \leq \gamma(v^*) = \beta(v^*)$, the right hand side of Eq. (72) can be bounded from above as

$$\begin{aligned}\text{right-hand} &\leq \varepsilon\left((1-\gamma_0)\ p + \left(\frac{(1-\gamma_0)\ v^*}{v_{\max}} - 1\right)\left(\int_0^1 I^2(\tau,\xi;v)\ d\xi\right)\right)\\ &\leq \varepsilon\ M\ p - \varepsilon \int_0^1 I^2(\tau,\xi;v)\ d\xi\\ &\leq \varepsilon\ M\ p - \varepsilon\left(\int_0^1 I(\tau,\xi;v)\ d\xi\right)^2 = \varepsilon\ M\ p - \varepsilon\ p^2\end{aligned} \tag{77}$$

where

$$M = (1-\gamma_0)\left(1 + \frac{v^*}{v_{\max}}\right) > 0\ . \tag{78}$$

Then, the solution to the equation

$$Q_\tau = \varepsilon\ M\ Q - \varepsilon\ Q^2 \tag{79}$$

with the initial condition $Q(0;v) = p_0(v)$ is a supersolution to Eqs. (72-73). Then,

$$p(\tau;v) \leq \frac{p_0(v)\ M}{p_0(v) - (p_0(v) - M)\ \exp(-\varepsilon\ M\ \tau)}\ . \tag{80}$$

Finally, we obtain

$$\begin{aligned}p_0(v)\ \exp\left(-\varepsilon\ \gamma(v)\ \tau\right) &\leq p(\tau;v)\\ &\leq \frac{p_0(v)\ M}{p_0(v) - (p_0(v) - M)\ \exp(-\varepsilon\ M\ \tau)}\ .\end{aligned} \tag{81}$$

Infortunately, Eq. (81) is not sufficient to describe the asymptotic behavior of the SA. To do so, let us define E and q by the following equations

$$I(\tau,\xi;v) = \frac{\sigma(v)}{\beta(v)} + E(\tau,\xi;v) \tag{82}$$

and

$$p(\tau;v) = \frac{\sigma(v)}{\beta(v)} + q(\tau;v) \tag{83}$$

where

$$q(\tau;v) = \int_0^1 E(\tau,\xi;v)\ d\xi\ . \tag{84}$$

Substituting Eq. (82) and Eq. (84) in Eq. (72) we obtain

$$\begin{aligned}\frac{d}{d\tau}q\,(\tau;v) &= -\varepsilon\left\{2\;\sigma\,(v)\;\;q\,(\tau;v)+\beta\,(v)\;\int_0^1 E^2(\tau,\xi;v)\;d\xi\right\} \\ &\leq -\varepsilon\left\{2\;\sigma\,(v)\;\;q\,(\tau;v)\right\}\end{aligned} \tag{85}$$

so, for any initial value $q\,(0;v)$ solution $q\,(\tau;v)$ to the above equation tends to zero as $\tau\to+\infty$. Hence, the SA verifies the relation

$$\lim_{\tau\to+\infty} p\,(\tau;v)=\frac{\sigma\,(v)}{\beta\,(v)}\;. \tag{86}$$

Similarly, we can conclude that

$$\lim_{\tau\to+\infty} I\,(\tau,\xi;v)=\frac{\sigma\,(v)}{\beta\,(v)}\;. \tag{87}$$

We resume the above relations in Eqs. (86-87) saying that the spatially homogeneous stationary solution $\overline{I}$ in Eq. (71) is orbitally stable. The quotient $\frac{\sigma(v)}{\beta(v)}$ represents, indistinctly, the *spatially homogeneous limit value* (**SHLV**) of the infected fraction and also the SHLV of the SA. Further, Eq. (87) ensures that the expected limit in Eq. (68) actually takes place.

3.4. *The fuzzy SHLV of the infected fraction*

As we have seen in the above section, an important role is played by the parameter

$$\widehat{I}\,(v)=\widehat{p}\,(v)=\frac{\sigma\,(v)}{\beta\,(v)} \tag{88}$$

which represents the SHLV when $\tau\to+\infty$. The set of values of v for which $\widehat{I}\,(v)>0$ and $\rho\,(v)>0$ is given in Eq. (26). The reader can check that $\widehat{I}\,(v)$ is a nonnegative strictly increasing function for VL values in Eq. (26) variying up to $(1-\gamma_0)$. Let us introduce a fuzzy set associated to the $\widehat{I}\,(v)$. Note that

$$\frac{\widehat{I}\,(v)}{1-\gamma_0} \tag{89}$$

is a fuzzy set, and we may introduce the fuzzy SHLV as

$$\widehat{I}^f=(1-\gamma_0)\;\;FEV\left(\frac{\widehat{I}\,(v)}{1-\gamma_0}\right) \tag{90}$$

with respect to the fuzzy measure in Eq. (20). Let us now determine the FEV in Eq. (90). To do so, we need first to introduce some notations. To simplify the formulae we will consider $v_{\max} = \overline{v} + \delta$. If it is not the case, in the formulae in this section the reader should change $v_{\max}$ by $\overline{v} + \delta$.

Let v_α the VL defined by the relation

$$\frac{\widehat{I}\left(v_\alpha\right)}{1-\gamma_0} = \alpha \tag{91}$$

and, the function

$$H\left(\alpha\right) = \mu\left\{ v \left| \frac{\widehat{I}\left(v\right)}{1-\gamma_0} \geq \alpha \right.\right\} = \sup_{v_\alpha \,\leq\, v \,\leq\, v_{\max}} \rho\left(v\right) \tag{92}$$

defined for $0 \leq \alpha \leq 1$. So, $H\left(0\right) = 1$ and $H\left(1\right) = \rho\left(v_{\max}\right)$. Let us select α such that

$$0 \leq \alpha \leq \frac{1}{1-\gamma_0}\widehat{I}\left(v_M\right) = \frac{v_M}{v_{\max}} \tag{93}$$

or, correspondingly,

$$v^* \leq v_\alpha \leq v_M\ . \tag{94}$$

For this range of values of the VL we have

$$\widehat{I}\left(v\right) = 1 - \frac{\left(v_{\max} - \left(1-\gamma_0\right)\ v\right)\ \left(v_M - v_{\min}\right)}{v_{\max}\left(v - v_{\min}\right)}\ . \tag{95}$$

Then, from Eqs. (2-3) and the definition of v_α in Eq. (91) follows

$$v_\alpha = v_{\max}\left\{\frac{v_M - \alpha\ \left(1-\gamma_0\right)\ v_{\min}}{\left(1-\gamma_0\right)\left(v_M - v_{\min} - \alpha\ v_{\max}\right)}\right\}. \tag{96}$$

Further, if

$$\frac{v_M}{v_{\max}} \leq \alpha \leq \frac{1}{1-\gamma_0}\widehat{I}\left(v_{\max}\right) = 1 \tag{97}$$

or, correspondingly,

$$v_M \leq v_\alpha \leq v_{\max} \tag{98}$$

then

$$\widehat{I}\left(v\right) = \frac{\left(1-\gamma_0\right)\ v}{v_{\max}} \tag{99}$$

and follows the expression

$$v_\alpha = \alpha\ v_{\max}. \tag{100}$$

We shall see one of the following situations, according to the relative positions of the distinguish values of the VL.

3.4.1. *Weak VL*

This is the case when

$$\overline{v} + \delta < v^*. \tag{101}$$

Then follows $H(\alpha) = 0$ for all α, and $\widehat{I}^f = 0$.

3.4.2. *Strong VL*

Now we consider the relations

$$v_M < \overline{v} - \delta < \overline{v} + \delta < v_{\max}. \tag{102}$$

Due to Eq. (99) we conclude:

- If
$$0 \leq \alpha \leq \frac{1}{1-\gamma_0}\widehat{I}(\overline{v}) = \frac{\overline{v}}{v_{\max}} \tag{103}$$
then $H(\alpha) = 1$.
- If
$$\frac{\overline{v}}{v_{\max}} < \alpha \leq \frac{1}{1-\gamma_0}\widehat{I}(\overline{v}+\delta) = \frac{\overline{v}+\delta}{v_{\max}} \tag{104}$$
then $H(\alpha) = \rho(v_\alpha)$.
- If
$$\frac{\overline{v}+\delta}{v_{\max}} < \alpha \leq \frac{1}{1-\gamma_0}\widehat{I}(v_{\max}) = 1 \tag{105}$$
then $H(\alpha) = 0$.

From the above relation follows a sandwich estimate for the FEV

$$\frac{\overline{v}}{v_{\max}} < FEV\left(\frac{\widehat{I}(v)}{1-\gamma_0}\right) < \frac{\overline{v}+\delta}{v_{\max}} \tag{106}$$

so,

$$(1-\gamma_0)\frac{\overline{v}}{v_{\max}} < \widehat{I}^f < (1-\gamma_0)\frac{\overline{v}+\delta}{v_{\max}}. \tag{107}$$

From the continuity of H follows that the solution to

$$H(\alpha) = \alpha \tag{108}$$

will be the FEV. From Eq. (92) and Eq. (100) follows,

$$\widehat{I}^f = (1-\gamma_0)\ FEV\left(\frac{\widehat{I}(v)}{1-\gamma_0}\right) = (1-\gamma_0)\ \frac{\overline{v}+\delta}{v_{\max}+\delta}\ . \tag{109}$$

3.4.3. *Middle VL*

In this case we are assuming

$$v^* < \overline{v} - \delta < \overline{v} + \delta < v_M. \tag{110}$$

From Eq. (95) follows:

- If

$$\begin{aligned} 0 \le \alpha &\le \frac{1}{1-\gamma_0}\widehat{I}\left(\overline{v}\right) \\ &= \frac{1}{(1-\gamma_0)}\left\{1 - \frac{\left(v_{\max} - (1-\gamma_0)\,\overline{v}\right)\,\left(v_M - v_{\min}\right)}{v_{\max}\left(\overline{v} - v_{\min}\right)}\right\} \end{aligned} \tag{111}$$

  then $H(\alpha) = 1$.
- If

$$\frac{1}{1-\gamma_0}\widehat{I}\left(\overline{v}\right) < \alpha$$

$$\begin{aligned} &\le \frac{1}{1-\gamma_0}\widehat{I}\left(\overline{v}+\delta\right) \\ &= \frac{1}{(1-\gamma_0)}\left\{1 - \frac{\left(v_{\max} - (1-\gamma_0)\,\left(\overline{v}+\delta\right)\right)\,\left(v_M - v_{\min}\right)}{v_{\max}\left(\overline{v}+\delta - v_{\min}\right)}\right\} \end{aligned} \tag{112}$$

  then $H(\alpha) = \rho(v_\alpha) = 1 - \frac{v_\alpha - \overline{v}}{\delta}$.
- If

$$\frac{1}{1-\gamma_0}\widehat{I}\left(\overline{v}+\delta\right) < \alpha \le \frac{1}{1-\gamma_0}\widehat{I}\left(v_{\max}\right) = 1 \tag{113}$$

  then $H(\alpha) = 0$.

From Eq. (111) and Eq. (112) follows a sandwich estimate for the FEV, so

$$1 - \frac{\left(v_{\max} - (1-\gamma_0)\,\overline{v}\right)\,\left(v_M - v_{\min}\right)}{v_{\max}\left(\overline{v} - v_{\min}\right)} < \widehat{I}^f$$

$$< 1 - \frac{\left(v_{\max} - (1-\gamma_0)\,\left(\overline{v}+\delta\right)\right)\,\left(v_M - v_{\min}\right)}{v_{\max}\left(\overline{v}+\delta - v_{\min}\right)}. \tag{114}$$

Considering the continuity of H follows that $FEV\left(\frac{\widehat{I}(v)}{1-\gamma_0}\right)$ is the fixed point

$$H(\alpha) = \alpha \tag{115}$$

hence, from Eq. (92) and Eq. (96) follows that a will be the positive root of the second degree equation

$$A\,\alpha^2 + B\,\alpha + C = 0 \tag{116}$$

where

$$A = -v_{\max}\,\delta\,\,(1-\gamma_0) < 0 \tag{117}$$

$$B = (1-\gamma_0)\,\,((v_M - v_{\min} + v_{\max})\,\,\delta + v_{\max}\,(\overline{v} - v_{\min})) > 0 \tag{118}$$

$$C = v_{\max}\,v_M - (1-\gamma_0)\,\,(v_M - v_{\min})\,\,(\overline{v} + \delta) > 0 \tag{119}$$

and

$$\widehat{I}^f = (1-\gamma_0)\,\,FEV\left(\frac{\widehat{I}\,(v)}{1-\gamma_0}\right)\,. \tag{120}$$

Note that the estimates for the FEV follow from the inequalities

$$\frac{1}{1-\gamma_0}\widehat{I}\,(\overline{v}) < \widehat{I}^f < \frac{1}{1-\gamma_0}\widehat{I}\,(\overline{v}+\delta) \tag{121}$$

so, from the continuity and injectiveness we conclude, in any case, the existence of a VL value $\widetilde{v}$,

$$\overline{v} < \widetilde{v} < \overline{v} + \delta \tag{122}$$

defined by the relation

$$\frac{1}{1-\gamma_0}\widehat{I}\,(\widetilde{v}) = \widehat{I}^f\,. \tag{123}$$

4. Conclusions

Here we consider a model with heterogeneity in the VL and with spatially dependent infected population. The role played by the threshold number $R_0\,(v)$ and its fuzzy integral R_0^f in [5] is now played by the number $\widehat{I}\,(v)$ and the corresponding $\widehat{I}^f$. Of course, there is an algebraic connection between $R_0\,(v)$ and $\widehat{I}\,(v)$, and the condition there $R_0\,(v) > 1$ is transformed here into the condition $\widehat{I}\,(v) > 0$. Nevertheless, there is no simple algebraic connection between the other two fuzzy integral numbers. One should note that if $\widehat{I}\,(v) > 0$ and $J\,(x;v) > 0$ for at least one value of x then, in spite of how less be the average $p_0\,(v) = \int_0^1 J(\xi;v)\,\,d\xi$ this average will growth up to its limit value $\widehat{I}\,(v)$. For a proper evaluation of the heterogeneity we shall substitute $\widehat{I}(v)$ by its fuzzy integral $\widehat{I}^f$. As the asymptotic limits were obtained for the solution and for the average, the reader can interpret the results even if the initial condition is not smooth but a generalized function, like for example a δ- function.

We remark that, due to the homogeneous Neumann conditions in Eq. (25) (or in Eq. (38)) which reflects an actual quarantine situation for the population inside the considered region, it is expected that the evolution of this population behaves like in the spatially homogeneous population model, at least for large values of time. But the actual situation is a little bit more complex, and might exist at the same time, spatial coordinates in which the infected population is locally very much greater, or locally very much lower, than the infected population predicted in the ODE model. Finally, we remark that, in the real situation, the spatial variable is mostly bidimensional, but in spite of this fact, it is not difficult to see that the main results about the fuzzy integral $\widehat{I}^f$ remains valid.

References

1. Georgescu, A.:"**Asymptotic Treatment of Differential Equation**". Chapman & Hall, London. (1995)
2. Kermack, W.O.; McKendrick, A.G.: Contribution to the mathematical theory of epidemics, Roy. Stat. Soc. J., 115, 700-721. (1927)
3. Edelstein-Keshet, L.:"**Mathematical Models in Biology**", Birkhauser, NY. (1988)
4. Murray, J.D.:"**Mathematical Biology**", Biomathematics Texts, Springer-Verlag, Berlin. (1993)
5. Barros, L.C.; Bassanezi, R.C.; Oliveira, R.Z.G.; Leite, M.B.F.: "A disease evolution model with uncertain parameters". Joint 9th IFSA World Congress and 20th NAFIPS International Conference- VIII IEEE International Conference of Fuzzy Systems, Vancouver, pg.1626-30. (August 2001)
6. Barros, L.C.; Leite, M.B.F.; Bassanezi, R.C.: "The SI Epidemiological Models with a Fuzzy Transmission Parameter", A International Journal Computers & Mathematics with Applications, Pergamon Press, ElsevierScience Ltd., pp 1619 -1628. (2003)
7. Sugeno M.: "Fuzzy measures and fuzzy integrals: A survey", in **Fuzzy Automata and Decission Processes**, Gupta, Saridis, Gaines (eds.), 89-102. (1977)
8. Sugeno M.: "**Theory of Fuzzy Integrals and Its Applications**", Doctoral Thesis, Tokyo Institute of Technology. (1974)
9. Shigesada, N.:"Spatial Distribution of Dispersing Animals in Heterogeneous Enviroments", Kyoto University, Japan. (1982)
10. Díaz, L.A.; Scapin, M.I.: "Dispersão Rápida/ Crescimento Lento em Dinâmica Populacional", Proceedings del Segundo Congreso Latinoameri-

cano de Biomatemática, Eds: R.C. Bassanezi, G.L. Diniz, Campinas, Brasil, Oct.29-Nov 2, pp203-208. (2002)

STRUCTURE PREDICTION OF ALPHA-HELICAL PROTEINS

SCOTT R. MCALLISTER AND CHRISTODOULOS A. FLOUDAS

Department of Chemical Engineering
Princeton University
Princeton, NJ 08544
E-mail: floudas@titan.princeton.edu

Within the field of protein structure prediction, the packing of α-helical proteins has been one of the more difficult problems. The use of distance constraints and topology predictions is shown to be highly useful for reducing the conformational space that must be searched by deterministic algorithms to find a protein structure of minimum conformational energy. We present a novel first principles framework to predict the structure of α-helical proteins that includes three main stages. These stages include a novel optimization model for generating interhelical distance restraints between α-helices, the analysis and clustering of loop structures with flexible stems, and finally the prediction of tertiary structure using a hybrid optimization algorithm. This approach does not assume the form of the helices, so it is applicable to all α-helical proteins, including helices with kinks and irregular helices. The interhelical contact prediction model was evaluated on 5 proteins, where it identified an average contact distance below 10.0 Å for the entire set. The proposed overall framework was applied to a 63 residue α-helical bundle (1r69) with 5 helices.

1. Introduction

The protein folding problem, in its most basic form, is the prediction of the three-dimensional structure of a protein from its primary amino acid sequence. Despite the existence of more than 30,000 experimentally-determined three-dimensional structures in the Protein Data Bank (PDB),[1] there are numerous protein sequences that have yet to be studied or cannot be studied due to limitations of the current experimental techniques. The computational task requires searching through a vast conformational space for the native structure. First principles protein folding prediction approaches are limited by the overwhelming number of possible conformations accessible to the polypeptide backbone and complex representations of interatomic interactions.[2] These challenges can only be met through the application of powerful algorithms, such as the αBB branch-and-bound method for optimization used in this work, [3,4,5] along with experimentally accurate information and models.

1.1. *α-Helical Protein Structure*

Helical proteins with no β-sheets are the focus of this chapter. The right-handed α-helix has 3.6 residues per turn and a translation per residue of 1.50 Å. Thus, there is a rotation of 100 degrees per residue in a perfect helix, and a helical wheel representation of the α-helix demonstrates that residues i, $(i + 3)$, $(i - 3)$, $(i - 4)$, and $(i + 4)$ will all be on the same side of the helical wheel. Many α-helices are amphipathic, meaning that they have mainly hydrophobic side chains along one side of the helical cylinder and polar residues along the remainder of the surface. The backbone carbonyl oxygen of each residue hydrogen bonds to the backbone amino group of the fourth residue along the chain, forming bonds that are about 2.86 Å long, quite straight, and nearly parallel to the helical axis in the classical α-helix. Yet, the exact hydrogen-bond geometry may vary in folded proteins, depending upon their environment.[6] While α-helices are very ordered structurally, isolated α-helices in aqueous solution are usually only marginally stable, if at all. Helices are formed very rapidly, within time spans of 10^5 to 10^7 seconds, but their unraveling is just as quick.[6] Once these helices form, it is their hydrophobic packing into a compact, globular shape that gives a helical protein its stability.

1.2. *Previous Research and the Proposed Method*

Ab initio methods have recently received increased attention in the prediction of loop structures. Loops exhibit greater structural variability than secondary structure elements due to their surface exposure and the relatively few contacts they have with the remainder of the structure. One approach built candidate loop structures from smaller fragments, then subjected them to minimization by simulated annealing with a heuristic scoring function combining sequence similarity, secondary structure similarity for loop stems, and geometric fit into the overall protein structure.[7] Deane and Blundell (2001) developed a consensus method that combines a database of loops from known structures with a database of representative computer-generated fragments to identify loop structures that fit into a given protein structured.[8] Fiser et al. (2000) used a combination of a physical energy function and a scoring function that takes statistical preferences for dihedral angles and non-bonded atomic contacts into account. The resulting energy function is minimized with a combination of local optimization, molecular dynamics and simulated annealing. Zhang et al. (2003) developed a new statistical potential that can quantitatively predict the likelihood of a residue to be buried.[9]

De Bakker and coworkers generated candidate loop geometries by sampling protein backbone angle distributions constructed from known loop structures.[10,11] After filtering steric clashes and poor fitting conformers, clustering approaches are used to classify candidate loops by similar geometry. Jacobson et al. (2004) developed a dihedral angle sampling approach where loop structures are discarded if they contain steric clashes, if insufficient space exists for side chains, if loops travel too far away from the remainder of the protein, or if loops ends do not fit into the remainder of the protein. A K-means clustering analysis[12] is then applied to pick representative conformers prior to side chain optimization. Forrset and Woolf (2004) combined Monte Carlo sampling and multi-temperature molecular dynamics to generate sets of conformations that are close to the native conformation.[13] Mönnigman and Floudas (2005) investigated the loop structure prediction problem with flexible stems where neither the loop anchor geometry nor the overall protein geometry are known.[14] They used a dihedral angle sampling approach to build up ensembles, which were subsequently optimized with an atomistic level energy function and iteratively clustered to identify and discard conformers far from the native structure.

Methods for the packing of α-helical proteins have also received increasing attention. Ortiz et al. (1998) have used multiple sequence alignments to derive distance constraints in order to guide Monte Carlo searches for the native structures of small proteins.[15] The PHD secondary structure prediction algorithm that incorporates multiple sequence information and employs neural network theory provided their secondary structure restraints. This method employed database-derived information in the form of secondary structure prediction and distance constraints. Predicted tertiary contacts were extracted on the basis of evolutionary information contained in the multiple sequence alignments. They demonstrated that the incorporation of distance restraints into a tertiary structure prediction algorithm could allow for the assembly of native-like structures using a lattice-based reduced protein model.

Predicting which residues in the helices of helical proteins contact one another was a strategy for structure prediction also pursued by Huang et al. (1999).[16] They divided the protein folding problem into two subproblems: the generation of a fold library of possible protein structures, and the selection of the best fold in the given library. A distance geometry procedure with input of generic distance restraints of 5.0 to 11.0 Å between contacting helical residues was used to generate a library of 500 possible structures for each of 11 small helical proteins. Even though they only specified that the predicted contacts lie between 5.0 and 11.0 Å, distances commonly found in small helical proteins, Huang et al. (1999) found that these distance constraints were sufficient to produce many native-like folds within their fold libraries for each protein.

Zhang, Hou, and Kim (2002) described one procedure for predicting the protein folding of α-helical proteins.[2] They combined a torsion angle dynamics (TAD) program with a secondary structure prediction program, whose output becomes the input for a contact prediction algorithm. The TAD program uses predicted tertiary contacts and secondary structural states in order to predict three-dimensional structures. Observing that helix packing interfaces in globular proteins consist of contact patches involving residues regularly spaced in the sequence, they aimed at identifying interhelical contact patches directly with a scoring scheme. The individual residue contacts are between two separate triangular patches on interacting helices. Zhang, Hou, and Kim (2002) were able to produce native-like folds for the majority of their 24 targets, with 14 of the 15 medium-sized proteins (80-100 residues long) having predicted structures within 6.5 Å root-mean-square deviation, RMSD. For small proteins, a coordinate RMSD of about 6 Å relative to the native structure has been suggested as a target value.[17] It is important to note that these RMSD values are based only upon the helical residues and not upon loop residue predictions.

Fain and Levitt (2001) proposed a method to sample the entire conformational space of helical proteins.[18] This approach relied on the enumeration of all the geometrically possible three dimensional arrangements of the helices subject to a set of rules extracted from a database of existing structures. A further analysis through a refinement procedure showed this algorithm had a small, but positive, effect on the number of near native conformers.

Besides these focused techniques, α-helical protein structures can also be predicted through more general protein structure prediction algorithms. These approaches are extensively outlined in two recent reviews.[19,20]

2. Theory and Modeling

The first principles folding prediction of helical proteins in the absence of *a priori* distance information is of interest in this paper. The interhelical contact prediction model of this chapter predicts interhelical hydrophobic-to-hydrophobic residue contacts between α-helices of globular proteins in order to generate bounds on these distances. A novel approach to clustering loops with flexible stems is used to select likely candidate loops from an ensemble generated by sampling techniques.[14] Together, these two sets of restrictions are used to narrow the conformational search space that the hybrid optimization algorithms must consider.

2.1. *Interhelical Contact Prediction Model*

Given the locations of helices in a protein's primary sequence, the first step in the proposed method is to predict the interhelical residue contacts. This is done in order to impose distance constraints upon such contacts in the tertiary structure prediction section of the model.

To accomplish this task, two mixed-integer linear programming (MILP) optimization problems were formulated. A MILP problem consists of a linear objective function to be maximized or minimized subject to linear constraints. Additionally, the objective function and the constraints are written in terms of binary variables only. The constructed MILP problems are guaranteed to have global optimal solutions in their feasible regions, but there may be multiple global optimum solutions.

The first MILP, denoted as the Level 1 Model, identifies a set of the most probable interhelical PRIMARY contacts. These hydrophobic contacts are expected to be the strongest interacting points between helices and be the driving force behind the assembly of α-helical proteins. The PRIMARY contacts selected by the Level 1 Model are stabilized by the presence of the most probable WHEEL contacts predicted in the Level 2 Model. This second model has the added benefit of distinguishing between equally likely results from the Level 1 model.

The following sections first cover the probability generation method, then explain the model in detail, presenting the objective functions or constraints following each of the appropriate discussions.

2.1.1. *Probability Generation and Probability Sets*

For the purpose of calculating the PRIMARY probabilities, two helices h_m and h_n of a given protein were considered to interact if they had a contact between a pair of residues i in h_m and j in h_n such that this PRIMARY distance was greater than or equal to 4.0 Å and less than 10.0 Å. Such contacts were designated PRIMARY to distinguish them from WHEEL contacts, which are discussed below. Once the PRIMARY probabilities were calculated, interactions or contacts were predicted by the model based on these PRIMARY probabilities, without knowledge of the actual distances between the residues i and j. In this way, the model remains a true first principles method.

For a parallel helix-to-helix interaction, possible WHEEL contacts include the following residue combinations if they fall within the boundaries of the two helices: $(i+3)$ to $(j+3)$; $(i+3)$ to $(j+4)$; $(i+4)$ to $(j+3)$; $(i+4)$ to $(j+4)$; $(i-3)$ to $(j-3)$; $(i-3)$ to $(j-4)$; $(i-4)$ to $(j-3)$; and $(i-4)$ to $(j-4)$. For an antiparallel helix-to-helix interaction, the following are the possible WHEEL contacts if they fall within the helical boundaries: $(i+3)$ to $(j-3)$, $(i+3)$ to $(j-4)$; $(i+4)$ to $(j-3)$; $(i+4)$ to $(j-4)$; $(i-3)$ to $(j+3)$; $(i-3)$ to $(j+4)$; $(i-4)$ to $(j+3)$; and $(i-4)$ to $(j+4)$. Figure 1 shows the PRIMARY contacts and one side of the WHEEL contacts for an antiparallel helical interaction. Similarly, two residues k and l that were WHEEL residues of i and j, respectively, were considered to interact if the WHEEL distance between them was greater than or equal to 4.0 and less than 12.0 Å. Once the WHEEL probabilities were calculated, the predictions of WHEEL contacts were made without reference to actual (k, l) distance pairs.

The formulation of the problem as a set of PRIMARY and WHEEL contacts provides a significant advantage over other methods. By selecting a set of interhelical point contacts within a specified distance range, there is no need to make assumptions about the form of the helices. Instead of representing a helix as a simple cylinder, this model allows for irregular helices, including those that bend or kink.

A C language program was written to calculate the probabilities that a given PRIMARY contact for the hydrophobic pair (i, j) was one of each of the unique combinations of hydrophobic pairs. These were segregated into parallel and antiparallel cases and constructed by counting the total number of times a specific hydrophobic-to-hydrophobic minimum interhelical distance occurred within the 4.0 to 10.0 Å distance range. This distance was the minimum distance for all hydrophobic pairs, but not necessarily the minimum distance between the two helices. Each PRIMARY probability equals the proportion of the entire number of helix-to-helix interactions that occurred for a given pair of hydrophobic residues in either the par-

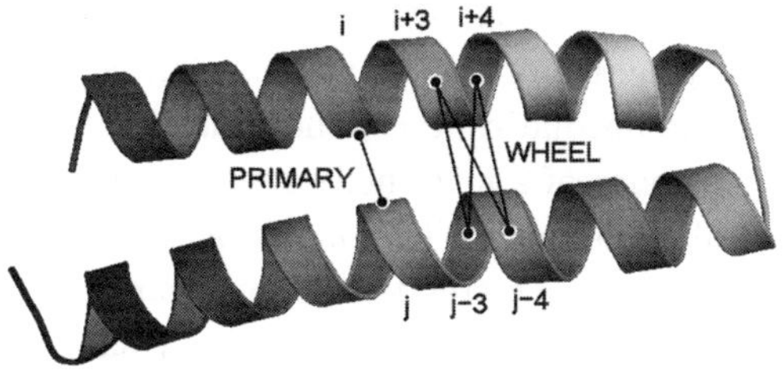

Figure 1. Two interacting α-helices in the example protein 1rop (PDB). The helices here interact in an antiparallel manner, where hydrophoibc residues i and j form a PRIMARY contact, and the residues $(i+3)$, $(i+4)$ can each interact with $(j-3)$, $(j-4)$ to form WHEEL contacts if both residues of a given pair are hydrophobic. The helices can also interact in a parallel manner, leading to the opposite potential WHEEL contacts. This protein figure was created with PyMol.[21]

allel or antiparallel cases. All the distances in this paper refer to C^{α}-C^{α} distances.

Given that two α-helices have either a parallel or an antiparallel interaction with a corresponding hydrophobic-to-hydrophobic minimum interaction distance within 4.0 to 10.0 Å between residues i and j, the conditional probability that the residues on the same side of the helical wheel as i and j and in their vicinity form any hydrophobic-to-hydrophobic contact within 4.0 to 12.0 Å was determined. The WHEEL probabilities were developed by considering the number of hydrophobic-to-hydrophobic WHEEL contacts and the total number of possible WHEEL contacts for every specific helix h_m to helix h_n interaction individually. The probability for each interhelical (i,j) residue contact was then calculated by averaging over the total number of such (i,j) contacts after the entire database PDB set has been considered.

2.1.2. *Level 1 Model*

In the Level 1 problem, the binary variables y^a_{mn} and y^p_{mn} are defined as whether or not the helices h_m and h_n of the same protein interact in an antiparallel or a parallel fashion, respectively. In addition, the binary variables w^{mn}_{ij} are defined as active when the hydrophobic residue pair (i,j) forms a

PRIMARY contact, where i is in helix h_m and j is in helix h_n. The objective function of Level 1, Equation 1, corresponds to maximizing the number of the most likely hydrophobic contacts (i, j) for the given primary sequence by considering the product of the binary variable w_{ij}^{mn}, representing the existence of a residue-residue contact, the binary variables y_{mn}^a and y_{mn}^p representing the existence of a helix-helix contact, and the probability of a parallel or antiparallel contact, $p^p(i, j; m, n)$ or $p^a(i, j; m, n)$, respectively. These probabilities are those estimated earlier from the database PDB set that the specific hydrophobic pair (i, j) forms a PRIMARY contact given that h_m and h_n have an antiparallel or parallel interaction, respectively.

$$\max \quad \sum_m \sum_n y_{mn}^a \cdot \sum_i \sum_j w_{ij}^{mn} \cdot p^a(i, j; m, n) + \sum_m \sum_n y_{mn}^p \cdot \sum_i \sum_j w_{ij}^{mn} \cdot p^p(i, j; m, n) \tag{1}$$

$$y_{mn}^a, y_{mn}^p, w_{ij}^{mn} = \{0, 1\} \tag{2}$$

This objective function is nonlinear due to the products of binary variables that result. Let us introduce the variables Z_{mn}^a and Z_{mn}^p as shown in Equations 3 and 4.

$$Z_{mn}^a = \sum_i \sum_j w_{ij}^{mn} \cdot p^a(i, j; m, n) \quad \forall (m, n) \tag{3}$$

$$Z_{mn}^p = \sum_i \sum_j w_{ij}^{mn} \cdot p^p(i, j; m, n) \quad \forall (m, n) \tag{4}$$

The objective function can then be linearized to the form of Equation 5 through standard optimization techniques by introducing a second pair of variables, U_{mn}^a and U_{mn}^p, as defined in Equations 6 and 7.

$$\max \sum_m \sum_n U_{mn}^a + \sum_m \sum_n U_{mn}^p \tag{5}$$

$$U_{mn}^a = Z_{mn}^a \cdot y_{mn}^a \tag{6}$$

$$U_{mn}^p = Z_{mn}^p \cdot y_{mn}^p \tag{7}$$

Equations 8 through 11 are equivalent to the definitions given by Equations 6 to 7 but used to keep the system linear.[22] The upper and lower bound constraints on U_{mn}^a and U_{mn}^p result from their definitions and are necessary in order to replace the nonlinear products of $Z_{mn}^a \cdot y_{mn}^a$ and $Z_{mn}^p \cdot y_{mn}^p$. When there is an antiparallel contact, $y_{mn}^a = 1$, and Equation 8 specifies that $U_{mn}^a = Z_{mn}^a$, in agreement with Equation 6. If no antiparallel contact is predicted, then $y^a_{mn} = 0$, and Equation 10 becomes active. Similarly, Equations 10 and 11 for parallel helical interactions are equivalent to the definition given by Equation 7.

$$Z^a_{mn} - (Z^a_{mn})^U \cdot (1 - y^a_{mn}) \leq U^a_{mn} \leq Z^a_{mn} \qquad \forall (m,n) \tag{8}$$

$$0 \leq U^a_{mn} \leq (Z^a_{mn})^U \cdot y^a_{mn} \quad \forall (m,n) \tag{9}$$

$$Z^p_{mn} - (Z^p_{mn})^U \cdot (1 - y^p_{mn}) \leq U^p_{mn} \leq Z^p_{mn} \qquad \forall (m,n) \tag{10}$$

$$0 \leq U^p_{mn} \leq (Z^p_{mn})^U \cdot y^p_{mn} \quad \forall (m,n) \tag{11}$$

Every hydrophobic amino acid of helix h_m, i, can have at most one PRIMARY contact with another hydrophobic amino acid of helix h_n, j, and this is given by Equation 12.

$$\sum_{j;j>i} w_{ij} + \sum_{j;j<i} w_{ij} \leq 1 \tag{12}$$

Equation 13 indicates an upper limit, *max_contact*, on the number of PRIMARY contacts that two interacting helices h_m and h_n can have specified, although more hydrophobic-to-hydrophobic contacts could physically exist in the solution structure.

$$\sum_i \sum_j w^{mn}_{ij} \leq max\_contact \cdot \left(y^{mn}_a + y^{mn}_p\right) \quad \forall (m,n) \tag{13}$$

For every helix h_m, Equation 14 establishes the maximum number of PRIMARY contacts that can be specified involving h_m. If helix h_m has either zero or one hydrophobic residues, then the parameter $counth(m)$ is set to zero or one respectively. On the other hand, if helix h_m has at least two hydrophobic residues that are not WHEEL residues to each other, then the parameter is set to two for these residues. Two was taken as an upper limit in order to require that the model choose the most critical PRIMARY contacts.

$$\sum_n (y^a_{mn} + y^p_{mn}) \leq counth(m) \quad \forall m \tag{14}$$

There is a minimum number of residues necessary to allow for the two turns required for a parallel interaction. The model ensures that any two helices that are less than six residues apart can only interact in an antiparallel fashion, and likewise an antiparallel interaction requires at least one intervening residue between the helices.

Equation 15 simply states that two helices h_m and h_n can either interact in a parallel or antiparallel fashion, but not both.

$$y^a_{mn} + y^p_{mn} \leq 1 \quad \forall (m,n) \tag{15}$$

A direct link between the w_{ij} PRIMARY contact variables and the y^a_{mn} and y^p_{mn} helical interaction variables is provided by Equations 16 and 17. The constraints are especially important in light of the integer cuts, since they prevent a helical interaction if all of the potential hydrophobic pairs between two helices have been disallowed from the solution space.

$$w_{ij}^{mn} \leq y^a_{mn} + y^p_{mn} \tag{16}$$

$$y^a_{mn} + y^p_{mn} - \sum_i \sum_j w_{ij}^{mn} \leq 0 \tag{17}$$

Equation 18 requires that (i', j') cannot be a WHEEL contact to the PRIMARY contact (i, j) and limits the size of kinks in the protein backbone that result from a differing separation between (i and i') and (j and j').

$$w_{ij}^{mn} + w_{i'j'}^{mn} \leq 1 \quad \forall (i, i', j, j') : \ |\text{diff}(i, i')| - |\text{diff}(j, j')| \leq 2 \\ \text{or either } |\text{diff}(i, i')| \leq 5 \text{ or } |\text{diff}(j, j')| \leq 5 \tag{18}$$

where $\text{diff}(i, i')$ refers to the difference in sequence numbering between i and i'.

Equation 19 states that if a PRIMARY contact (i, j) occurs, then none of the WHEEL residues for i can also be part of a PRIMARY contact themselves.

$$w_{kl}^{mn} + w_{ij}^{mn} \leq 1 \quad \forall (i, j, k, l) : \ i, k \in m \text{ and} \\ k \text{ is in a WHEEL position of } i \tag{19}$$

Equations 20 and 21 indicate that if a contact is predicted to be parallel with PRIMARY contact (i, j), then both w_{ij}^{mn} and $w_{i'j'}^{mn}$ are allowed to be active. Residues i' and j' must satisfy $i' > i$ and $j' > j$ in the sequence numbering and be in the same helices as i and j, respectively. However, for residue $j'' < j$ in sequence numbering and in the same helix, only one of the variables w_{ij}^{mn} or $w_{ij''}^{mn}$ is allowed to be active.

$$w_{ij}^{mn} + w_{i'j'}^{mn} + y^a_{mn} \leq 2 \quad \forall (i, j, i', j') : \ i' > i, j' > j \text{ and} \\ |i' - i| < |j' - j| + 3 \text{ or} \\ |i' - i| > |j' - j| - 3 \tag{20}$$

$$w_{ij}^{mn} + w_{i'j'}^{mn} + y^p_{mn} \leq 2 \quad \forall (i, j, i', j') : \ i' > i, j > j' \text{ and} \\ |i' - i| < |j' - j| + 3 \text{ or} \\ |i' - i| > |j' - j| - 3 \tag{21}$$

Equations 22 and 23 specify that an overlap between two helices of at least two-thirds the length of the short helix is required for a PRIMARY contact pair (i, j) to be predicted.

$$w_{ij}^{mn} + y_{mn}^{a} \leq 1 \quad \text{if } m, n \text{ overlap} < 2/3 \text{ of shorter helix} \tag{22}$$

$$w_{ij}^{mn} + y_{mn}^{p} \leq 1 \quad \text{if } m, n \text{ overlap} < 2/3 \text{ of shorter helix} \tag{23}$$

Additional constraints are based upon transitive rules for the topology of three interacting helices. Equations 24 through 25 give these restrictions.

$$y_{mn}^{a} + y_{np}^{a} + y_{mp}^{a} \leq 2 \quad \forall (m, n, p) : m \neq n \neq p, nhel \geq 3 \tag{24}$$

$$y_{mn}^{p} + y_{np}^{a} + y_{mp}^{p} \leq 2 \quad \forall (m, n, p) : m \neq n \neq p, nhel \geq 3 \tag{25}$$

$$y_{mn}^{a} + y_{np}^{p} + y_{mp}^{p} \leq 2 \quad \forall (m, n, p) : m \neq n \neq p, nhel \geq 3 \tag{26}$$

$$y_{mn}^{p} + y_{np}^{p} + y_{mp}^{p} \leq 2 \quad \forall (m, n, p) : m \neq n \neq p, nhel \geq 3 \tag{27}$$

Equation 28 eliminates a number of helical interactions from the Level 1 solutions equal to the value of the parameter *subtract*. The *subtract* parameter effectively loosens the helical packing, which may be desired in predicting only the most essential and hopefully smallest distance contacts. In the general blind prediction method, the top ranking solutions should be compiled for each possible *subtract* parameter value from 0 to (*nhel*-1) and the structure refinement run on all these solutions to find the minimum energy structure.

$$\sum_{m} \sum_{n} (y_{m}^{a} n + y_{m}^{p} n) \leq \left(\sum_{m} counth(m)/2 \right) - subtract \tag{28}$$

Equations 1-28 result in a mixed-integer linear optimization problem (MILP), representing the Level 1 mathematical model.

2.1.3. *Level 2 Model*

The Level 2 MILP problem serves as a check on the ordering of the solutions found in Level 1. The Level 2 objective function, Equation 29, maximizes the most probable hydrophobic (k, l) WHEEL contacts based on the database PDB set. The parameters $p^{a}(k, l; i, j; m, n)$ and $p^{p}(k, l; i, j; m, n)$ give the probabilities that any hydrophobic (k, l) pair will occur on the same side of the helical wheel as a specific PRIMARY pair (i, j) for antiparallel or parallel orientations, respectively. The

Level 2 objective function, however, must be based upon not only the $p^a(k,l;i,j;m,n)$ and $p^p(k,l;i,j;m,n)$ probabilities, but also upon the probabilities $p^a(i,j;k,l;m,n)$ and $p^p(i,j;k,l;m,n)$ as shown in Equations 30 through 35. These latter probabilities treat the (k,l) WHEEL contacts as PRIMARY contacts themselves, and the values reflect the relative weights of the different (i,j) WHEEL contacts to these (k,l) PRIMARY contacts. This approximation allows for distinguishing between the possible (k,l) hydrophobic contacts that may be specified when there is a choice of more than one pair.

$$\max \sum_m \sum_n \sum_i \sum_j X^a_{ijmn} + \sum_m \sum_n \sum_i \sum_j X^p_{ijmn} \tag{29}$$

$$X^a_{ijmn} = \zeta^a_{mn;ij} \cdot y^a_{ijmn} \tag{30}$$
$$X^p_{ijmn} = \zeta^p_{mn;ij} \cdot y^p_{ijmn} \tag{31}$$
$$\zeta^a_{mn;ij} = \sum_k \sum_l w^{mn}_{kl;ij} \cdot [p^a(k,l;i,j;m,n) + p^a(i,j;k,l;m,n)] \tag{32}$$
$$\zeta^p_{mn;ij} = \sum_k \sum_l w^{mn}_{kl;ij} \cdot [p^p(k,l;i,j;m,n) + p^p(i,j;k,l;m,n)] \tag{33}$$
$$y^a_{ijmn} = w^{mn}_{ij} \cdot y^a_{mn} \tag{34}$$
$$y^p_{ijmn} = w^{mn}_{ij} \cdot y^p_{mn} \tag{35}$$

$$y^a_{mn}, y^p_{mn}, w^{mn}_{ij}, w^{mn}_{kl;ij}, y^a_{ijmn}, y^p_{ijmn} = \{0,1\} \tag{36}$$

The upper and lower bound constraint Equations 37 through 44 are similar to the constraints used in Level 1 to keep the problem linear.[22] Equations 45 and 46 state that a maximum of one WHEEL contact is allowed to be specified per primary contact.

$$w^{mn}_{ij} + y^a_{mn} - 1 \leq y^a_{ijmn} \leq w^{mn}_{ij} \quad \forall(m,n,i,j) \tag{37}$$
$$0 \leq y^a_{ijmn} \leq y^a_{mn} \quad \forall(m,n,i,j) \tag{38}$$
$$w^{mn}_{ij} + y^p_{mn} - 1 \leq y^p_{ijmn} \leq w^{mn}_{ij} \quad \forall(m,n,i,j) \tag{39}$$
$$0 \leq y^p_{ijmn} \leq y^p_{mn} \quad \forall(m,n,i,j) \tag{40}$$

$$\zeta^a_{mn;ij} - (p^a(k,l;i,j;m,n))^U \cdot (1 - y^a_{ijmn}) \leq X^a_{ijmn} \leq \zeta^a_{mn;ij} \tag{41}$$
$$0 \leq X^a_{ijmn} \leq y^a_{mn;ij} \tag{42}$$
$$\zeta^p_{mn;ij} - (p^p(k,l;i,j;m,n))^U \cdot (1 - y^p_{ijmn}) \leq X^p_{ijmn} \leq \zeta^p_{mn;ij} \tag{43}$$
$$0 \leq X^p_{ijmn} \leq y^p_{mn;ij} \tag{44}$$

$$\sum_k \sum_l w_{kl;ij}^{mn} \leq w_{ij}^{mn} \quad \forall (m,n,i,j) : y_{mn}^a = 1 \tag{45}$$

$$\sum_k \sum_l w_{kl;ij}^{mn} \leq w_{ij}^{mn} \quad \forall (m,n,i,j) : y_{mn}^p = 1 \tag{46}$$

Equations 29-46 represent the Level 2 mathematical model, a mixed-integer linear optimization problem (MILP).

2.2. *Loop Prediction*

As part of the improved ASTRO-FOLD framework detailed in Figure 2, the structure of the loop regions should be predicted prior to the tertiary structure prediction stage. Mönnigman and Floudas (2005) investigated the loop structure prediction problem with flexible stems, where neither the loop anchor geometry nor the overall protein geometry are known.[14] The first step of this approach uses dihedral angle sampling based on probability bins derived from a set of nonhomologous protein structures. After sampling, each of the loops is subject to a local optimization of the energy based on an atomistic level force field. Once a large ensemble of loop conformers (including information on the stem residues) has been generated, a new iterative approach to protein clustering is applied. After calculating the pairwise rmsd among all the ensemble of conformers, a threshold rmsd value is selected. The cluster size for each conformer is formed by determining the number of conformers that fall below the threshold rmsd relative to the selected conformer. Any conformer that does not meet the cluster size threshold is removed from the ensemble. This aim of this approach, unlike most clustering algorithms, is to identify and remove conformers that are likely to be far from the native structure, thus improving the ensemble. Once the clustering algorithm has completed, a hybrid selection strategy based upon the minimum energy, the minimum colony energy, and the largest cluster size at three stages of the clustering method is applied to the remaining ensemble. Information from these conformers is then included in the tertiary protein structure prediction stage.

2.3. *Tertiary Structure Prediction*

A key aspect of protein secondary structure is the patterned formations of local bonding. As a result, bounds can be set on the dihedral angles and distances within the protein. The helical residues are constrained with an i, $(i+4)$ C^α-C^α distance of 5.5 to 6.5 Å where i is a given helical residue

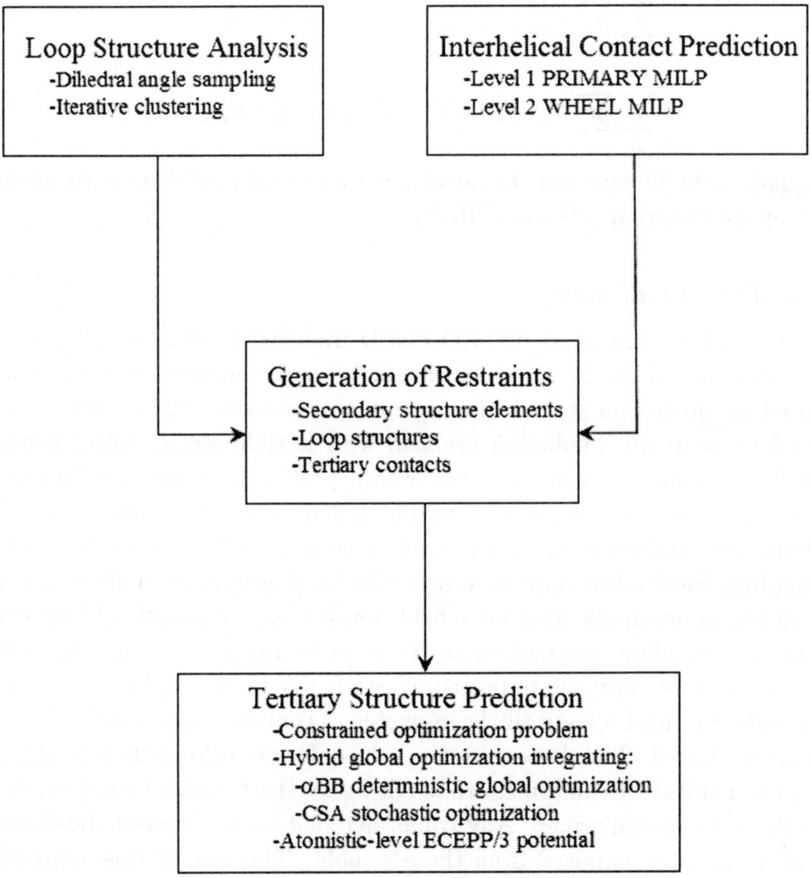

Figure 2. The proposed framework for the prediction of α-helical proteins. The loop structure analysis is combined with the interhelical contact model to develop restraints for the protein. These restraints are then input to a tertiary structure prediction algorithm to identify low-energy conformations.

and $(i+4)$ is the residue four places away in the primary sequence. This restraint represents the hydrogen bond that results from helix formation. For the residues that are predicted to be helical, the dihedral angles are restricted to [-85,-55] for ϕ and [-50,-10] for ψ.

The distance constraints for the predicted interhelical contacts bound the C^α-C^α distance to between 5.0 Å and 10.0 Å for the PRIMARY contacts. Similarly constraints are imposed on the predicted WHEEL contacts, requiring a distance in the range of 5.0 Å to 12.0 Å.

Finally, it is necessary to address the loop regions between secondary structure elements to develop additional restraints for use in the tertiary structure prediction. Once a loop conformer is selected for inclusion in the tertiary structure analysis, the dihedral angles for the loop residues of the predicted protein are restricted to fall with $\pm 20°$ of the dihedral angles of that conformer.

After the distance constraints and dihedral angle bounds have been established, the goal is to minimize the potential energy of the tertiary structure while satisfying all the constraints. The ASTRO-FOLD approach combines the deterministic αBB global optimization algorithm with a stochastic global optimization algorithm and a molecular dynamics approach in torsion angle space.[5,23] The basic formulation is the minimization of the force field energy over torsion angle space, subject to upper and lowering bounding constraints on these angles. Representing the model in torsion angles offers the benefit of significantly decreasing the size of the independent variable set, while only modestly increasing the model complexity.

The use of the αBB global optimization algorithm [24,25,26,27,28,29] guarantees convergence to the global minimum solution by a convergence of upper and lower bounds on the potential energy minimum. The upper bounds of this model result from local minimizations of the original nonconvex problem. Adding separable quadratic terms to the objective and constraint functions then produces a convex lower bounding function. This bounded problem can be branched over the variable space iteratively, fathoming portions of this space when the lower bound rises above the best upper bound.

The deterministic approach is exceptionally difficult due to the highly nonlinear force field. The application of torsion-angle dynamics methods can be used to rapidly identify feasible low energy conformers, significantly improving the performance of the αBB method. In addition, stochastic optimization methods may further improve the performance of the upper bounding approach of the formulation. One such hybrid global optimization method, described as Alternating Hybrids, has been recently introduced.[30,31] It combines the deterministic αBB approach with the stochastic approach of conformational space annealing.[32,33,34,35,36] Conformational space annealing balances genetic algorithm approaches of mutations and crossovers with simulated annealing to identify low energy conformers. The use of this Alternating Hybrid approach yields a much more efficient search for the native state, while still retaining the deterministic guarantees of convergence.[31]

The ASTRO-FOLD methodology has been successfully applied to a varied set of proteins throughout the range of small to medium-sized proteins.[23] The recent success of the ASTRO-FOLD method in a double blind prediction of a four-helix bundle reinforces the value of the approach.[37]

3. Results and Discussion

3.1. *Interhelical Contact Prediction Results*

The Level 1 and Level 2 MILP optimization problems were applied to 5 target proteins with known structures in the PDB, termed the test PDB set. For each of these proteins, only the primary amino acid sequence and the experimentally-determined locations of the helices were presented to the model. The model predicted the interhelical hydrophobic residue contacts between such helices using the PRIMARY and WHEEL probabilities developed from globular helical proteins. Therefore, only globular helical proteins with no β-sheets and two or more helices, each with at least two hydrophobic residues per helix, were tested. An upper limit of 14 Å was identified as a goal for the average distance corresponding to the contact predictions, since such a distance constraint would significantly improve the structure refinement.

The test PDB set has been selected such that there is no overlap with the database PDB set used to develop the probabilities. 1p68 is a representative left-turning four-helix bundle and 1r69 has 5 α-helices that form a compact hydrophobic interior. Both have been previously studied by Floudas and co-workers.[38,37] The proteins 1a1w (6 helices), 1b4f (5 helices), and 1hta (3 helices) were also selected due to their varied structures and biological importance.

For the reported best contact distance average, only the predictions that satisfied:

$$\text{number of contacts} \geq \text{number of helices} - 1 \tag{47}$$

were considered to ensure a reasonable number of contacts would be enforced for proteins with many helices.

The usefulness of the parametric analysis via the *subtract* parameter for smaller systems was also apparent for the protein 1hta. 1hta consists of a long central helix that has two short helices on either end. The original predictions for 1hta included interactions between helices 1 and 3, which corresponded to long-range contacts clearly not present in the experimental structure. The relaxation of helical packing dramatically improved the results (Table 1). The best average contact distance fell from 14.12 Å ± 8.36 Å SD for a *subtract* value of 0 to 9.72 Å ± 2.12 Å SD for a *subtract* value of 1 and a *max_contact* value of 2. However, even though setting the *subtract* value to 2 results in the lowest average contact distance, only one contact is predicted in this case versus 4 and 5 when *subtract* values of 1 and 0 were used, respectively. This demonstrates that increasing the *subtract* parameter does not always result in better solutions. The model should be run for all *subtract* values possible in the general case, allowing for multiple degrees of helical packing in finding the most appropriate contacts.

The best contact distance average solution for the protein 1p68 is given in Table 2. It is interesting to compare several of the top ranking solutions by iteration to the solution with the best average contact distance. As seen in Table 3, the first solution predicts antiparallel interactions between helices (1 and 2), (1 and 3), (2 and 4), and (3 and 4). Although this first solution is inconsistent with the experimental topology, the other three highlighted iterations in Table 3 predict helical interactions between (1 and 2), (1 and 4), (2 and 3), and (3 and 4). These prediction sets not only identify the correct experimental topology but additionally yield 15-16 interhelical contacts all within the desired 14 Å upper limit.

The protein 1r69 was tested in similar detail to the 4-helix bundle, and the best average contact distance prediction as well as the top-ranking solutions are given in Tables 2 and 4. The interaction of helices 4 and 5 is predicted by the three highest ranking solutions for a *subtract* value of 1. This is also reflected in the best average distance solution. The strong agreement across parameter values indicates that this experimentally-consistent helical interaction should be emphasized during the subsequent structure refinement. Both of the best predictions for 1r69 also indicate a definitive contact between helices 1 and 5, leading to average contact distances of 8.45 Å ± 2.38 SD and 8.24 ± 0.93 SD for *subtract* values of 1 and 0 respectively.

The proteins 1b4f-A and 1a1w are similar in size, with the former containing 5 helices in 74 residues and the latter 6 helices in 83 residues. Although their best distance averages are similar as shown in Table 1 and the specific contacts in Table 2, the model predicts many more contacts for the 1a1w system than for the 1b4f-A system. For example, for a *subtract* value of 2, the model specifies 12 contacts for 1a1w distance constraints, while it only predicts 4 for 1b4f-A. An examination of the helical sequences reveals the reasoning for this difference. The helices of 1a1w are rich with leucine, containing a total of 19 leucine residues. The helices of 1b4f-A contain only

Table 1. Summary of the best average contact distance predictions for each parametrization of the five proteins that were the focus of the test set.

PDB Name	subtract, max_contact	Number of Contacts	Distance Average	Distance Std. Dev.	Iteration Rank
	0 , 1	4	13.15	7.90	6
	1 , 1	2	8.75	0.49	11
1hta	2 , 1	1	8.40	0.00	1
	0 , 2	5	14.12	8.36	13
	1 , 2	4	9.72	2.12	6
	2 , 2	1	8.40	0.00	9
	0 , 1	8	8.96	1.34	11
	1 , 1	6	8.62	1.27	17
	2 , 1	4	8.28	1.38	19
1p68	3 , 1	2	7.95	1.91	5
	0 , 2	13	8.11	0.76	16
	1 , 2	12	9.21	2.24	15
	2 , 2	8	9.01	1.15	19
	3 , 2	4	8.03	1.27	15
	0 , 1	5	8.24	0.93	3
	1 , 1	4	8.45	2.38	7
1r69	2 , 1	1	7.20	0.00	18
	0 , 2	5	8.24	0.93	3
	1 , 2	4	8.45	2.38	7
	2 , 2	1	7.20	0.00	17
	0 , 2	6	11.42	2.34	8
1b4f	1 , 2	5	10.80	1.32	3
	2 , 2	4	9.72	1.72	12
	3 , 2	2	9.55	2.05	3
	0 , 1	10	10.24	2.21	6
	1 , 1	8	9.39	1.75	9
	2 , 1	5	8.56	1.28	19
	3 , 1	4	9.58	2.36	1
1a1w	4 , 1	2	7.55	0.92	17
	0 , 2	16	10.11	3.03	6
	1 , 2	14	10.80	4.40	1
	2 , 2	12	10.71	3.20	8
	3 , 2	7	8.90	1.97	19
	4 , 2	4	8.62	2.33	4

3 leucine residues. This difference is significant, due to leucine's prevalence in interhelical interactions. Thus, while the model is able to match a large number of leucine contacts in 1a1w, it is constrained to only a few possible leucine contacts in 1b4f-A. These few contacts are selectively picked over other interactions that could possibly provide a larger number of contacts. This difference emphasizes the model's sensitivity to the amino acid sequence.

Table 2. Best average contact distance predictions given a prediction with at least as many helical contacts as α-helical pairs.

PDB Name	subtract, max_contact	PRIMARY Contact	PRIMARY Distance	WHEEL Contact	WHEEL Distance	Helix-Helix Interaction
1bha	0 , 2	13L-61L	6.7	-	-	1-2 A
		25L-48L	8.6	22L-52I	11.7	1-2 A
1p68	3 , 2	30L-71L	8.0	33V-68L	8.0	2-3 A
		44I-57L	6.5	41I-61M	9.5	2-3 A
		2I-34L	7.5	6V-31I	9.2	1-5 A
1r69	0 , 2	6V-59L	7.2	-	-	1-5 A
		45L-60L	9.2	48L-56V	8.1	4-5 A
1b4f	2 , 2	18L-63I	8.4	14V-67I	11.4	1-5 A
		49I-64L	8.1	46M-67I	11.0	4-5 A
		5L-50L	8.2	8L-46F	6.9	1-4 A
1alw	3 , 2	11V-43L	8.8	7L-46F	10.3	1-4 A
		20L-70L	10.7	23L-66L	6.1	2-5 A
		26L-62L	11.3	-	-	2-5 A

3.2. *Loop Structure Prediction Results*

The protein 1r69 was selected as an initial validation for the α-helical prediction framework. It has four loop regions, all of the form helix-loop-helix. These loop structures were all subject to the dihedral angle sampling approach, and the iterative clustering approach was used to refine the ensemble. The final selection of the loop was based upon the largest cluster after two rounds of iterative clustering–the same selection that would be made in a blind structure prediction approach. The loop conformers selected for constraining the tertiary structure prediction algorithm are detailed in Table 5.

3.3. *Tertiary Structure Prediction Results*

Given the interhelical contact prediction and loop structure prediction results from previous sections, a set of restraints was introduced to the final stage of the framework. The input to this tertiary structure prediction algorithm consisted of 43 distance bounds due to the local helical structure, 5 distance bounds for the predicted contacts from the Level 1 and Level 2 MILP models, a reduced ϕ, ψ dihedral angle space for the helical residues, and a $\pm 20°$ ϕ, ψ angle variation around the modeled structures for the loop regions.

The lowest energy structure identified from this analysis had an energy of -358.766 kcal/mol and an overall rmsd to the native 1r69 structure of 6.05 Å. The top five low-energy structures as well as the structure with

Table 3. The first three iterations of the contact distance prediction for protein 1p68 compared to the best average contact distance prediction for *subtract* 0, *max_contact* 2.

Iteration Rank	PRIMARY Contact	PRIMARY Distance	WHEEL Contact	WHEEL Distance	Helix-Helix Interaction	Distance Average
	5L-71L	24.7	9L-68L	14.9	1-3 A	
	19L-57L	21.9	16L-60M	16.8	1-3 A	
	8L-44I	11.5	5L-48M	7.5	1-2 A	
1	16L-37L	8.7	12L-41I	9.1	1-2 A	14.14
	30L-100V	29.0	33V-96L	20.4	2-4 A	
	40V-92I	12.6	37L-96L	13.9	2-4 A	
	68L-89I	8.4	64F-93F	9.0	3-4 A	
	75L-82V	7.6	71L-85I	10.3	3-4 A	
	5L-99L	7.2	9L-96L	7.7	1-4 A	
	12L-92I	7.0	-	-	1-4 A	
	8L-40V	14.0	12L-37L	8.8	1-2 A	
2	16L-30L	8.8	12L-33V	13.6	1-2 A	8.79
	37L-68L	9.3	40V-64F	6.6	2-3 A	
	44I-60M	7.8	48M-57L	8.4	2-3 A	
	57L-96L	7.4	61M-92I	9.4	3-4 A	
	71L-82V	7.6	68L-85I	8.2	3-4 A	
	5L-100V	9.9	8L-96L	8.8	1-4 A	
	12L-92I	7.0	9L-96L	7.7	1-4 A	
	9L-37L	10.7	5L-41I	10.7	1-2 A	
3	16L-30L	8.8	12L-33V	13.6	1-2 A	9.35
	33V-71L	8.3	30L-75L	12.0	2-3 A	
	48M-57L	8.4	44I-61M	8.4	2-3 A	
	68L-89I	8.4	64F-93F	9.0	3-4 A	
	75L-82V	7.6	71L-85I	10.3	3-4 A	
	5L-99L	7.2	8L-96L	8.8	1-4 A	
	12L-92I	7.0	-	-	1-4 A	
	9L-44I	8.5	5L-48M	7.5	1-2 A	
16	16L-37L	8.7	12L-41I	9.1	1-2 A	8.11
	30L-71L	8.0	33V-68L	8.0	2-3 A	
	57L-96L	7.4	61M-92I	9.4	3-4 A	
	68L-85I	8.2	71L-82V	7.6	3-4 A	

Table 4. The first three iterations of the contact distance prediction for protein 1r69 for *subtract* 1, *max_contact* 2. compared to the best average contact distance for these parameter values.

Iteration Rank	PRIMARY Contact	PRIMARY Distance	WHEEL Contact	WHEEL Distance	Helix-Helix Interaction	Distance Average
1	34L-59L	11.4	-	-	3-5 A	9.57
	45L-60L	9.2	48L-56V	8.1	4-5 A	
2	34L-59L	11.4	-	-	3-5 A	11.30
	48L-60L	11.2	-	-	4-5 A	
3	2I-34L	7.5	6V-31I	9.2	1-3 A	8.50
	45L-60L	9.2	48L-56V	8.1	4-5 A	
7	2I-59L	5.4	6V-56V	11.1	1-5 A	8.45
	45L-60L	9.2	48L-56V	8.1	4-5 A	

Table 5. Loop structures selected for restraining the tertiary structure prediction algorithm. The residues in the table indicate the loop regions of 1r69, but the RMSD values refer to the secondary structure stems as well as the loop region.

Begin Residue	End Residue	RMSD to Native (Å)	Energy (kcal/mol)	Cluster Size (k=2)	Lowest RMSD of Sampling Run (Å)
13	16	3.01	-69.93	513	0.42
23	28	2.59	-44.22	513	0.99
36	44	2.95	-129.65	824	2.25
51	55	3.76	-18.39	341	0.47

the lowest RMSD to the native structure are presented in Table 6. This stage of the analysis required approximately 1 wall clock day on 40 3.0 GHz Intel processors in a Beowulf cluster and was the rate-limiting step in the framework illustrated in Figure 2.

Table 6. The top structures identified using dihedral angle bounds from the loop structure prediction analysis

Energy (kcal/mol)	RMSD to Native (Å)
-358.77	6.05
-351.92	7.95
-347.71	6.72
-337.80	5.88
-337.52	8.04
-210.13	4.68

A second analysis was conducted to determine the value of high quality loop modeling efforts. If the loop regions are instead modeled with a dihedral angle restriction of $\pm 10^\circ$ ϕ, ψ around the native loop regions, a much improved structure results. This new lowest potential energy structure had an energy value of -381.543 kcal/mol and an overall rmsd to the native structure of 3.72 Å. The top five low-energy structures, which in this case also includes the conformer closest to the native structure, are listed in Table 7.

3.4. *Discussion*

The larger protein systems introduce more energetic complexity, and contacts corresponding to small experimentally-determined distances are more difficult to forecast. However, the number of contacts that can be predicted by the inter-helical contact prediction model when a small *subtract* value is used increases. These larger numbers of contacts for the protein sys-

Table 7. The top five low-energy structures using dihedral angle bounds from the loops of the native structure. The conformer with the lowest RMSD in the ensemble is the fourth structure listed below.

Energy (kcal/mol)	RMSD to Native (Å)
-381.54	3.72
-376.10	3.40
-375.17	4.45
-363.77	2.03
-356.85	4.91

tems allow for more accurate tertiary structure refinement by more strictly limiting the feasible conformational space that must be searched.

A general goal of 5.0 to 14.0 Å for the actual distance range of contacts predicted by the model was set, since such a distance range would significantly improve the structure refinement. This goal was attained and surpassed for the entire set of the test proteins. The averages for the target proteins fall far below 14.0 Å, suggesting that lower distance restraints such as 12.0 Å or even 10.0 Å and below, may be appropriate in some or even most cases.

Table 1, also referred to earlier, summarizes the best combined PRIMARY and WHEEL distance averages for every model run for the 5 proteins in the test set. It provides the standard deviations of these best solution distances as well as the number of unique PRIMARY and WHEEL contacts that were predicted by the model for later ASTRO-FOLD structure refinement. It is promising to note that even the parameter values that did not lead to the best overall average contact distance still produced high quality predictions. Of the 5 proteins, only 1hta had any average contact distance beyond 12 Å.

The results achieved from the tertiary structure analysis underline the importance of accurately modeling the loop regions of proteins. High-quality loop structure prediction methods that address the problem of flexible stems are needed to further bridge the gap between the medium-resolution structure obtained and the high-resolution structure predictions that are possible.

4. Conclusions

A novel optimization model for predicting interhelical contacts and generating interhelical distance restraints in α-helical globular proteins was successfully introduced. The model was tested on 5 α-helical proteins ranging from 3-6 helices and 63-102 residues, and the best average contact distances found for the systems fell below 10.0 Å in all 5 of these systems. For each protein system, several rank-ordered lists of solutions were constructed from the optimization model using different values of a parameter that varied the degree of helical packing imposed on the system.

The process of going from a list of predicted contacts between helices to a final tertiary structure model was illustrated for the protein system 1r69. Combining the contact prediction results and a loop structure analysis produced a set of restraints for use in the tertiary structure optimization. This tertiary structure prediction stage was able to produce a number of near-native conformers while illustrating the importance of accurate loop modeling efforts.

Acknowledgements

CAF gratefully acknowledges financially support from the National Institutes of Health (R01 GM52032).

References

1. H. M. Berman, J. Westbrook, Z. Feng, G. Gilliland, T. N. Bhat, H. Weissig, I. N. Shindyalov, and P. E. Bourne, *Nucleic Acids Res.* **28**, 235 (2000).
2. C. Zhang, J. T. Hou, and S. H. Kim, *Proc. Natl. Acad. Sci. USA.* **99**, 3581 (2002).
3. C. S. Adjiman, S. Dallwig, C. A. Floudas, and A. Neumaier, *Comp. Chem. Eng.* **22**, 1137 (1998a).
4. C. S. Adjiman, I. P. Androulakis, and C. A. Floudas, *Comp. Chem. Eng.* **22**, 1159 (1998b).
5. J. L. Klepeis and C. A. Floudas, *J. Global. Optim.* **25**, 113 (2003b).
6. T. E. Creighton. *Proteins: Structures and Molecular Properties.* W. H. Freeman and Company: New York, 1993.
7. C. A. Rohl, C. E. M. Strauss, D. Chivian, and D. Baker, *Prot. Struct. Funct. Bioinf.* **55**, 656 (2004).
8. C. M. Deane and T. L. Blundell, *Prot. Sci.* **10**, 599 (2001).
9. C. Zhang, S. Liu, and Y. Zhou, *Prot. Sci.* **13**, 391 (2003).
10. P. I. W. de Bakker, M. A. DePristo, D. F. Burke, and T. L. Blundell, *Prot. Struct. Funct. Bioinf.* **51**, 21 (2003).
11. M. A. DePristo, P. I. W. de Bakker, S. C. Lovell, and T. L. Blundell, *Prot. Struct. Funct. Bioinf.* **51**, 41 (2003).
12. J. A. Hartigan and M. A. Wong, *Appl. Stat.* **28**, 100 (1979).
13. L. R. Forrest and T. B. Woolf, *Prot. Struct. Funct. Bioinf.* **52**, 491 (2003).
14. M. Mönnigmann and C. A. Floudas, *Prot. Struct. Funct. Bioinf.* **61**, 748 (2005).

15. A. R. Ortiz, A. Kolinski, and J. Skolnick, *J. Mol. Biol.* **277**, 419 (1998).
16. E. S. Huang, R. Samudrala, and J. W. Ponder, *J. Mol. Biol.* **290**, 267 (1999).
17. B. A. Reva, A. V. Finkelstein, and J. Skolnick, *Fold Des.* **3**, 141 (1998).
18. B. Fain and M. Levitt, *J. Mol. Biol.* **305**, 191 (2001).
19. C. A. Floudas, *AIChE J.* **51**, 1872 (2005).
20. C. A. Floudas, H. K. Fung, S. R. McAllister, M. Mönnigmann, and R. Rajgaria, *Chem. Eng. Sci.* **61**, 966 (2006).
21. W. L. DeLano. *The PyMol Molecular Graphics System.* DeLano Scientific, San Carlos, CA USA, 2002. http://www.pymol.org.
22. C. A. Floudas. *Nonlinear and Mixed-integer Optimization: Fundamentals and Applications.* Oxford University Press, 1995.
23. J. L. Klepeis and C. A. Floudas, *Biophys. J.* **85**, 2119 (2003c).
24. C. S. Adjiman, I. P. Androulakis, C. D. Maranas, and C. A. Floudas, *Comp. Chem. Eng.* **20**, S419 (1996).
25. C. S. Adjiman, I. P. Androulakis, and C. A. Floudas, *Comp. Chem. Eng.* **21**, S445 (1997).
26. C. S. Adjiman, S. Dallwig, C. A. Floudas, and A. Neumaier, *Comp. Chem. Eng.* **22**, 1137 (1998a).
27. C. S. Adjiman, I. P. Androulakis, and C. A. Floudas, *Comp. Chem. Eng.* **22**, 1159 (1998b).
28. I. P. Androulakis, C. D. Maranas, and C. A. Floudas, *J. Global. Optim.* **7**, 337 (1995).
29. C. A. Floudas. *Deterministic Global Optimization : Theory, Methods and Applications.* Nonconvex Optimization and its Applications. Kluwer Academic Publishers, 2000.
30. J. L. Klepeis, M. T. Pieja, and C. A. Floudas, *Comp. Phys. Comm.* **151**, 121 (2003a).
31. J. L. Klepeis, M. T. Pieja, and C. A. Floudas, *Biophys. J.* **84**, 869 (2003b).
32. J. Lee, H. A. Scheraga, and S. Rackovsky, *Biopolymers* **46**, 103 (1998).
33. J. Lee, H. A. Scheraga, and S. Rackovsky, *J. Comput. Chem.* **18**, 1222 (1997).
34. J. Lee and H. A. Scheraga, *Int. J. Quantum Chem.* **75**, 255 (1999).
35. D. Ripoll, A. Liwo, and H. A. Scheraga, *Biopolymers* **46**, 117 (1998).
36. J. Lee, J. Pillardy, C. Czaplewski, Y. Arnautova, D. R. Ripoll, A. Liwo, K. D. Gibson, R. J. Wawak, and H. A. Scheraga, *Comp. Phys. Comm.* **128**, 399 (2000).
37. J. L. Klepeis, Y. N. Wei, M. H. Hecht, and C. A. Floudas, *Prot. Struct. Funct. Bioinf.* **58**, 560 (2005).
38. J. L. Klepeis and C. A. Floudas, *J. Comput. Chem.* **23**, 245 (2002).

QUALITY AND EFFECTIVENESS OF PROTEIN STRUCTURE COMPARATIVE MODELS

DOMENICO RAIMONDO[1], ALEJANDRO GIORGETTI[1], DOMENICO COZZETTO[1], ANNA TRAMONTANO[1,2]

[1] *Department of Biochemical Sciences, University of Rome "La Sapienza".*
[2] *Istituto Pasteur Fondazione Cenci Bolognetti, University of Rome "La Sapienza",P.le A. Moro, 5 - 00185 Rome, Italy.*
anna.tramontano@uniroma1.it

Since decades, the problem of deciphering the code that relates the amino acid sequence of a protein and its native three-dimensional structure has been the subject of innumerable investigations and, in spite of the many frustrations caused by its elusiveness, the interest in the problem is not fading away, on the contrary. What stands in our way, notwithstanding all these efforts, is the complexity of protein structures. In a three dimensional protein structure thousands of atoms are held together by weak forces and give rise to a conformation which is only marginally stable. The consequence of this, as we will discuss, is that it is very unlikely that we can use our understanding of the laws of physics to compute the native functional structure of a protein in the foreseeable future. However, we have at our disposal the experimentally solved structures of a reasonable number of proteins, a few thousands as of today. They represent solved instances of our problem and their knowledge has allowed us to devise methods to predict the structure of their evolutionary neighbours. Here we describe how we can evaluate and estimate the quality of a protein model and analyze how this impacts on their usefulness.

1. Introduction

The biochemical function of a protein is determined, by and large, by its three-dimensional structure. In turn, the structure of a protein is mainly dictated by the specific linear sequence of its amino acids as first demonstrated by Anfinsen in a historical experiment[3]. He showed that a protein (in its particular case Ribonuclease A) once denatured in vitro, recovers its native conformation when the denaturing agents are removed from the test tube. The same experiment can be done using a chemically synthesized protein. This implies that the information about the three-dimensional structure of a protein is contained in its amino acid sequence. Subsequently, it was discovered that there exist cellular mechanisms that catalyze folding of some proteins; however these systems accelerate the folding process, but do not affect the structure of the final native state[8].

The structure of a protein can be experimentally determined with techniques such as x-ray crystallography or Nuclear Magnetic Resonance, but they are time and labour consuming and it is impossible to imagine that the structure of all the proteins of the universe can be experimentally determined. Therefore we would like to infer (or predict, as we usually say) the three-dimensional structure of a protein given its amino acid sequence.

Proteins are only marginally stable, with a stability of the order of a few Kcal/mol, and therefore, although the free energy of the unfolded state and that of the native state are rather large terms, their difference is a very small number. In order to estimate them with a reasonable accuracy, we would need to estimate the energy of the folded and unfolded state with a very high precision that is not achievable today. This implies that the route of calculating the energy of every possible conformation of a protein chain to select the native low energy state is not only impossible because of the enormous number of conformations that we should explore, but also because of the more fundamental problem of the precision of our energetic calculations.

However, the large set of available protein structures gives us a quite important handle to try and find a solution to the problem on the basis of heuristic rules. These can be learnt from the set of available solved instances, i.e. of proteins for which both the sequence and the structure are known.

Evolution provides us with many examples of proteins that descend from a common ancestral protein whose sequence has been modified via a process of residue substitutions and, albeit less frequently, of insertions and deletions of amino acids. We know that these proteins are functional, or at least not deleterious, as they have been accepted in the population, therefore they are expected to have a stable native structure. Because of the requirement that the evolutionary changes preserve function, we also expect their structure to be similar. Clearly as mutations accumulate, local rearrangements add up so that the longer the evolutionary distance between the two proteins, the less conserved will their structures be[4].

We have methods to detect evolutionarily related proteins from their sequences and this implies that we can detect proteins very likely to have a similar structure[11]. The structure of one protein therefore can represent an approximate model for the structure of all the proteins of its evolutionary family. The closer the evolutionary distance between the target protein (of unknown structure) and the template one (with known structure), the better the approximation.

This is the basis of the most used method for protein structure prediction: homology or comparative modelling[14].

The question that arises, and that will be discussed here, is how good are the models that we can build and which is their expected usefulness.

2. How good are comparative models? the CASP experiment

In 1994 John Moult proposed a world wide experiment named CASP (Critical Assessment of Techniques for Protein Structure Prediction) aimed at establishing the current state of the art in protein structure prediction, identifying what progress has been made, and highlighting where future effort may be most productively focused[10,9].

This is how the experiment is organized:

Crystallographers and NMR spectroscopists who are about to solve a protein structure are asked to make the sequence of the protein available together with a tentative date for the release of the final coordinates.

Predictors produce and deposit models for these proteins (the CASP targets) before the structures are made available. CASP also tests publicly available servers on the same set of targets providing a unique opportunity to verify the effectiveness of human intervention in the modeling procedure.

A panel of three assessors compares the models with the structures as soon as they are available and tries to evaluate the quality of the models and to draw some conclusions about the state of the art of the different methods. The task is divided among assessors in such a way that one looks at models of proteins that share significant sequence similarity with a protein of known structure, one to those sharing a significant structural similarity, but no clearly detectable sequence similarity with proteins of known structure, while the third evaluates all the remaining ones. It is expected that the first set of targets is predicted using comparative modelling and the second using fold recognition methods, however since the experiment is run blindly, i.e. the assessors do not know who the predictors are until the very end of the experiment, it is entirely possible that different techniques are used by different groups for the same target.

The results of the comparison between the models and the target structures are discussed in a meeting where assessors and predictors convene. The conclusions are made available to the whole scientific community via the World Wide Web and via the publication of a special issue of the journal Proteins: Structure, Function, and Bioinformatics.

The assessors change every time, with rare exceptions, and they are free to analyse the data in different ways and draw their own conclusions, although a set of measures are provided by the CASP organizers and they are available for all the targets of all the editions. The most important are the GDT-TS values defined as the weighted sum of the fraction of atoms within 1Å, 2Å, 4Å and 8Å from the corresponding atoms in the experimental structure, r.m.s.d. values for several subsets of superimposed atoms, the number of correctly aligned residues between target and model (a residue is considered correctly aligned if, after superposition of the experimental and modelled structure, its Cα atom falls within 3.8Å of the corresponding experimental atom, and there is no other Cα atom of the experimental structure that is nearer).

Figure 1. Example of a comparative model submitted to the CASP experiment (in green) compared to the experimental structure (in blue).

The results of the experiment have shown that models with very respectable accuracy can be built for proteins sharing a significant similarity with proteins of known structure[5] (Figure 1). A sequence identity between target and template of above 30% essentially guarantees that the overall structure is correctly predicted by most, if not all, techniques (Figure 2).

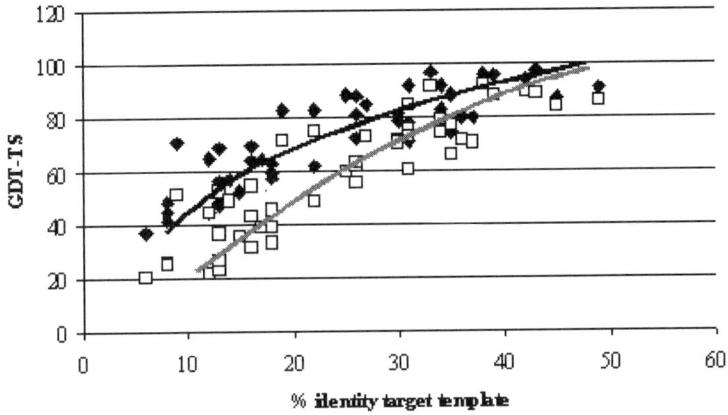

Figure 2. Quality of the best (black diamonds, black line) and average (white squares, gray line) models submitted to the CASP experiments as a function of the sequence identity between the target protein and its closest homologue of known structure.

3. Can we evaluate the quality of a protein comparative model a priori?

By and large, the quality of a comparative model depends upon two factors: the extent of structural divergence between the target and the template and the quality of the sequence alignment between the two protein sequences. The latter is usually derived from a multiple sequence alignment of as many proteins of the family as possible and its accuracy will depend upon the number and similarity distribution of the sequences of the protein family.

The distance between two sequences can be estimated by counting the number of different amino acids in corresponding positions, i.e. in structurally equivalent positions. In a pioneering work, Chothia and Lesk[4] analysed pairs of proteins of known structure, in which case the correct identification of the pairs of corresponding positions (alignment) in any two structures is almost straightforward.

Intuitively, the expected accuracy of an alignment between two protein sequences depends upon their sequence similarity. However present model building procedures rarely make use of two sequences alone. Homology is transitive, therefore if two proteins are evolutionarily related to a third protein, they are also evolutionarily related. This can be used to detect more distant evolutionary relationships in database searching strategies, by "hopping" in sequence space from one homologous protein to the next thus increasing the number of proteins that we can include in the family[13,1].

Making use of sequences of proteins that are intermediate between target and template not only allows more distant evolutionary relationships to be detected, but also improves our chances of obtaining a correct alignment. The alignment of two closely related sequences is less prone to errors and, once obtained, can be used to guide the alignment of a third, more distant, sequence and so on iteratively until all known members of the protein family are aligned. This is the method of choice in all present comparative modeling experiments and its advantages have been clearly demonstrated in many cases.

However, the use of many intermediate sequences between target and template makes it incorrect to evaluate the expected accuracy of a model solely on the basis of the sequence identity or similarity between target and template. The quality of the alignment between these two sequences will also depend upon the number and the similarity distribution of all the sequences of the multiple sequence alignment.

This issue has several implications, for example that the "quality" of the multiple sequence alignment needs to be taken into account when evaluating a priori the expected quality of a comparative model, an important practical issue since it permits to decide whether the model will be of sufficient quality for the desired applications.

It also has important implications for the evaluation of prediction methods. There are several world-wide initiatives, the most popular being CASP, for assessing the state of the art in the field. Better results in subsequent CASP experiments can be due to genuine method improvements but also to differences in the difficulty of the targets between the two sets. A larger protein family may facilitate obtaining a correct alignment, therefore targets of the same apparent difficulty, as judged by their sequence similarity with the template, can become progressively easier as sequence databases continue to grow. Pair-wise sequence similarity between target and template is therefore not a good parameter to estimate the difficulty of the targets.

We have developed a method to evaluate the difficulty of the pair-wise alignment between two sequences implied by a given multiple sequence alignment[6].

For each of the CASP targets, we collected all sequences in the PSI-BLAST[2] output whose length spanned at least 80% of the region of superposition between target and template in the multiple sequence alignment and constructed a graph similar to that shown in Figure 3. We subsequently calculated the parameter μ, that is the maximum distance (i.e. the minimum percent sequence identity) between two sequences of the alignment that had to be included to connect the target to the template (Figure 3).

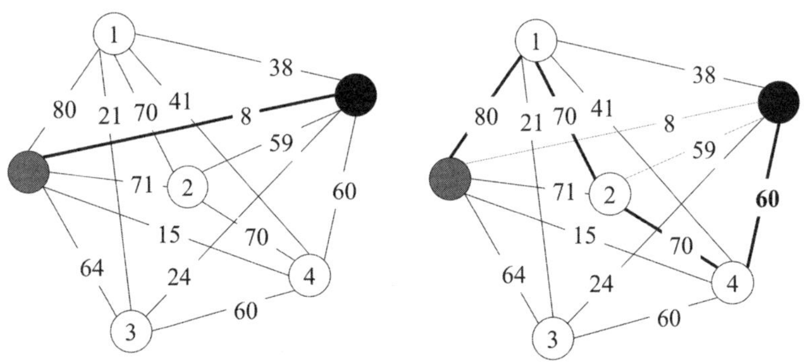

Figure 3. An illustration of the method used to compute the difficulty of aligning a target protein (gray) with a template protein (black). Even if their sequence identity is only 8%, the use of intermediate sequences implies that we only need to align protein sequences sharing, at worst, 60% sequence identity.

Using this approach, we were able to show that the apparent improvement in the alignment quality in subsequent CASP experiments is due to the availability of larger databases, rather than to better methods (Figure 4).

4. Useful applications of comparative models: molecular replacement in x-ray crystallography

Let us recall here that, in a diffraction experiment, a crystal is irradiated with a particular X-ray wavelength and the resulting diffracted waves are collected on physical or electronic devices. However, in this passage from 3D to 2D all the information on the phases of the diffracted waves is lost and this is one of the fundamental problems of structural science.

There are three approaches to solve the phase problem: direct methods, interference based methods and molecular replacement (MR) methods. The latter methods are based on the fact that the prior knowledge of a protein

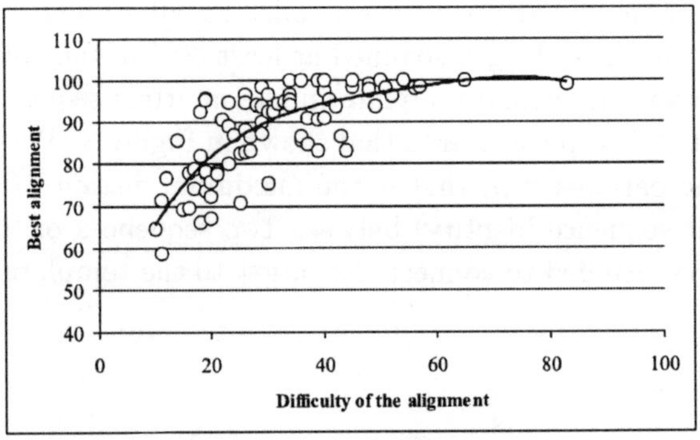

Figure 4. There is a clear relationship between the difficulty of the alignment and the quality of CASP models.

structure simplifies the solution of a different crystal form of the same molecule[12]. In some cases, the structure of a homologous protein or a model of the target protein can be sufficient to approximate the relative position of the atoms in the structure and allow the structure factors to be computed. Historically, it has been very difficult to decide a priori the quality of a model that is required for a successful molecular replacement experiment. A generally accepted "rule of thumb" is that molecular replacement is effective if the model is reasonably complete and shares at least 40%-50% sequence identity with the unknown structure.

However, we have shown that matters are more complex than this.

We used a set of models deposited in the CASP database as input for a completely automatic molecular replacement procedure and recorded in which cases the model was sufficient to obtain the phases and, therefore, to solve the structure[7].

The conclusions of the work can be summarized as follows:

In this specific application, what really counts is the overall quality of the model rather than the details of the less well predicted parts.

In all cases, a GDT-TS above 84 is sufficient to guarantee the success of the procedure regardless of the sequence identity between the target and template structure, of the method used for producing the model and of the structural class of the protein under examination. In the automatic procedure, models with GDT-TS below 80 were never successful.

For models of intermediate quality, the results vary. Most of the times a large fraction of the structure can be automatically built with respectable quality and it is likely that, in these cases, more iterations and, most of all, manual intervention can lead to success. Even limited improvements in the quality of a model can be instrumental for the success of an MR experiment. This observation can explain why it has been so difficult so far to predict beforehand when a model can be successfully used in MR solely on the basis of the sequence identity between a model and the structural template used to build it.

5. Conclusions

The knowledge, even approximate, of the three-dimensional structure of a protein is essential for understanding the details of its molecular function and gives valuable insights for the development of effective rational strategies for experiments such as studies of disease related mutations, site directed mutagenesis, or structure based drug design.

The number of known protein structures is an order of magnitude smaller than the number of protein sequences that can be deduced from genome data. The only way to bridge this gap is to recur to computational methods such as comparative modelling.

These methods have matured, and information from three-dimensional protein models has been successfully used in a wide variety of biomedical applications. There is no doubt that structure prediction methods are an essential part of the scientific background and of the toolbox of life scientists. The need for integrating experimental knowledge with theoretical hypotheses will only grow in the future: it has been estimated that every new experimental structure carries information about the structure of at least a hundred other proteins.

It is therefore important to understand not only the methods and advantages of these methods, but also their limitations.

Acknowledgments

The work described here is supported by the EU funded BioSapiens Network of Excellence, contract number LHSG-CT-203-503265, the AIRC funded BICG Project "Bioinformatics tools for identifying, understanding and attacking targets in cancer" and by the Institute Pasteur, Fondazione Cenci Bolognetti.

References

1. Altschul, Stephen F.; Koonin, Eugene V. Iterated profile searches with PSI-BLAST - a tool for discovery in protein databases. Trends Biochem Sci, v. 23,

n. 11, p. 444-447, 1998.

2. Altschul Sthephen F.; Madden, Thomas L.; Schaffer, Alejandro A.; Zhang, Jinghui; Zhang, Zheng; Miller, Webb; Lipman, David J. Gapped BLAST and PSI-BLAST: a new generation of protein database search programs. Nucl Acids Res, v. 25, n. 17, p. 3389-3402, 1997.
3. Anfinsen, Christian B.; Haber, Edgar; Sela, Michael; White Jr, Frederick. The kinetics of formation of native ribonuclease during oxidation of the reduced polypeptide chain. Proc Natl Acad Sci U S A, v. 47, p. 1309-1314, 1961.
4. Chothia, Cyrus; Lesk, Arthur M. The relation between the divergence of sequence and structure in proteins. EMBO J., v.5, n. 4, p. 823-826, 1986.
5. Cozzetto, Domenico; Di Matteo, Adele; Tramontano, Anna. Ten years of prediction ... and counting. FEBS Journal, v. 272, n. 4, p. 881-882, 2005.
6. Cozzetto, Domenico; Tramontano, Anna. The relationship between multiple sequence alignment and the quality of protein comparative models. Proteins, v.58, n. 1, p. 151-157, 2005.
7. Giorgetti, Alejandro; Raimondo, Domenico; Miele, Adriana E.; Tramontano, Anna. Evaluating the usefulness of protein structure models for molecular replacement. Bioinformatics, v. 21 Suppl. 2, p.72-76, 2005.
8. Hartl, F. Ulrich. Molecular chaperones in cellular protein folding. Nature, v. 381, n. 6583, p. 571-579, 1996.
9. Moult, John. A decade of CASP: progress, bottlenecks and prognosis in protein structure prediction. Curr Opin Struct Biol., v. 15, n. 3, p. 285-289, 2005.
10. Moult, John; Pedersen, Jan T.; Judson, Richard; Fidelis, Krzysztof. A largescale experiment to assess protein structure prediction methods. Proteins, v. 23, n. 3, p. ii-v, 1995.
11. Pearson, William R.; Sierk, Michael L. The limits of protein sequence comparison? Curr Opin Struct Biol., v. 15, n. 3 p. 254-260, 2005.
12. Rossman, Michael G.; Blow, David M. (1962) The detection of sub-units within the crystallographic asymmetric unit. Acta Cryst., v. 15, p. 24-31, 1962.
13. Teichmann, Sarah A.; Chothia, Cyrus; Church, George M.; Park Jong. Fast assignment of protein structures to sequences using the intermediate sequence library PDB-ISL. Bioinformatics, v. 16, n. 2, p. 117-124, 2000.
14. Tramontano, Anna. Homology modeling with low sequence identity. Methods v. 14, n. 3, p. 293-300, 1998.

STEINER MINIMAL TREES AND TWIST ANGLES IN FOLDED PROTEIN STRUCTURES

JAMES MACGREGOR SMITH

Department of Mechanical and Industrial Engineering University of Massachusetts Amherst MA 01003, e-mail: jmsmith@ecs.umass.edu

The Steiner Minimal Tree (SMT) problem determines the minimal length network for connecting a given set of vertices in 3-dimensional space. SMTs are also useful in modelling minimum energy configurations such as those in proteins. With the Steiner tree topologies, we can define planes within the amino acids that have a surprising regularity property for the twist (dihedral) angles of the planes. This angular property is quantified for amino acids with the optimal Steiner tree topology structure. The angular properties are then examined in some detail in fibroin structures where Ala-Gly and Gly-Ala dipeptides are evaluated for their dihedral angular properties as well as other geometric properties.

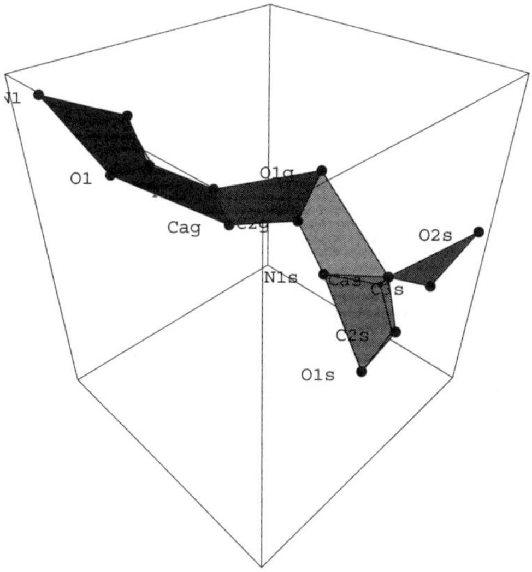

Figure 1. Gly-Gly-Ser Folded Steiner Structures

1. Motivation

The protein folding problem remains one of the most important and unsolved problems in the biological sciences. Developing a computer program to generate the 3-D structure of a protein from its amino acid sequence remains an important computational and biological objective. It is felt that SMTs may play an important part in the development of such a computer program. Figure 1 illustrates a tri-peptide planar structure comprised of the two amino acids of Glycine and one amino acid Serine of a protein and its underlying Steiner minimal tree structure. The planar folds linking the atoms as defined by the Steiner tree gives one a sense of the overall folding problem. In effect, the problem boils down to determining the twist(dihedral) angles between planes of the atoms as they fold up in space. The fundamental properties of Steiner trees are felt to be important for modelling proteins in the quest to better understand how they fold up in space.

1.1. *Purpose and Objectives*

Steiner minimal trees represent the shortest network interconnecting a given set of terminals $V = \{v_1, v_2, \ldots, v_N\}$ in E^d. The Steiner points arise from the fact that additional points from a set $S = \{s_1, s_2, \ldots, s_M\}$ may be necessary in order for the network interconnecting V to be as short as possible. The main interest of this paper is to further explore and articulate the relationship between Steiner Minimal Trees and protein structures.

1.2. *Outline of Paper*

In §2 we describe the background the protein folding problem and of the Steiner Minimal Tree problem along with a brief literature review of some of the important papers related to their study. In §3, we describe the mathematical model surrounding the energy function for the protein folding problem as well as the mathematical model of the SMT problem and its link to the protein folding problem. We also discuss the notion of thermodynamic independence (TI) which is an essential assumption for the energy function as well as an important concept for the use of SMTs in protein modelling. In §4, we summarize some of the important results we have generated for the amino acids in proteins as well as the important lower bounds they represent on the topological structure that the amino acids can assume. In §5 we examine the fibroin structure of silk and see how the dihedral angular properties as well as other geometric properties occur in the folded structure. §6 rounds out the paper.

2. Problem Background

We provide a brief overview of the problem background for the protein folding problem and the Steiner minimal tree problem. One of the major overriding goals of the paper is that we are seeking to examine how these two research problems overlap and influence each other.

2.1. *Protein Folding Problem*

The protein folding problem may be conceptualized from the way biologists and chemists normally view the fundamental structure of a protein. There are basically four classification schemes: i) primary; ii) secondary; iii) tertiary; and iv) quarternary, see Brandon and Tooze (1991). Brandon and Tooze provides a good overview of the folding problem and they postulate that there must be a set of rules that govern the way a protein folds up in space.

The primary structure refers to the linear sequence of amino acids in the chain. Secondary refers to the helices and sheets in which the acids form from the primary sequence of amino acids along the backbone chain. Tertiary refers to the complete 3-D structure of the protein and quarternary refers to the spatial relationships between the different polypeptides or subunits of the overall globular protein. It is felt that the 3-D structure is largely determined by the primary composition level of the amino acids as argued in Voet and Voet (1991). We shall largely focus on this primary and secondary level and examine what Steiner Minimal trees offer here. Some of the key 3-D rules are felt to be contained in this primary structure. We shall take a theoretical rather than an empirical approach to determining the 3-D structure since the formalism of Steiner Minimal trees allows us to do so.

The link between Steiner trees and proteins as we shall argue is through the potential energy function of a protein and the geometric way in which the forces are resolved in 3-D. The geometric properties of the Steiner trees are felt to be central to the proteins structure.

3. Mathematical Models and Properties

In protein modelling, the potential-energy objective function often used to measure the Minimum Energy Configuration is the following, where the parameters $K_{b_i}, K_{\theta_i}, A_{ij}, B_{ij}$, and ϵ are adjustable weights; see Cohen (1995). This potential-energy objective function is based on the theoretical molecular mechanical force field model used to model most molecular structures; see Leach (1996). It is interesting to observe that the objective function is a sum of nonlinear terms with little interaction between the terms.

$$\begin{aligned}
E_{tot} &= \sum_i K_{b_i}(b_i - b_0)^2 \text{ bond lengths } [E_{bs}] \\
&+ \sum_i K_{\theta_i}(\theta_i - \theta_0)^2 \text{ bond angles } [E_{ab}] \\
&+ \sum_i K_{\tau_i}(\cos(3\tau_i - \gamma_0)) \text{ torsion angles } [E_{tor}] \\
&+ \sum_i \sum_j A_{ij} d_{ij}^{-6} + B_{ij} d_{ij}^{-12} \text{ van der Walls } [E_{14vdW}] \\
&+ \sum_i \sum_j V_i V_j / \epsilon d_{ij} \text{ electrostatic interactions } [E_{14e}]
\end{aligned}$$

where:

E_{bs}: The first term is the sum of energies arising from bond stretching or compression beyond the optimum bond length, and b_i, b_0 are the actual equilibrium bond lengths.

E_{ab}: The second term is the sum of energies for angles that are distorted from their optimum values, and θ_i, θ_0 are the equilibrium bond angles.

E_{tor}: The third term is the sum of the torsional energies that arise from rotations about each respective dihedral angle, and τ_i, γ_0 are the torsion angles.

E_{14vdW}: The sum of energies due to Van der Walls interactions.

E_{14e}: The final term is for the electrostatic interactions.

3.1. *Thermodynamic Independence (TI)*

A crucial assumption made in most molecular modelling is the notion that the total free energy in a protein can be expressed as a sum of individual energy components, thus, the total system can be separated into a series of independent subsystems Mark and van Gunsteren (1994). Thus, the energy function of a protein is separable. This Thermodynamic Independence (TI) is mathematically equivalent to the concepts of joint probabilistic independence and additive independence in utility theory. This is an important mathematical property in optimization and plays a significant role in protein modelling as we shall see.

Not everyone agrees that the energy function is separable, for example see Dill (1997), since there may be interaction between the terms in the objective function, but if one makes this assumption, it can yield a simplifying tool for analysis. The separability is also crucial for the objective function in the Steiner Minimal Tree Problem which we will elaborate upon in the next section.

3.2. *Notation*

We will approach the problem of protein structure from a Steiner Minimal Tree perspective. As such, we need to lay down some of the basic definitions and notation of SMTs. The following is a list of useful notation and definitions:

$M :=$ number of Steiner vertices from point set S.
$N :=$ number of given terminal vertices from the set V.
$FST :=$ Full Steiner tree with the maximum number of Steiner points $M = N - 2$.
$MST :=$ Minimum spanning tree with the number of Steiner points $M = 0$.
$\rho_3(V) :=$ Steiner ratio of a given terminal vertex set V, i.e. $\rho_3(V) = SMT/MST$ in E^3.

3.3. *Steiner Trees*

For the most part in this paper, we are interested in FSTs. MSTs are important and they represent upper bounds on the length of the SMTs. The identification of effective lower bounds for the SMT problem remains elusive in 3-D.

One important aspect of SMTs is the ratio $\rho_3(V)$ of the SMT to an MST for a given terminal vertex set V. Since we will be working primarily in 3-D, this ratio $\rho_3(V)$ helps us determine whether the optimal SMT configuration has been achieved for a particular given terminal vertex set V.

There are certain elemental facts that are important in E^3 and higher dimensions. They are:

- $M \leq N - 2 \ \forall E^d$, Gilbert and Pollak (1968),
- angles subtended at each Steiner point are equal to $2\pi/3 \ \forall E^d$, Gilbert and Pollak (1968), and

- the ratio in the plane of dimension two $\rho_2(V) = \sqrt{3}/2$ is attained for equilateral triangles, ladders and lattice configurations, see Du et al. (1982) and Du and Hwang (1992).

Of special importance linking Steiner trees and proteins is the fact that the solutions of Steiner trees in E^3 are characterized by planes joining the terminal vertices together. In fact, one key to finding the optimal Steiner tree structure is to know what the twist angles are between the various planes. We illustrate the Steiner solution for $N = 4$ for an equilateral tetrahedron. The given terminal vertices for this point set are $\{V_i, V_j, V_k, V_\ell\}$ which are from an equilateral tetrahedron. The larger nodes in the diagrams which follow represent the terminal vertices while the smaller nodes either represent equilateral points or Steiner points.

What we seek to do is to find the maximum distance between the equilateral vertices of the edge pairs or, in essence, find the orientation of the circles through the equilateral reflection points that are furthest apart. We have the following optimization problem involving the equilateral reflection points $T_{ij} - T_{k\ell}$ of the Melzak Circles with the constraints ensuring the equilateral edges are satisfied:

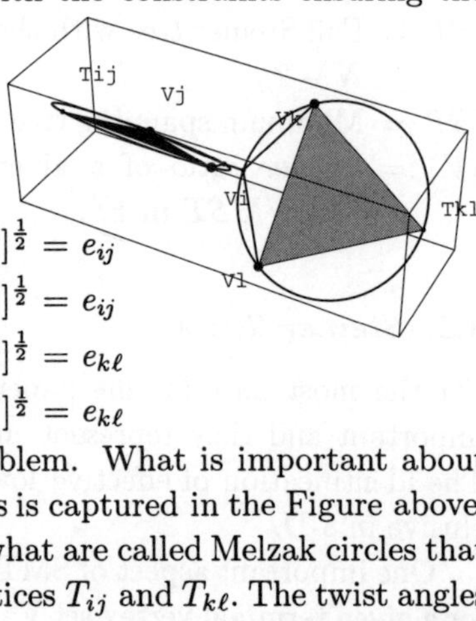

$$\text{Maximize } \Phi = \|T_{ij} - T_{k\ell}\|$$

subject to:

$$[(x_{ij} - x_i)^2 + (y_{ij} - y_i)^2 + (z_{ij} - z_i)^2]^{\frac{1}{2}} = e_{ij}$$
$$[(x_{ij} - x_j)^2 + (y_{ij} - y_j)^2 + (z_{ij} - z_j)^2]^{\frac{1}{2}} = e_{ij}$$
$$[(x_{k\ell} - x_k)^2 + (y_{k\ell} - y_k)^2 + (z_{k\ell} - z_k)^2]^{\frac{1}{2}} = e_{k\ell}$$
$$[(x_{k\ell} - x_\ell)^2 + (y_{k\ell} - y_\ell)^2 + (z_{k\ell} - z_\ell)^2]^{\frac{1}{2}} = e_{k\ell}$$

This is a non-trivial optimization problem. What is important about the problem as it relates to all that follows is captured in the Figure above. In the above Figure, the circles represent what are called Melzak circles that link pairs of terminals with equilateral vertices T_{ij} and $T_{k\ell}$. The twist angles between the planes defined by the Melzak circles are crucial to the SMT topology. For the $N = 4$ regular tetrahedron case, the twist angles between the planes are $\pi/2$. These twist angles define the Steiner tree topology for this point set. When one has point sets of $N = 4$ which deviate from the regular tetrahedron, then the twist angles also change and, in general, are very difficult to compute.

In the diagrams for $N = 5, 6$, the t_i, t_j, t_k, t_ℓ represent equilateral points of the Melzak circles in the solution process of constructing the Steiner

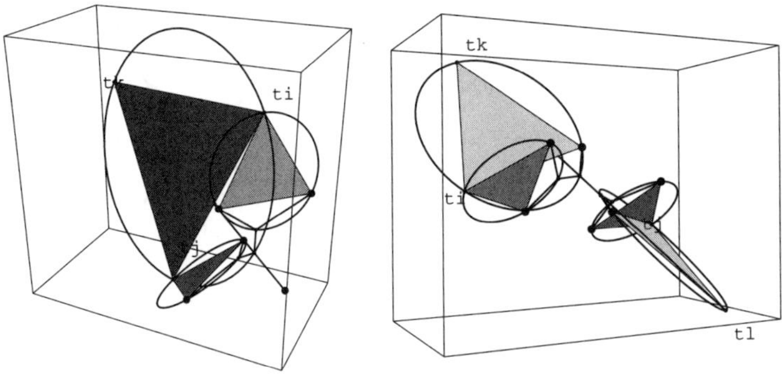

Figure 2. SMT solution for $N = 5, 6$ (top to bottom)

trees. These latter two points sets are also regular structures. Again, the twist angles between the planes are essential for determining the Steiner tree topology. It is actually the fact that there are so many more planes in space with unknown twist angles rather than the single embedded plane for E^2 problems that makes computational complexity of the Steiner problem in E^3 so much more difficult.

3.4. *Steiner Trees and MECs*

We now need to examine what is the crucial link between Steiner trees and MECs. This relationship between SMTs and MECs is described by Maxwell's theorem; see Gilbert and Pollak (1968) for a proof.

Let F_1, F_2, F_3, F_4 be unit forces acting at fixed terminals $\{v_1, v_2, v_3, v_4\}$ respectively. If one designs a network linking these terminals with Steiner points $\{s_1, s_2\}$ that can be moved into position, then one seeks to find the location of the Steiner points and the network where these forces will be in equilibrium. Figure 3 illustrates the MEC.

Theorem 1. *If one draws unit vectors from a Steiner tree in the direction of each of the lines incident to* $v_1, v_2, \ldots, v_N$*, and letting* F_i *denote the sum of the unit vectors at* v_i*, then in mechanical terms,* F_i *is the external force needed at* v_i *to hold the tree in equilibrium. The length of the tree SMT has the simple formula*

$$SMT = \sum_{i=1}^{N} v_i F_i$$

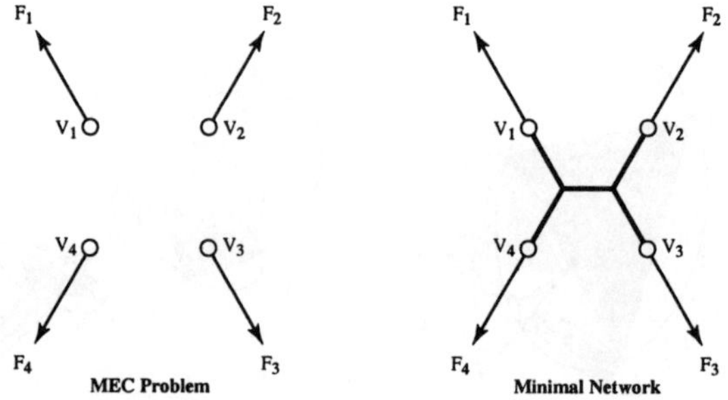

Figure 3. Maxwell's Theorem, Given V (Left) SMT≡MEC (Right)

If the forces at the vertices are not all uniform, then the SMT acts as only a lower bound. This was discussed in some detail in our previous paper, see Stanton and Smith(2004). Notice that in Maxwell's Theorem that the function is separable in the force components. Thus, the FST can be decomposed into its FST components. One way to construct a FST is to identify its FST components. Thus, if we can somehow subdivide the overall terminal point set into its FST components, then we can compute the overall SMT by constructing the SMT for its FST components.

3.5. *Minimum Energy Configurations MECs*

One of the surprising features of the link between Steiner trees and proteins is seen in Figure 4 on the left where the trans-peptide group illustrates the angles of the carbon and nitrogen atoms in the amide plane. They are very close to 120°. This is apparently due to the partial double-bond nature of the backbone plane (Dickerson and Geis (1969). The fact that the peptide groups with few exceptions assume the trans-peptide conformation as opposed to the cis-peptide conformation Voet and Voet (1991) is considered also to be important.

Figure 4 on the right illustrates the relationship between the peptide backbone planes since the peptide bond is planar there are only two degrees of freedom permitted per residue: the twist angle about the $C_\alpha - N$ bond ϕ and the twist angle along the $C_\alpha - C$ bond ψ (Dickerson and Geis (1969). Extensive studies have been carried out examining the acceptable values of the twist angles of the peptide bonds for many protein conformations, see (Voet and Voet (1991). One hypothesis which will emerge in this paper is whether or not additional information about the relationship of the twist (dihedral) angles of those atoms within the side chains may also be an important factor in the protein folding problem.

The overall network of peptide bonds is normally configured as in Figure 5 as a linear polymer rather than a branched polymer chain (Voet and Voet (1991). With this linear chain of peptide bond planes, the resulting geometry of the secondary structures can either be an alpha helix, a beta sheet, or combinations of these structures as they fold up into space. Of course the complicating feature of the network structures that result is the geometry of the side chains and how the entire structure packs together.

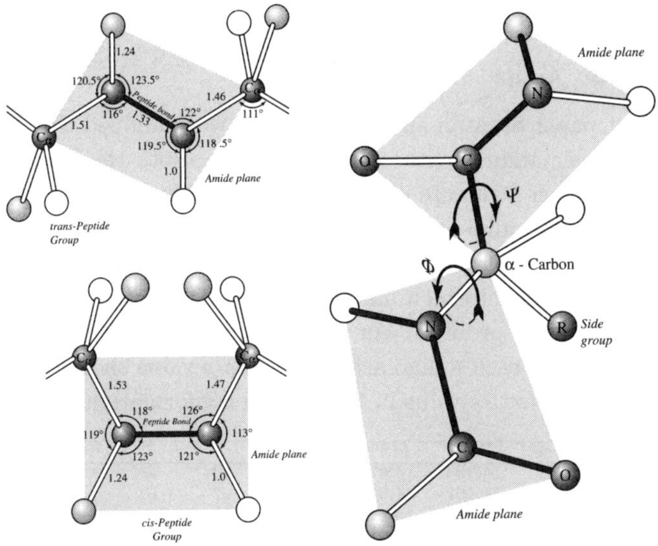

Figure 4. Possible Amide Plane Conformations

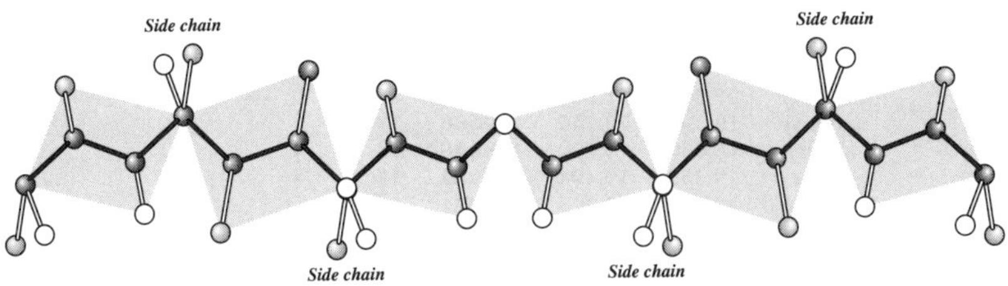

Figure 5. Network of Amide Planes in a Protein

Since the protein structure in space is uniquely determined by the interaction of the amino acids in 3-D, let us further examine the relationship of Steiner trees to the amino acids themselves.

4. Amino Acid Results

The SMT structure for each amino acid is felt to be an important part of the 3-D structure of proteins since everything must be a MEC. This amounts to the influence of local structure on the overall structure of the protein, yet if the amino acids themselves are MECs, then they also must bear some relationship to Steiner trees.

4.1. *Amino Acids*

In a recent paper Stanton and Smith(2004), we examined the Steiner ratio for the 20 basic amino acids found in proteins. Both the Steiner ratio and the actual Steiner structure of the individual acids are felt to be significant features of the overall 3-D structure of a protein. Since the Steiner ratio is a dimensionless quantity that signifies the geometry and the energy measure for a MEC, it is felt to be a unique and valuable performance measure in the protein folding problem. As we shall argue, if one knows the optimal SMT structure of each amino acid, then this ρ value should be an optimal scoring function for its value in the protein conformation.

Name	SMT	MST	ρ	V	$S' \subseteq S$	$\mathbf{C}+\mathbf{N}$
(Ala)	14.394	14.478	**0.9942**[a]	13	4	4
(Arg)	31.162	31.357	**0.9938**	27	10	10
(Asn)	19.382	19.481	**0.9949**	17	6	6
(Asp)	17.292	17.391	**0.9943**[a]	15	5	5
(Cys)	16.395	16.541	**0.9912**[a]	14	4	4
(Gln)	23.071	23.186	**0.9950**	20	7	7
(Glu)	20.991	21.096	**0.9950**	18	6	6
(Gly)	10.720	10.775	**0.9949**[a]	10	3	3
(His)	23.291	23.518	**0.9904**[b]	20	8	9
(Ile)	25.417	25.586	**0.9934**	22	7	7
(Leu)	25.447	25.588	**0.9945**	22	7	7
(Lys)	28.912	29.102	**0.9935**	25	8	8
(Met)	23.801	23.990	**0.9918**	20	6	6
(Phe)	27.222	27.300	**0.9972**[b]	23	9	10
(Pro)	19.491	19.756	**0.9866**	17	6	6
(Ser)	15.696	15.792	**0.9940**[a]	14	4	4
(Thr)	19.359	19.494	**0.9931**	17	5	5
(Trp)	32.253	32.478	**0.9931**[b]	27	12	13
(Tyr)	28.473	28.562	**0.9969**	24	10	10
(Val)	21.752	21.885	**0.9939**	19	6	6

[a] Those acids marked with *a* are optimal structures
[b] Those acids where some of the **C**, **N** atoms are leaf vertices

In Table 1 we have extended the optimal results of our previous experiments to include the fact that we have optimal SMTs for the acids Asp,

Cys, and Ser. Computational results attempting to identify the optimal SMT structure of the remaining amino acids have not been successful at this point in time, yet we feel that we are very close to the optimal SMT structure for each amino acid as we shall demonstrate. His, Phe, and Trp which have unique SMT topologies in that in each of these acids, certain of the carbon and nitrogen atoms act as terminal vertices due to the atomic structures involved.

4.1.1. *Glycine*

The first acid we will examine in some detail is glycine (Gly). The reason we will examine this one first is that it has the smallest number of atoms, $N = 10$ and its SMT structure is easy to compute.

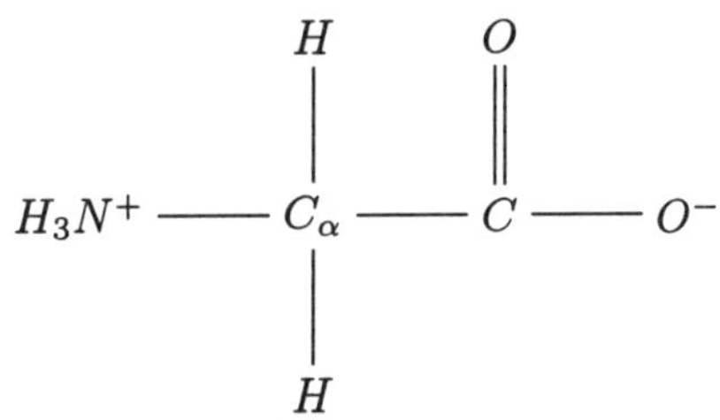

Gly Chemical Structure

Gly has a single H atom as a side chain. It is classified as a nonpolar amino acid and was first found (1820) as a component of gelatin, but it is also prevalent in fibroin, the protein of silk which we will examine in some detail in this paper. Its average occurrence in proteins is around 7.2% (Voet and Voet (1991).

Below is the optimal Steiner tree topology for Gly.

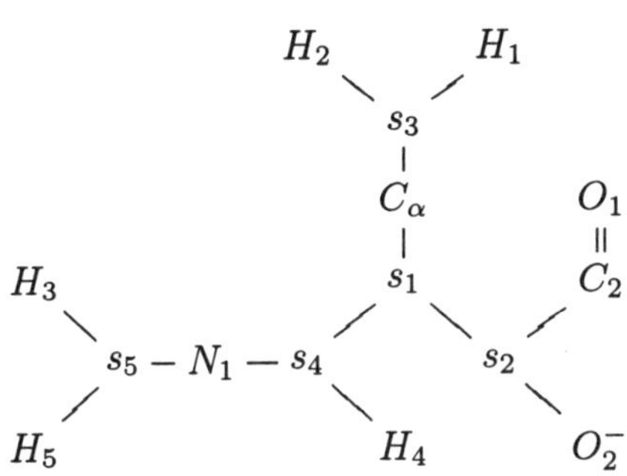

GLY Steiner Structure

The optimal 3-d SMT structure is depicted in Figure 6 where we have identified the clusters of the FST components of the given terminals via the Steiner tree topology which interconnects them. Two different viewpoints are provided. The SMT topology provides more information about the geometric structure of the amino acid than just the chemical diagram. The Steiner points are felt to be essential to define how the forces are transmitted through the molecular structure of the acid. Whether the Steiner points relate to some physical entity remains an open question. Defining the planes of interconnected atoms seems to be their primary role (*i.e.*) to define the conduit for the forces in E^3. The tetrahedral-planes isolate the FST components of the SMT acid structure. This helps explain why certain of the carbon and nitrogen atoms are Steiner points in the SMT structure.

In Figure 6, we have not shown the Melzak Circles, but they are implicit in the diagrams. For example, one Melzak Circle in either the diagrams of Figure 6 would pass through H_3, H_5 and the Steiner point linking them.

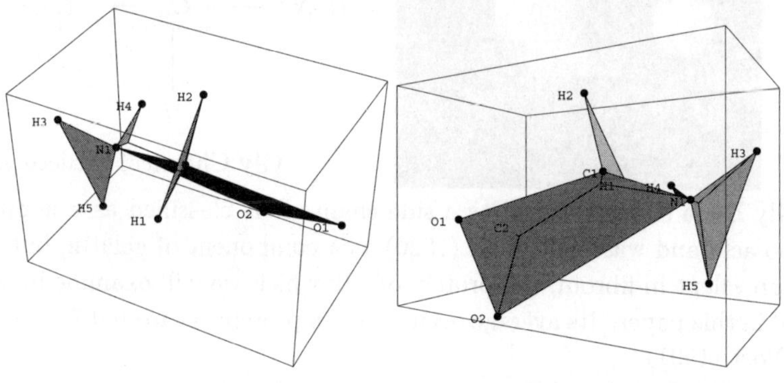

Figure 6. Glycine Optimal Structure Two Viewpoints

If we examine more carefully the two tetrahedral-planes within glycine which are incident to the C_α we see that the planes are essentially squashed tetrahedra with almost zero volume. The tetrahedra constructed are ones which are comprised of $\mathbf{T}_1 = \{C_\alpha, O_1, O_2, N_1\}$ and the other one which is comprised of $\mathbf{T}_2 = \{H_1, H_2, C_\alpha, s_3\}$. In $\mathbf{T}_1$, H_4 was excluded since it did not lie in the plane of $\mathbf{T}_1$. Notice that C_2 lies inside $\mathbf{T}_1$ as a Steiner point. It appears that $\mathbf{T}_1$ lies in a plane due to the double bond nature of the carbon-oxygen pair.

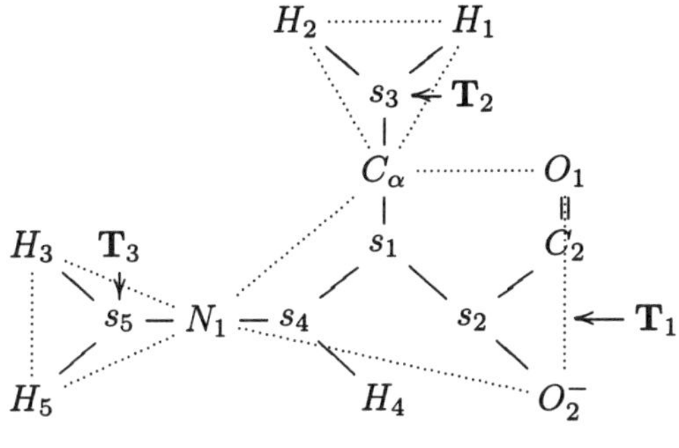

GLY Planar Structures

The volume of the $\mathbf{T}_1$ is 0.0993 while $\mathbf{T}_2$ is $0.55616x10^{-7}$. If we define planes from these structures, $\mathbf{P}_1 = \{C_\alpha, O_1, O_2\}$ and $\mathbf{P}_2 = \{C_\alpha, H_1, H_2\}$, the angle between their normals in the planes is 1.57055 radians which is essentially $\pi/2$. We can define another plane from Tetrahedron $\mathbf{T}_3$ as $\mathbf{P}_3 = \{N_1, H_3, H_5\}$ and it is incident to the plane $\mathbf{P}_4 = \{C_\alpha, N_1, O_2\}$ which has an angle of 1.57072 which is also essentially $\pi/2$. What is also interesting is that the angle between $\mathbf{P}_3$ and $\mathbf{P}_4$ is 1.04784 radians and $\mathbf{P}_2$ $\mathbf{P}_3$ is 1.048164 radians which are both essentially $\pi/3$ or 60°. Thus to summarize, the tables below array the pairs of twist angles of the above planes, the first matrix is output from MAPLE in degrees (to show the numerical accuracy involved), then the suggested rounded values in radians are included in the second matrix:

[0.0]	[89.99]	[60.05]	[0.0]
[89.99]	[0.0]	[59.90]	[89.99]
[60.06]	[59.90]	[0.0]	[60.04]
[0.0]	[89.99]	[60.04]	[0.0]

P	1	2	3	4
1	–	$\pi/2$	$\pi/3$	0
2		–	$\pi/3$	$\pi/2$
3			–	$\pi/3$
4				–

It is felt that one wouldn't necessarily know the results about the twist angles within the amino acid unless one first computed the SMT structure, so it appears that these results are novel. The force planes as defined by the SMT components appear to be based on the covalent bonds of the atoms. What is important to realize here is that the computer program locates the Steiner points coincident with the carbon and nitrogen atoms without any prior information or understanding that these would be Steiner points. Finally, what appears to be happening here is that the Steiner structure defines 3-D planes involving the atoms and these planes have certain regular dihedral angular relationships much as the Ramachandran angles Voet and

Voet (1991) in the peptide backbone.

4.1.2. *Alanine*

The second amino acid we shall examine is alanine (Ala) which has the next smallest number of atoms $N = 13$. Ala is also classified as a nonpolar amino acid and its side chain is made up of a methyl group CH_3. It is considered to be non-essential protein for mammals, yet its average occurrence in proteins is around 7.8%, (Voet and Voet (1991). It is also prevalent in silk which we will examine later. The chemical diagram of Ala follows. Below that is the optimal Steiner topology for Ala.

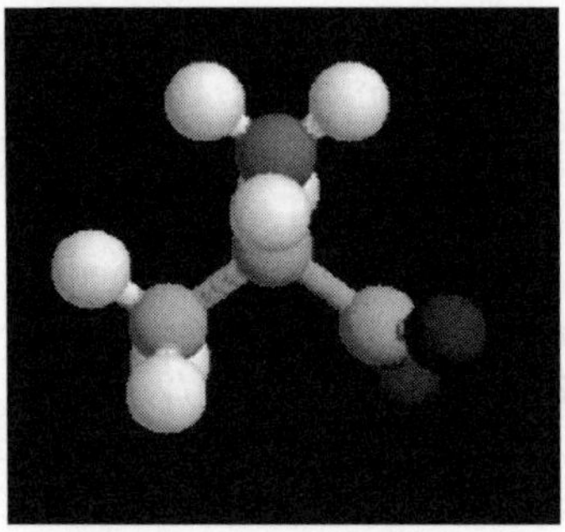

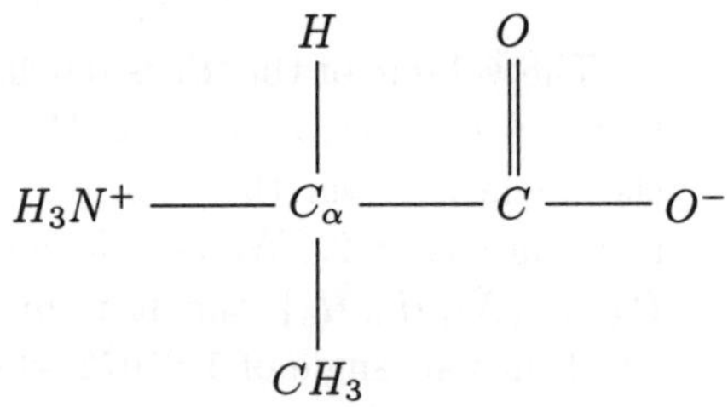

Ala Chemical Structure

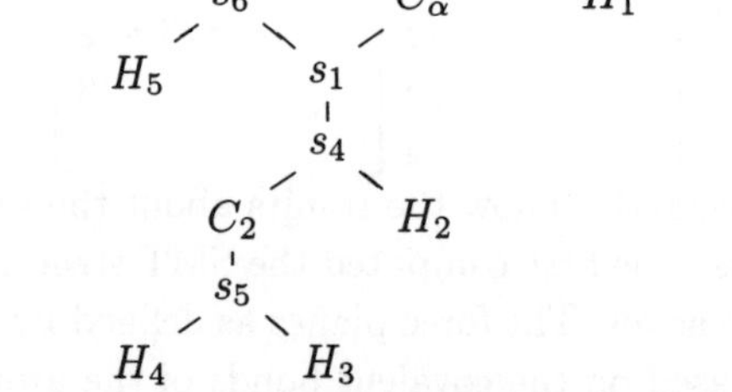

Ala Steiner Structure

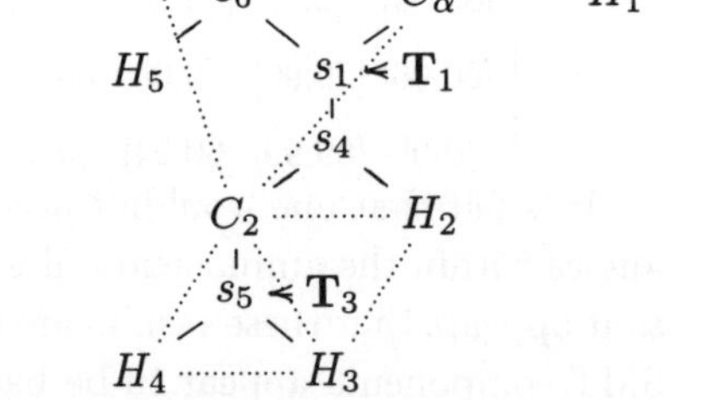

Ala Planar Structures

Figure 7 depicts the optimal SMT structure of Ala from two different viewpoints. Ala again has an interesting collection of FST tetrahedra and corresponding planes which form the SMT MEC structure as depicted in Figure 7. Figure 7 also illustrates the two central planar structures incident again to the C_α.

Let's again examine some of the tetrahedral structures within alanine. Please examine the decomposition of the Steiner tree topology above. The first tetrahedron is $\mathbf{T}_1 = \{C_\alpha, C_2, N_1, s_1\}$ while the tetrahedron orthogonal to it $\mathbf{T}_2 = \{O_1, O_2, H_1, C_\alpha\}$. Within $\mathbf{T}_2$, C_3 lies inside also as a Steiner point.

The volume of $\mathbf{T}_1 = 0.0003704$ while of $\mathbf{T}_2 = 0.01563$. They are squashed tetrahedra with essentially zero volume and almost orthogonal to each other. Further, if we examine some of the planes where $\mathbf{P}_1 = \{C_\alpha, N_1, C_2\}$ and $\mathbf{P}_2 = \{C_\alpha, H_1, O_2\}$, the angle between the normals of the planes is 1.57073 radians or essentially $\pi/2$. Also, some additional planes are $\mathbf{P}_3 = \{C_2, H_3, H_4\}$, $\mathbf{P}_4 = \{N_1, H_6, H_7\}$, and $\mathbf{P}_5 = \{C_2, H_2, H_3\}$ for the other tetrahedra. The planes and the atoms contained within defined for Ala are slightly different than that for Gly but what follows appears to be a regular property of the twist angles of these two acids. Again, the first matrix is output from MAPLE in degrees, while the second is the radian approximations:

$$\begin{bmatrix} [0.0] & [90.0] & [60.0] & [60.0] & [60.0] \\ [90.0] & [0.0] & [60.0] & [60.0] & [60.0] \\ [60.0] & [60.0] & [0.0] & [60.0] & [0.0] \\ [60.0] & [60.0] & [60.0] & [0.0] & [60.0] \\ [60.0] & [60.0] & [0.0] & [60.0] & [0.0] \end{bmatrix} \qquad \begin{array}{c} \mathbf{P} \\ 1 \\ 2 \\ 3 \\ 4 \\ 5 \end{array} \begin{array}{c} \begin{array}{ccccc} 1 & 2 & 3 & 4 & 5 \end{array} \\ \begin{bmatrix} - & \pi/2 & \pi/3 & \pi/3 & \pi/3 \\ & - & \pi/3 & \pi/3 & \pi/3 \\ & & - & \pi/3 & 0 \\ & & & - & \pi/3 \\ & & & & - \end{bmatrix} \end{array}$$

This appears to be a remarkably consistent result in that the twist angles are so regular and only from a small subset of possible angular values.

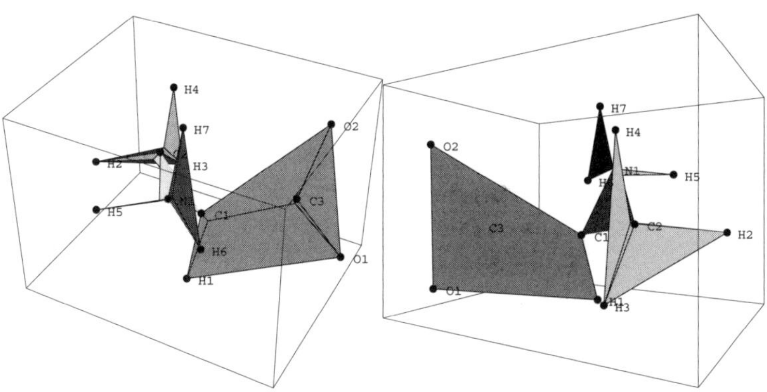

Figure 7. Alanine Optimal Structures

4.2. *Aspartic Acid*

The next acid we shall examine for its optimal structure is Aspartic acid. It has a total of 15 atoms. The chemical diagram of Asp appears below and just after it the optimal SMT topology appears.

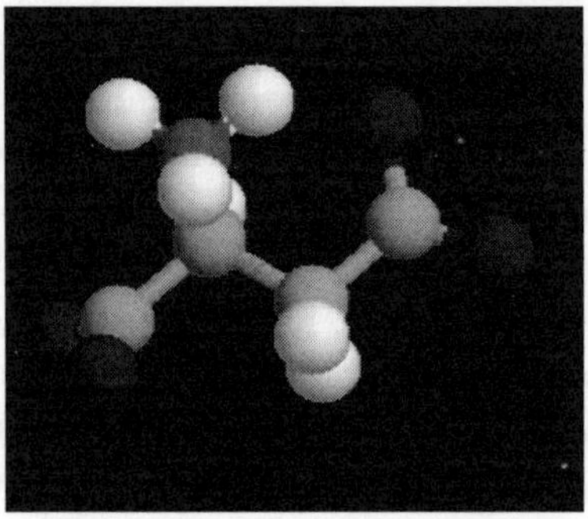

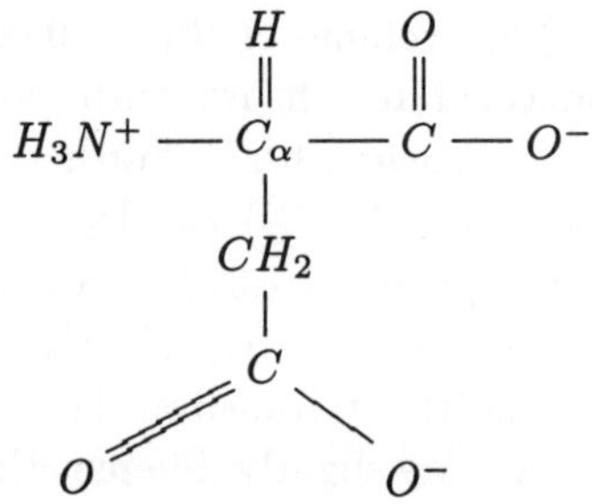

Asp Chemical Structure

Asp has a charged polar side chain but it is not essential for mammals. Its average occurrence in proteins is $\approx 5.3\%$ (Voet and Voet (1991). The side chain has a carboxylic acid component. The acid aids in the expulsion of ammonia from the body and apparently increases resistance to fatigue and increases endurance.

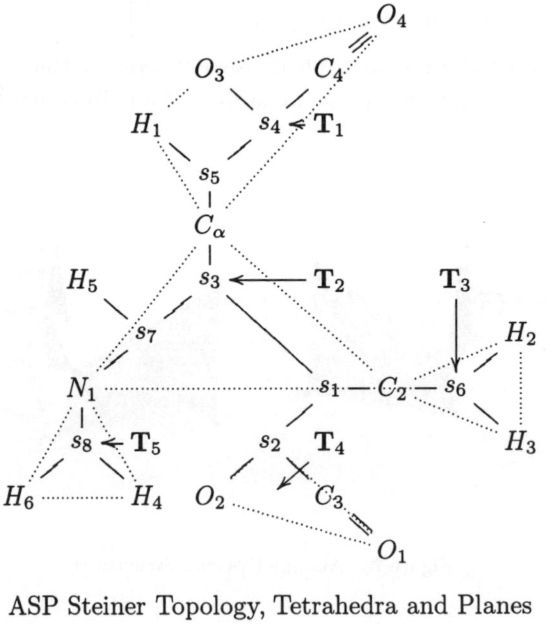

ASP Steiner Topology, Tetrahedra and Planes

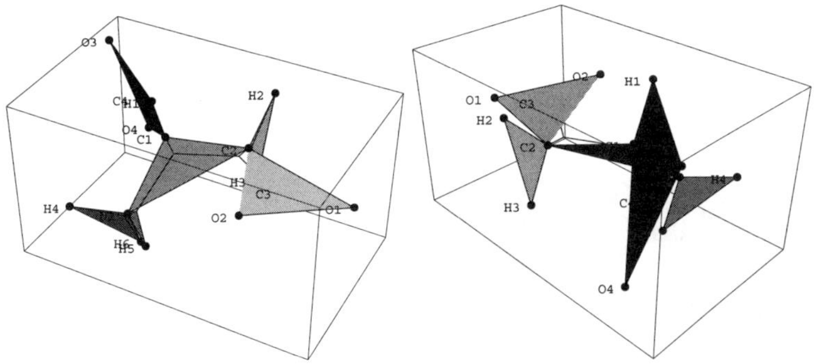

Figure 8. Aspartic Acid Optimal Structures

The optimal Steiner structure is indicated in Figure 8 from two different viewpoints. If we define planes and tetrahedra in the following manner where we are trying to maximize the number of atoms in a FST component of Asp, then we have identified the following subsets of atoms:-

$$\mathbf{T}_1 := \{O_3, O_4, C_\alpha, H_1\}; \mathbf{P}_1 := \{C_\alpha, H_1, O_3, O_4\}$$
$$\mathbf{P}_2 := \{C_\alpha, C_2, N_1\}; \mathbf{P}_3 := \{C_2, H_2, H_3\}$$
$$\mathbf{P}_4 := \{O_1, O_2, C_3\}; \mathbf{P}_5 := \{N_1, H_6, H_4\}$$

One sees again that C_4 lies within $\mathbf{T}_1$ as a Steiner point.

The first matrix is the MAPLE calculation of the twist angles in degrees, while the second matrix is a radian approximation to these angles:

$$\begin{bmatrix} [0.0] & [89.98] & [60.03] & [59.98] & [59.99] \\ [89.98] & [0.0] & [60.04] & [59.99] & [60.03] \\ [60.03] & [60.04] & [0.0] & [89.99] & [59.99] \\ [59.98] & [59.99] & [89.99] & [0.0] & [59.99] \\ [59.99] & [60.03] & [59.99] & [59.99] & [0.0] \end{bmatrix} \qquad \begin{array}{c} \mathbf{P} \\ 1 \\ 2 \\ 3 \\ 4 \\ 5 \end{array} \begin{array}{c} \begin{array}{ccccc} 1 & 2 & 3 & 4 & 5 \end{array} \\ \begin{bmatrix} - & \pi/2 & \pi/3 & \pi/3 & \pi/3 \\ & - & \pi/3 & \pi/3 & \pi/3 \\ & & - & \pi/2 & \pi/3 \\ & & & - & \pi/3 \\ & & & & - \end{bmatrix} \end{array}$$

Again, it is quite remarkable that the twist angles are all $\pi/2$ or else $\pi/3$ but one can additionally verify that this is the case upon inspecting Figure 8.

4.3. *Cysteine*

Cysteine is an amino acid with an uncharged polar side chain and has a thiol group which is unique among the 20 amino acids. Cysteine has a total of $N = 14$ atoms. It is particularly abundant in the proteins of hair, hooves, and the keratin of skin. It average occurrence in proteins is $\approx 1.9\%$ (Voet and Voet (1991). Below is the chemical structure of Cysteine and following that the chemical diagram is the optimal Steiner tree topology of Cys.

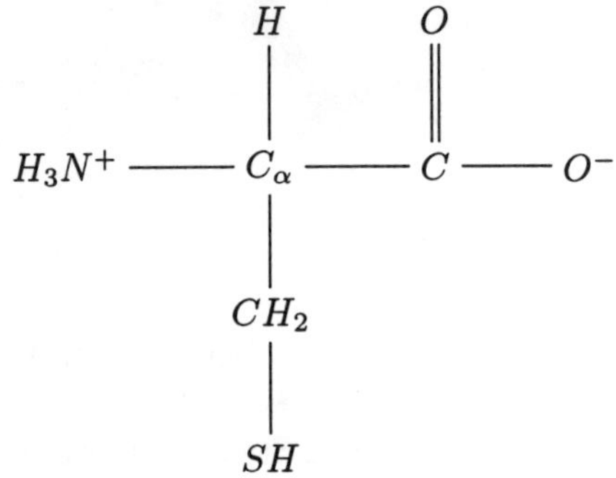

Cys Chemical Structure

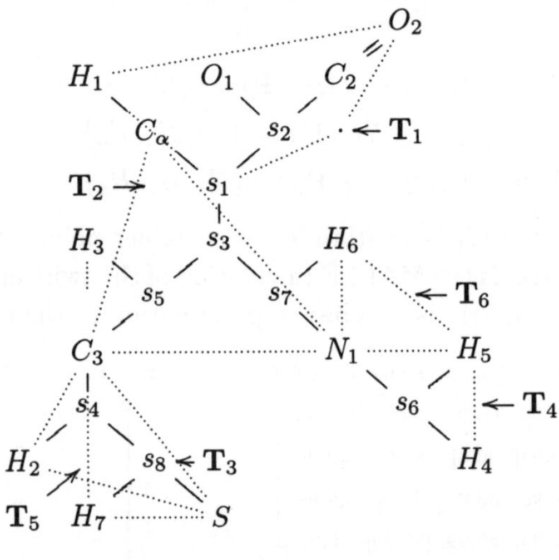

Cys Steiner Topology and Planes

The optimal Steiner structure is depicted in Figure 9 from two different viewpoints.

The first polyhedron of Cys we shall examine is comprised of $\mathbf{T}_1 = \{C_\alpha, H_1, O_1, O_2, C_2\}$ which actually has 5 vertices including the Steiner points, but you can see they all essentially lie in the same squashed plane as seen in the top diagram of Figure 9. C_2 lies in $\mathbf{T}_1$ as a Steiner point.

The tetrahedra and planes we have defined for Cys are as follows:

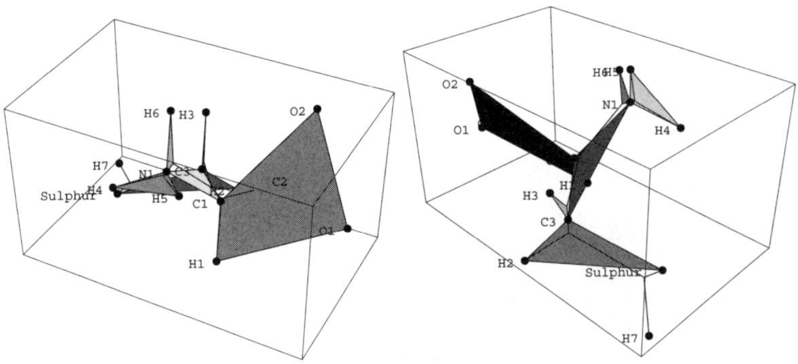

Figure 9. Cysteine Optimal Structure

$$\begin{aligned}
\mathbf{T}_1 &:=\{O_3, O_2, C_\alpha, H_1\}; \mathbf{P}_1 := \{H_1, O_2, C_\alpha\};\\
\mathbf{P}_2 &:=\{C_\alpha, C_3, N_1\}; \mathbf{P}_3 := \{S, C_3, H_2\};\\
\mathbf{P}_4 &:=\{N_1, H_4, H_5\}; \mathbf{P}_5 := \{N_1, H_5, H_6\};
\end{aligned}$$

Below is the MAPLE output of the planes we have defined, followed by our approximation of the twist angles:

$$\begin{bmatrix} [0.0] & [90.0] & [59.94] & [60.0] & [60.03] \\ [90.0] & [0.0] & [60.03] & [60.03] & [59.94] \\ [59.94] & [60.0] & [0.0] & [60.03] & [89.95] \\ [60.0] & [60.10] & [60.03] & [0.0] & [60.0] \\ [60.03] & [59.94] & [89.95] & [60.0] & [0.0] \end{bmatrix} \qquad \begin{array}{c|ccccc} \mathbf{P} & 1 & 2 & 3 & 4 & 5 \\ \hline 1 & - & \pi/2 & \pi/3 & \pi/3 & \pi/3 \\ 2 & & - & \pi/3 & \pi/3 & \pi/3 \\ 3 & & & - & \pi/3 & \pi/2 \\ 4 & & & & - & \pi/3 \\ 5 & & & & & - \end{array}$$

One caveat here is that because the Sulphur atom is unique to this acid, we see that one of the FST components includes $\{C_3, H_2, H_7, S\}$ so that the Steiner structure is a tetrahedron and not planar. This is clearly displayed in the lower drawing of Figure 9. As we examine more of the amino acids, the angles will not likely be restricted to the ones found in the previous structures.

4.4. *Serine*

The final optimal Steiner structure is an amino acid with an uncharged polar side chain. Ser has 14 atoms. It is not essential for the human diet but is important in metabolism. Serine was first obtained from silk protein

which we will examine in more detail in later sections of the paper. The average occurrence of Ser in proteins is ≈ 6.8% (Voet and Voet (1991). Below is the chemical structure of Serine, and following that is the optimal Steiner tree topology of Ser:

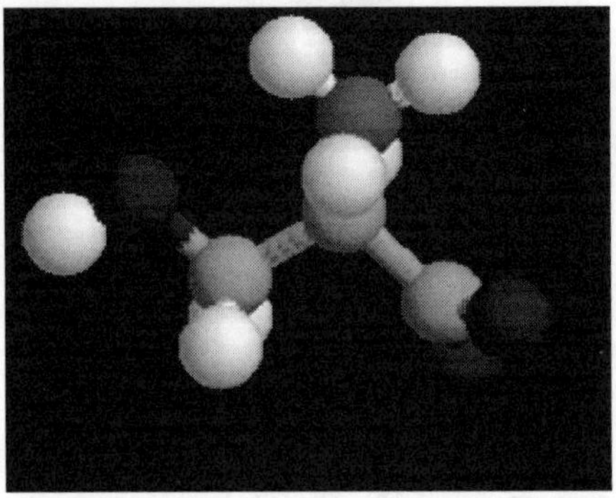

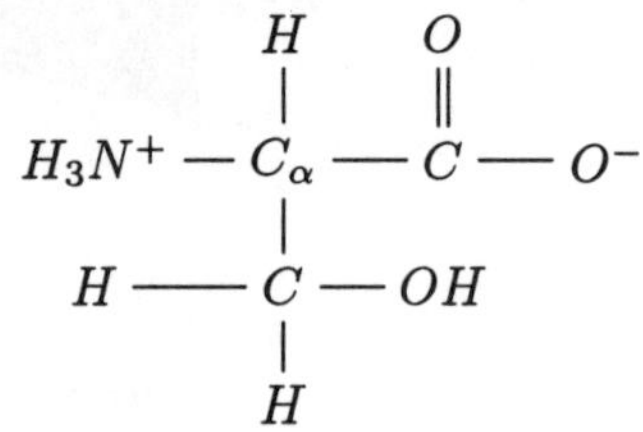

Ser Chemical Structure

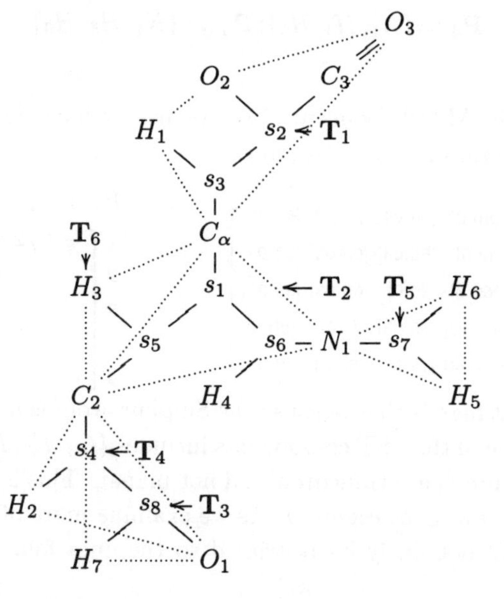

Ser Steiner Topology and Planes

Figure 10 illustrates the optimal Steiner structure of Ser again from two different viewpoints.

The planes we have defined for Ser are as follows:

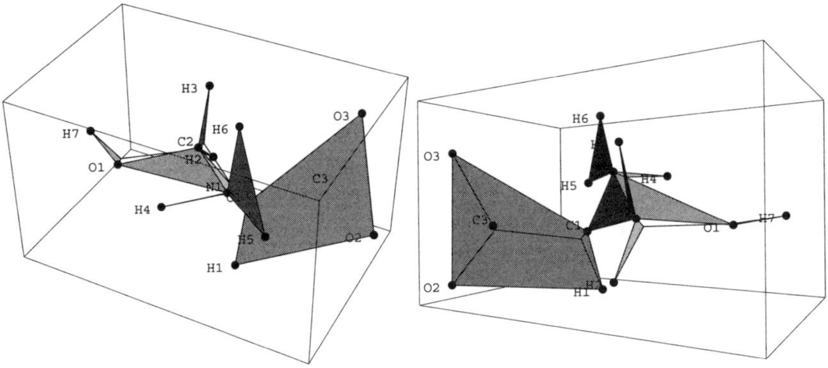

Figure 10. Serine Optimal Structure

$$
\begin{aligned}
&\mathbf{T}_1 := \{O_3, O_2, C_\alpha, H_1\}; \mathbf{P}_1 := \{O_2, C_\alpha, H_1\};\\
&\mathbf{T}_2 := \{C_\alpha, C_2, N_1, H_3\}; \mathbf{P}_2 := \{C_\alpha, C_2, N_1\};\\
&\mathbf{P}_3 := \{O_1, C_2, H_7\}; \mathbf{P}_4 := \{C_2, O_1, H_2\};\\
&\mathbf{P}_5 := \{N_1, H_5, H_6\}; \mathbf{P}_6 := \{C_\alpha, C_2, H_3\};
\end{aligned}
$$

Here again is the MAPLE output for the twist angles along with the approximation matrix:

$$
\begin{bmatrix}
[0.0] & [89.95] & [59.94] & [60.0] & [60.03] & [60.0] \\
[89.95] & [0.0] & [60.0] & [60.0] & [59.94] & [59.94] \\
[60.0] & [60.0] & [0.0] & [60.0] & [60.0] & [60.0] \\
[60.0] & [60.0] & [60.0] & [0.0] & [90.0] & [90.02] \\
[60.0] & [59.94] & [60.0] & [90.0] & [0.0] & [0.0] \\
[60.0] & [59.94] & [60.0] & [90.0] & [0.0] & [0.0]
\end{bmatrix}
\qquad
\begin{array}{c|cccccc}
\mathbf{P} & 1 & 2 & 3 & 4 & 5 & 6 \\
\hline
1 & - & \pi/2 & \pi/3 & \pi/3 & \pi/3 & \pi/3 \\
2 & & - & \pi/3 & \pi/3 & \pi/3 & \pi/3 \\
3 & & & - & \pi/3 & \pi/3 & \pi/3 \\
4 & & & & - & \pi/2 & \pi/2 \\
5 & & & & & & 0 \\
6 & & & & & & -
\end{array}
$$

Therefore, in all the optimal SMT topologies we have seen so far, there are many different planes defined by the Steiner trees. These planes often appear to be making dihedral angles which are either $0, \pi/2$ or $\pi/3$ which is rather surprising although when we begin to examine the secondary structures of silk and some of the other proteins, this regularity seems critical to the overall structure. Especially in silk, its orthogonal nature together with its primary composition of Gly and Ala makes sense.

5. Dipeptide Structures

What we wish to do in this part of the paper is examine the secondary structure of folded proteins to see how the Steiner ratio and dihedral properties occur in these structures and what the Steiner properties might have to say with regards to the rules involved in the folding problem as well as perhaps explain the differences in the energy levels of the protein structures. We shall assume the Thermodynamic Independence property for our analysis and first examine individual amino acids in the proteins, then pairs of acids, *i.e.* dipeptides.

5.1. *Silk Protein*

Silk is a protein produced in the posterior silk glands of the larva of the cultivated silk worm *Bombyx mori* for the construction of cocoons and it occurs in the webs of a number of various spiders (Voet and Voet (1991). The silk fibers are comprised of a protein called *fibroin* which is a theoretical model of the protein.

The protein is constructed from layers of antiparallel beta pleated sheets which run parallel to the fibre axis (Voet and Voet (1991). While each chain is comprised of multiple repeats of the sequence $(Gly - Ser - Gly - Ala - Gly - Ala)_n$ the protein is often approximated by a repeating units of $(Gly - Ala)_n$. It is this type of protein structure which we will analyze in some detail. We would like to understand the alignment and the twist angles of the Steiner planes in these two acids as perhaps a key to the protein structure of fibroin.

5.1.1. *Experiments for Individual Acids*

The two protein structures of fibroin that we will examine are from the paper by (Fossey et al. (1991). A sample of the ρ values from the amino acids is given in Table 2. It is surprising how consistent the ρ values are when compared with the "optimal" ρ value computed in Table 1 for these acid values.

Silk 1			Silk 2		Optimal
Atoms	Acid	ρ	Atoms	ρ	ρ
7-16	Ala	0.9939	7-16	0.9938	**0.9942**
17-23[a]	Gly	0.9957	17-23	0.9951	**0.9949**
24-33	Ala	0.9939	24-33	0.9938	**0.9942**
34-40[b]	Gly	0.9957	34-40	0.9951	**0.9949**
41-50	Ala	0.9940	41-50	0.9938	**0.9942**
51-57[c]	Gly	0.9956	51-57	0.9951	**0.9949**
71-80	Ala	0.9939	78-87	0.9940	**0.9942**
81-87[d]	Gly	0.9957	88-94	0.9953	**0.9949**
88-97	Ala	0.9940	95-104	0.9940	**0.9942**
98-104[e]	Gly	0.9957	105-111	0.9953	**0.9949**

[a] First dipeptitde pair
[b] Second pair
[c] Third pair
[d] Fourth pair
[e] Fifth pair

Within this Table 2, we have defined 5-pairs of dipeptides as indicated by the horizontal lines separating the pairs of Ala-Gly data sets. Fossey et. al showed that Silk2 was 1-kcal/mol per residue more stable than Silk1, thus the Silk2 conformation is more energy efficient than Silk1. It is difficult to see how this difference in energy is possible from the ρ values. What we wish to do is further examine the twist angles as well as the volume of the tetrahedra within the acids to see if we can measure the difference in the energy level from the geometry of the protein structures.

5.1.2. *Ala-Gly Dipepetides*

It is important to look at the overall geometric structure of the silk data sets. Figure 11 illustrates the total protein for the two different data sets wherein Silk2 is obviously more regular and symmetric than Silk1. But how can one account for this quantitatively? We must look more carefully at the details of the structures.

Figure 12 illustrates two adjacent amino acids, a dipepetide of Ala-Gly, called alanylglycine, from the data sets in Silk1 and Silk2, the first dipeptide pair in Table 2. The left figure is for Silk1 while the right one is for Silk2. The figures represent the planes of selected atoms in the same structures. Clearly Silk1 does not have as a definitive planar structure as Silk2 does.

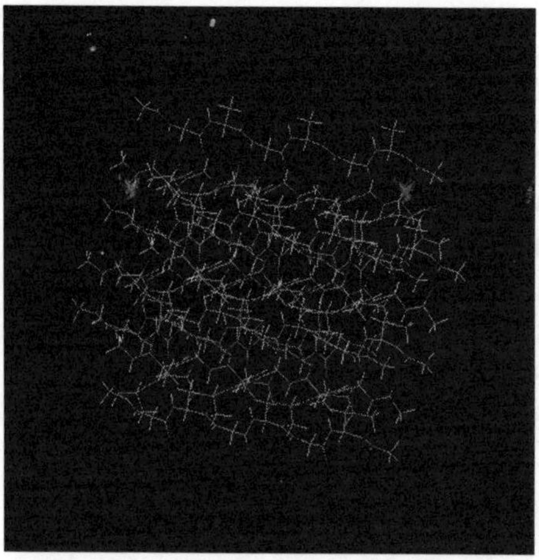

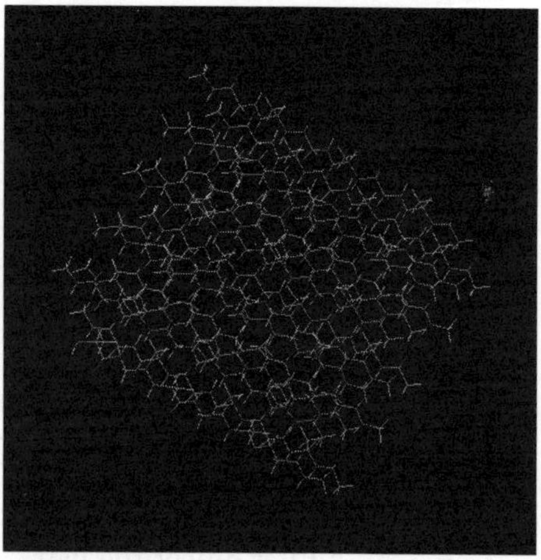

Figure 11. Silk1 (left) vs. Silk2 (right)

In both proteins we have identified the following six sets of atoms for analysis where after the subscript the atom from the Gly is denoted by little"g" and the atoms from Ala by little "a":

$$\mathbf{T}_1 : \{O_{1g}, C_{\alpha g}, N_{1g}, C_{2g}\}; \mathbf{P}_1 : \{O_{1g}, C_{\alpha g}, N_{1g}\}$$
$$\mathbf{T}_2 : \{O_{1a}, N_{1g}, C_{\alpha a}, N_{1a}\}; \mathbf{P}_2 : \{O_{1a}, N_{1g}, C_{\alpha a}\}$$
$$\mathbf{P}_3 : \{H_{2g}, H_{3g}, C_{\alpha g}\} \mathbf{P}_4 : \{H_{3a}, H_{4a}, C_{3a}\}$$
$$\mathbf{P}_5 : \{N_{1a}, C_{\alpha a}, C_{3a}\} \mathbf{T}_3 : \{H_{1a}, N_{1a}, C_{3a}, H_{5a}\}; \mathbf{P}_6 : \{H_{1a}, N_{1a}, C_{3a}\};$$

As one measure of detail, this can be clearly seen in the Figure 12, the planes for the atoms in $\mathbf{T}_2 : \{O_{1a}, N_{1g}, C_{\alpha a}, N_{1a}\}$ of Silk1 and Silk2 are clearly demarcated. The results of our experiments here are unfortunately not optimal, since $N = 17$. In the results which follow, the same amount of computing time was expended for both data sets seeking the optimal SMT topology.

Below we array the matrix of twist angles for the planes of the first of the two data sets. There is remarkable similarity between the values of the angles although there are more varied numerically than we had before for the individual isolated acids.

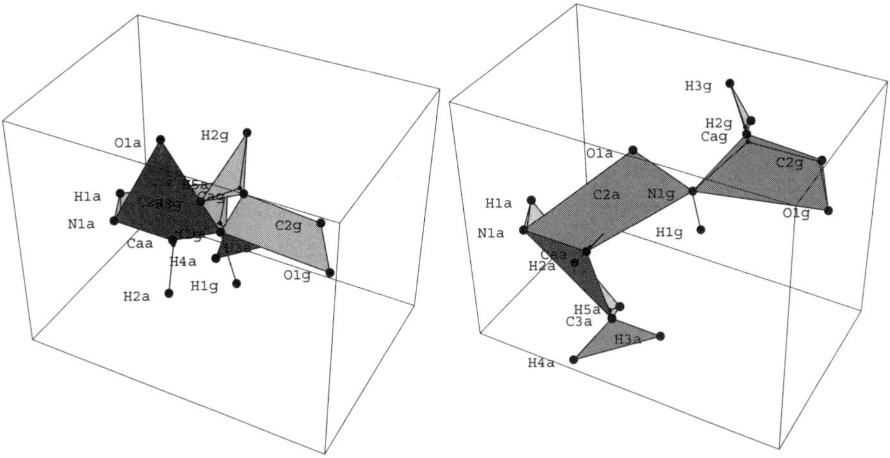

Figure 12. Silk1 (above) vs. Silk2 (below) 1st Dipeptide

$$\begin{bmatrix} [0.0] & [32.43] & [90.02] & [3.438] & [56.98] & [71.74] \\ [32.43] & [0.0] & [82.38] & [30.46] & [82.44] & [88.07] \\ [90.02] & [82.38] & [0.0] & [88.46] & [66.17] & [52.55] \\ [3.533] & [30.47] & [88.46] & [0.0] & [60.0] & [74.71] \\ [56.94] & [82.44] & [66.17] & [60.0] & [0.0] & [17.81] \\ [71.74] & [88.07] & [52.55] & [74.71] & [17.77] & [0.0] \end{bmatrix} \quad \begin{bmatrix} [0.0] & [29.67] & [90.02] & [8.028] & [51.76] & [62.61] \\ [29.66] & [0.0] & [86.51] & [22.87] & [76.33] & [60.51] \\ [90.02] & [86.51] & [0.0] & [86.74] & [65.25] & [33.04] \\ [8.148] & [22.90] & [86.74] & [0.0] & [59.87] & [57.26] \\ [51.76] & [76.33] & [65.25] & [59.87] & [0.0] & [74.90] \\ [62.61] & [60.51] & [33.04] & [57.26] & [74.90] & [0.0] \end{bmatrix}$$

Ala1-Gly1 (Silk1) Matrix Ala1b-Gly1b (Silk2) Matrix

The matrix of twist angles between the planes $1-5$ are fairly consistent while the twist angles in the last column #6 are not consistent between the two structures.

Now rather than examining each of the five dipeptides data sets from Table 2, let's examine the first 10 dipeptides of the two data sets, and average the dihedral angles, we achieve the following two matrices:

$$\begin{bmatrix} 0.0 & [33.48] & [90.0] & [5.579] & [55.95] & [72.06] \\ [33.49] & 0.0 & [85.56] & [32.72] & [84.33] & [85.36] \\ [90.0] & [85.56] & 0.0 & [84.89] & [67.15] & [50.80] \\ [5.714] & [32.72] & [84.91] & 0.0 & [59.96] & [76.79] \\ [55.94] & [84.33] & [67.15] & [59.97] & 0.0 & [20.24] \\ [72.06] & [85.36] & [50.80] & [76.79] & [20.24] & 0.0 \end{bmatrix} \quad \begin{bmatrix} 0.0 & [31.83] & [90.0] & [8.181] & [55.31] & [61.89] \\ [31.84] & 0.0 & [86.91] & [31.12] & [82.84] & [61.57] \\ [90.0] & [86.91] & 0.0 & [82.80] & [67.20] & [31.69] \\ [8.246] & [31.12] & [82.80] & 0.0 & [60.75] & [54.34] \\ [55.31] & [82.84] & [67.20] & [60.75] & 0.0 & [76.88] \\ [61.89] & [61.57] & [31.69] & [54.34] & [76.88] & 0.0 \end{bmatrix}$$

Silk 1 (n=10) Data Sets Matrix Silk 2 (n=10) Data Sets Matrix

While this is conjecture, it appears that the dihedral angle matrix of

Silk2 is tending towards the following ideal values which are also based upon the previous results for the angles found in Ala and Gly and would tend to make the most sense:

$$\begin{bmatrix} 0.0 & [30.00] & [90.00] & [0.00] & [55.00] & [60.00] \\ [30.00] & 0.0 & [90.00] & [30.00] & [90.00] & [60.00] \\ [90.0] & [90.00] & 0.0 & [90.00] & [68.00] & [30.00] \\ [0.00] & [30.00] & [90.00] & 0.0 & [60.00] & [55.00] \\ [60.00] & [90.00] & [60.00] & [60.00] & 0.0 & [75.00] \\ [60.00] & [60.00] & [30.00] & [60.00] & [75.00] & 0.0 \end{bmatrix}$$

There are three new angles in this sets of $55.00, 68.00, 75.00$ simply because they seem to be occurring in the resulting data set. Why they occur seems to be due to the way the acids in the dipeptide function together. For the average volume of the tetrahedra in the two proteins (sample size N=10) dipeptides, we have for Silk1 in the first column and following it Silk2 in the second column:

$$\mathbf{T}_1 : \{O_{1g}, C_{\alpha g}, N_{1g}, C_{2g}\} = 0.36062 \quad \mathbf{T}_1 : \{O_{1g}, C_{\alpha g}, N_{1g}, C_{2g}\} = 0.20010$$
$$\mathbf{T}_2 : \{O_{1a}, N_{1g}, C_{\alpha a}, N_{1a}\} = 0.58918 \quad \mathbf{T}_2 : \{O_{1a}, N_{1g}, C_{\alpha a}, N_{1a}\} = 0.55934$$
$$\mathbf{T}_3 : \{H_{1a}, N_{1a}, C_{3a}, H_{5a}\} = 0.24802 \quad \mathbf{T}_3 : \{H_{1a}, N_{1a}, C_{3a}, H_{5a}\} = 0.16627$$

Thus, we can see that the tetrahedra in Silk2 are closer to being in the plane which is what we should expect. They are not exactly planar, but closer than the conformation for Silk1. Thus, from the twist angle results and the tetrahedra as defined by the Steiner trees, Silk2's smaller energy seems to go hand in hand with the more regular twist angles and the more regular planar structure.

5.2. *Gly-Ala Dipeptide*

To balance our previous discussion, it is worth examining how the Gly-Ala dipeptide sequence operates and what angles are found in this sequence. The angles in this dipeptide direction were expected to be different.

Again, this is conjecture, for the Silk2 protein, that the twist angles seem to be converging towards these estimated values which are similar to the values found in the Ala-Gly dipeptide:

$$\begin{bmatrix} 0.0 & [55.00] & [68.00] & [30.00] & [30.00] & [90.00] & [55.00] \\ [55.00] & 0.0 & [60.00] & [30.00] & [30.00] & [90.0] & [0.00] \\ [68.00] & [60.00] & 0.0 & [75.00] & [75.00] & [68.00] & [60.00] \\ [30.00] & [30.00] & [75.00] & 0.0 & [0.00] & [75.00] & [30.00] \\ [30.00] & [30.00] & [75.00] & [0.00] & 0.0 & [75.00] & [30.00] \\ [90.00] & [90.0] & [68.00] & [75.00] & [75.00] & 0.0 & [90.00] \\ [55.00] & [0.00] & [60.00] & [30.00] & [30.00] & [90.00] & 0.0 \end{bmatrix}$$

It is also interesting when we examine the Gly-Ala dipeptide that the three new angles that we found in the Ala-Gly dipeptide $55.00, 68.00, 75.00$ over and above the $\{0, 30, 60, 90\}$ also occur in the Gly-Ala case. It is expected that the set of angles $\{0, 30, 55, 60, 68, 75, 90\}$ will also appear in the other amino acids when they are examined.

6. Summary and Conclusions

We have demonstrated a property of the twist angles of planes identified by the Steiner trees of certain amino acids where we have the optimal topology of the amino acid. We have taken a subset of the 20 amino acids and shown that there is a regular angular structure comprised of twist angles from the set $\{0, \frac{\pi}{3}, \frac{\pi}{2}\}$ for planes of the atoms defined through the Steiner tree of the acids. This regular twist angular structure also appears to be very consistent and regular within the fibroin structure involving Ala-Gly and Gly-Ala dipeptides. Whether this regular twist angular pattern carries over to the other remaining amino acids and their dipeptide pairs is a subject of future research.

References

Brandon, C., J. Tooze. 1991. *Introduction to Protein Structure.* Gabriel Publishing, New York.

Cohen, F.E. 1995. Folding the sheets: using computational methods to predict the structure of proteins. E. Lander, ed. *Calculating the Secrets of Life.* National Academy of Sciences, Washington D.C. 236–271.

Dickerson, R.E., I. Geis. 1969. *The Structure and Action of Proteins.* Harper and Row, New York.

Dill, K.A. 1997. Additivity principles in biochemistry. *J. Biological Chemistry* **272** 701–704.

Dijkstra, B.W., K.H. Kalk, W.G.J Hol, and J. Drenth, 1981 "Structure of Bovine Pancreatic Phospholipase a2 at 1.7 angstroms resolution." *J. Mol. Biol.* **147**, 97-123.

Du, D.Z. 1991. On Steiner ratio conjectures, *Annals of Operations Research* **33** 437–449.

Du, D.Z., F.K. Hwang. 1992. A proof of the Gilbert-Pollak conjecture on the Steiner ratio. *Algorithmica* **7** 121–135.

Du, D.Z., F.K. Hwang, J.F. Weng. 1982. Steiner minimal trees on zig-zag lines. *Trans. Amer. Math Soc.* **278** 149–156.

Fossey, S.A., G. Nemethy, K.D. Gibson, H.A. Scheraga. 1991. Conformational energy studies of beta-sheets of model silk fibroin peptides. I. sheets of poly(Ala-Gly) chains. *Biopolymers* **31** 1529.

Gilbert, E.N., H.O. Pollak. 1968. Steiner minimal trees. *Siam J. Appl. Math* **16** 1–29.

Holm, L. and C. Sander, 1992. "Evaluation of protein models by atomic solvation preference." *J. Mol. Biol.***225**, 93-105.

Lazardis, T. and M. Karplus, 1998. "Discrimination of the Native nad Misfolded Protein Models with an Energy Function Including Implicit Solvation," *J. Mol. Biol.***288**, 477-487.

Leach, A.R. 1996. *Molecular Modeling — Principles and Applications*, 2nd. ed. Prentice-Hall, New York.

Mark, A.E. and W. F van Gunsteren, 1994. Decomposition of the Free Energy of a System in Terms of Specific Interactions." *J. Mol. Biol.* **240** 167-176.

Ryan-Vollmar, S. *Natural Health*, January 1, 1999, 1 page.

Stanton, C. and J. MacGregor Smith, "Steiner Trees and 3-D Macromolecular Conformation," *Informs Journal of Computing, 16* (4), 470-485, 2004.

Voet and Voet. Biochemistry. 2nd edition. Wiley: New York, 1995.

STEINER TREES AS INTRAMOLECULAR NETWORKS OF THE BIOMACROMOLECULAR STRUCTURES

RUBEM P. MONDAINI

Federal University of Rio de Janeiro - UFRJ
Technology Centre, COPPE
21941-972, Rio de Janeiro, RJ, Brazil, P.O. Box 68511
E-mail: mondaini@cos.ufrj.br, rpmondaini@gmail.com

Some recent results on the mathematical modelling of biomacromolecular structures are reported. The usefulness of the concept of Intramolecular Networks as the physical scaffolds to keep these structures stable is emphasized. The guidelines of an analytical description are then given in order to derive a Steiner Ratio Function to study these structures and their stability instead of the usual numerical algorithms for Potential Energy minimization.

Our fundamental aim in the present work is to show some proposals which were made to unveil Nature's building code of molecular organization. This should be done by accommodating the phenomenological approach of potential energy description with its diverse contributions according the several codes already proposed[3] in an unified idea. We think that the consideration of Steiner Trees as a fundamental part of intramolecular structure[9] can shed light upon the phenomena related to biomacromolecular organization like chirality and an unified energy description in a Statistical Mechanics context. Recently[10,15,13,14] some new results were obtained on Euclidean Steiner Trees. These results have introduced for the first time the possibility of a full description of intramolecular interactions in an unified scheme. They give also the possibility of analytical description by the definition of a new Constrained Optimization Problem. In other words, the problem of Potential Energy minimization of a biomacromolecular structure[9] is going to be solved by the problem of length minimization of a Steiner Tree if the plausible assumption of the equality of interaction strength of an atom with its nearest neighbors is introduced. We hope to contribute in the future derivation of the solution of an optimization problem. Namely, the construction of a Potential Energy function from the concepts of Steiner points and Steiner trees instead the usual terms of phenomenological origin from the literature on molecular dynamics[2]. The robustness of the approach which we propose in this work seems to point out that we are in the right track of a new successful theory.

1. An Analytical approach to the search of Minimal Potential Energy configurations

This section is based on the work which was done with helical point sets where the points are evenly spaced along a right circular helix. Before deriving the fundamental formulae, we shall introduce the concepts of spanning trees and Steiner Trees.

Let A be a finite set of points in a metric manifold M. Let us consider all the possible ways of connecting pairs of points on each set of the manifold M through edges of minimal length. The resulting edges are geodesic arcs of the manifold. We shall collect only the subsets of A and their associated edges which form trees. A tree of the manifold which interconnects all the points of a given subset, it is a spanning tree for this subset. Among all the possible spanning trees s of the set A, with length $l_{SP}(s, A)$, there is one whose overall length is minimum. This is the Minimum Spanning Tree of set A, MST(A). Its length is

$$l_{MST}(A) = \min_{(s-\text{trees})} l_{SP}(s, A). \tag{1}$$

Let us now allow for the introduction of additional points on each set A of the manifold M in order to have spanning trees of smaller overall length. If we make the additional requirement of tangent lines to three geodesic edges meeting at an angle of 120° on each additional (Steiner) point we get a Steiner Tree (ST). Among all the Steiner trees t of the set A with length $l_{ST}(t, A)$, there is one whose overall length is minimum. We call it the Steiner Minimal Tree of the set A, SMT(A). Its length is

$$l_{SMT}(A) = \min_{(t-\text{trees})} l_{ST}(t, A). \tag{2}$$

The Minimum Spanning tree MST(A) is considered as the worst approximation to the Steiner Minimal Tree, SMT(A), or the "worst cut" for each set $A \subset M$. An useful measure of this approximation is the Steiner Ratio of the set $A \subset M$,

$$\rho(A) = \frac{l_{SMT}(A)}{l_{MST}(A)}. \tag{3}$$

The Steiner Ratio of the manifold ρ_M is then defined as the infimum of all the values $\rho(A)$ for all the sets A, or,

$$\rho_M = \inf_{A \subset M} \rho(A). \tag{4}$$

We henceforth take the 3-dimensional Euclidean Space E^3 as the metric manifold M. Our fundamental pattern of input points will be given by sets of evenly spaced points along a right circular helix of unit radius. We have,

$$P_j = (\cos j\omega, \sin j\omega, \alpha j\omega); \quad 0 \leq j \leq n-1, \tag{5}$$

where ω is the angular coordinate and $2\pi\alpha$ stands for the pitch of the helix.

After exhaustive computational experiments, we got the result that the corresponding Steiner points belong to another helix of the same pitch and smaller radius or,

$$S_k = (r(\omega,\alpha)\cos k\omega, r(\omega,\alpha)\sin k\omega, \alpha k\omega); \quad 1 \leq k \leq n-2, \tag{6}$$

where,

$$r(\omega,\alpha) = \frac{\alpha\omega}{\sqrt{A_1(A_1+1)}}; \quad A_1 = 1 - 2\cos\omega. \tag{7}$$

The function $r(\omega,\alpha)$ above is easily obtained from the requirement of meeting edges at an angle of 120° on each Steiner point[9,11,8]. To be rigorous, we should write,

$$r(\omega,\alpha) = \mathrm{Max}\left\{1, \frac{\alpha\omega}{\sqrt{A_1(A_1+1)}}\right\} \tag{8}$$

We now introduce a generalization of the formulae above by thinking on subsequences of input and Steiner points, corresponding to non-consecutive points. These subsequences are of the form:

$$(P_j)_{m,\,l_{Pmax}} : P_j, P_{j+m}, P_{j+2m}, \ldots, P_{j+l_{Pmax}.m} \tag{9}$$

$$(S_k)_{m,\,l_{Smax}} : S_k, S_{k+m}, S_{k+2m}, \ldots, S_{k+l_{Smax}.m} \tag{10}$$

where l_P, l_S are the number of intervals of skipped points before the present point on each subsequence and $(m-1)$ is the number of skipped points.

We also have:

$$0 \leq l_P \leq l_{Pmax} = \left[\frac{n-j-1}{m}\right]; \quad 1 \leq l_S \leq l_{Smax} = \left[\frac{n-k-2}{m}\right] \tag{11}$$
$$0 \leq j \leq m-1, \quad 1 \leq k \leq m$$

and the square brackets $[x]$ stand for the greatest integer value $\leq x$.

The sequences corresponding to Eqs. (5) and (6) are of course included in the scheme above. They are $(P_0)_{1,n-1}$ and $(S_1)_{1,n-2}$, respectively. In the general case, we can define new sequences of n and $n-2$ points instead of those given by Eqs. (5) and (6). We shall have respectively,

$$\mathbb{P}_m = \bigcup_{j=0}^{m-1} (P_j)_{m,l_{jmax}}; \qquad \mathbb{S}_m = \bigcup_{k=1}^{m} (S_k)_{m,l_{kmax}}. \tag{12}$$

The present development is independent of a specific coordinate representation of the points. If we now assume helical point sets whose points are evenly spaced along right circular helices, we get,

$$P_{j+l_P m} = (\cos(j+l_P m)\omega, \sin(j+l_P m)\omega, \alpha(j+l_P m)\omega), \tag{13}$$

$$S_{k+l_S m} = (r_m(\omega,\alpha)\cos(k+l_S m)\omega, r_m(\omega,\alpha)\sin(k+l_S m)\omega, \alpha(k+l_S m)\omega). \tag{14}$$

The function $r_m(\omega,\alpha)$ is obtained through the same requirement of meeting edges at 120° on each Steiner point. We have, analogously,

$$r_m(\omega,\alpha) = \mathrm{Max}\left\{1, \frac{m\alpha\omega}{\sqrt{A_m(A_m+1)}}\right\} \tag{15}$$

where,

$$A_m = 1 - 2\cos(m\omega). \tag{16}$$

In Figure 1, we show some sequences of input points for $n = 23$.

From Eqs. (13) and (14) and Figure 1, we can write for the length of the spanning trees,

$$\begin{aligned} l_{SP}(m,\omega,\alpha) = &\sqrt{m^2\alpha^2\omega^2 + A_m + 1}\, \textstyle\sum_{j=0}^{m-1} \left[\frac{n-j-1}{m}\right] \\ &+ (m-1)\sqrt{\alpha^2\omega^2 + A_1 + 1} \end{aligned} \tag{17}$$

The length of the Steiner Trees is then,

$$\begin{aligned} l_{ST}(m,\omega,\alpha) = &(1 - r_m(\omega,\alpha))\left(m + \sum_{k=1}^{m}\left[\frac{n-k-2}{m}\right]\right) \\ &+ \sqrt{m^2\alpha^2\omega^2 + r_m(\omega,\alpha)(A_m+1)}\ .\ \sum_{k=1}^{m}\left[\frac{n-k-2}{m}\right] \\ &+ 2\sqrt{m^2\alpha^2\omega^2 + (1-r_m(\omega,\alpha))^2 + r_m(\omega,\alpha)(A_m+1)}. \end{aligned} \tag{18}$$

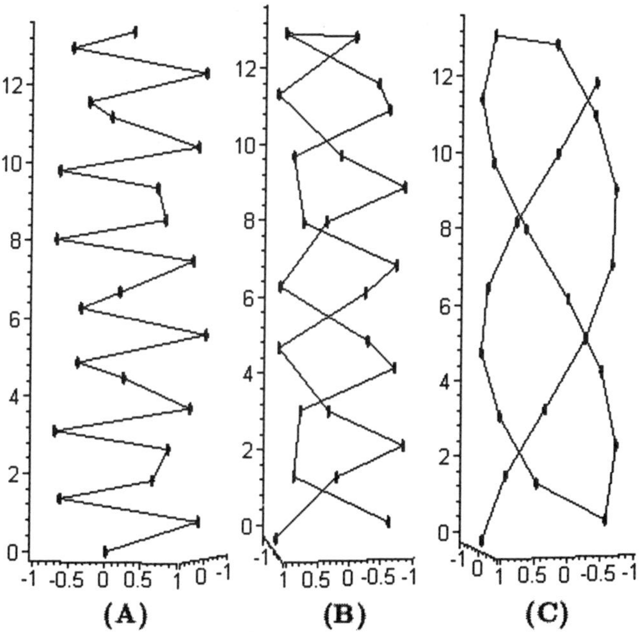

Figure 1. **(A)** The sequence $n = 23$, $m = 1$, $j = 0$; **(B)** The union of the subsequences $n = 23$, $m = 2$, $j = 0$ and $n = 23$, $m = 2$, $j = 1$; **(C)** The union of the subsequences $n = 23$, $m = 3$, $j = 0$; $n = 23$, $m = 3$, $j = 1$ and $n = 23$, $m = 3$, $j = 2$.

After using some useful relations like,

$$\sum_{j=0}^{m-1} \left[\frac{n-j-1}{m}\right] = n - m \tag{19}$$

$$\sum_{k=1}^{m} \left[\frac{n-k-2}{m}\right] = n - m - 2 \tag{20}$$

and taking the limit for $n \gg m$, we get,

$$l_{SP}(m, \omega, \alpha) = n\sqrt{m^2\alpha^2\omega^2 + A_m + 1} \tag{21}$$

$$l_{ST}(m, \omega, \alpha) = n\left(1 + m\alpha\omega\sqrt{\frac{A_m}{A_m + 1}}\right) \tag{22}$$

Following the prescriptions for writing the Steiner Ratio, we can write for the Steiner Ratio Function of very large helical point sets with points evenly spaced along right circular helices,

$$\rho(\omega,\alpha) = \frac{\min_{(m)}\left(1 + m\alpha\omega\sqrt{\frac{A_m}{A_m+1}}\right)}{\min_{(m)}\left(\sqrt{m^2\alpha^2\omega^2 + A_m + 1}\right)} \tag{23}$$

where the min process above should be understood in the sense of a piecewise function formed by the functions corresponding to the values $m = 1, 2, 3, \cdots$.

Eq. (23) is our proposal for a Steiner Ratio Function[13,14] (SRF). It allows for an analytic formulation of the search of the Steiner Ratio which is then defined as the minimum of the SRF function, Eq. (23). Actually, there is a further restriction to be imposed on function (23) in order to characterize it as an useful SRF function. This restriction is that we should consider only full Steiner Trees. This restriction can be imposed on the spanning trees, by requesting that the angle $\theta_m(\omega,\alpha)$ between consecutive edges formed with the points P_{j+l_Pm} as vertices should be lesser than 120°. We have,

$$-\frac{1}{2} \leq \cos\theta_m(\omega,\alpha) = -1 + \frac{(A_m+1)^2}{2(m^2\alpha^2\omega^2 + A_m + 1)}. \tag{24}$$

In Figure 2, we can see the restrictions corresponding to Eq. (24), for $m = 1, 2, 3$. The horizontal line is $\cos\theta_m = -1/2$.

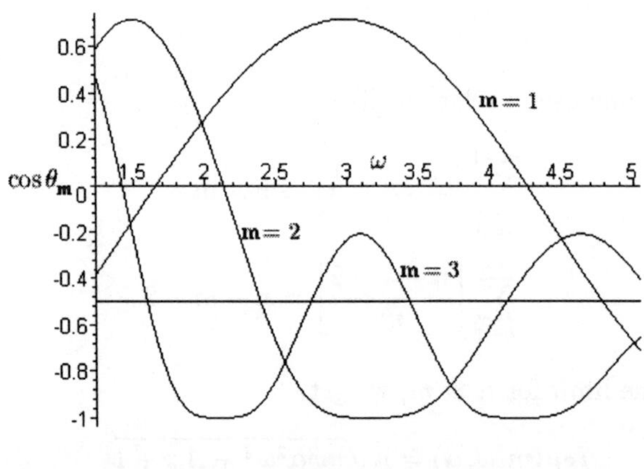

Figure 2. The restriction to Full Steiner Trees. The figure is a section $\alpha = \alpha_R$ (Eq. (27)) of the surfaces given by Eq. (24) corresponding to $m = 1, 2, 3$.

The $m = 1$ spanning tree is the only one which corresponds to Full Steiner trees in a large region of the ω-interval convenient for our work.

The other trees, $m = 2, 3$ correspond to forbidden regions in the same ω-interval. The corresponding Steiner trees to be obtained from the positions of the points $S_{k+l_{S}m}$ and $P_{j+l_{P}m}$ are necessarily degenerate and should not be taken into consideration. Thus, the prescription (23) for the SRF function turns into,

$$\rho(\omega,\alpha) = \frac{1+\alpha\omega\sqrt{\frac{A_1}{A_1+1}}}{\min_{(m)}\left(\sqrt{m^2\alpha^2\omega^2 + A_m + 1}\right)} = \mathrm{Max}_{(m)}\ \rho_m(\omega,\alpha) \tag{25}$$

where,

$$\rho_m(\omega,\alpha) = \frac{1+\alpha\omega\sqrt{\frac{A_1}{A_1+1}}}{\sqrt{m^2\alpha^2\omega^2 + A_m + 1}} \tag{26}$$

The function (25) has a global minimum at the point

$$(\omega_R, \alpha_R) = \left(\pi - \arccos\frac{2}{3}, \frac{\sqrt{30}}{9\left(\pi - \arccos\frac{2}{3}\right)}\right) \tag{27}$$

and

$$\rho(\omega_R,\alpha_R) = \frac{1}{10}(3\sqrt{3}+\sqrt{7}) = 0.78419037337\ldots \tag{28}$$

For a proof see Mondaini and Oliveira[13].

The last value corresponds to the main conjecture of Smith and MacGregor Smith[15] about the value of the Steiner Ratio in 3-dimensional Euclidean Space. It lead us also to think that Nature has solved the problem of energy minimization in the organization of intramolecular structure by choosing Steiner Trees as an intrinsic part of this structure[7].

2. The Stability of Networks under Elastic Force Deformations

In the following we continue to work in a $\mathbb{R}^3$ manifold with an Euclidean distance definition. Let us now introduce a tree as that of Figure 3.

There are n input points (position vectors $\vec{r_j}$) and $q = (n-2)/(p-2)$ Steiner points (position vectors $\vec{S_k}$). If q is not and integer number, there is not a tree with these n, p values[8]. In Figure 3 with $p = 5$, we assume n to be a feasible value. The knowledge of the Steiner Problem tell us that this tree structure is not stable since its total length can be reduced by decreasing the number p[5]. The usual Steiner problem corresponds to $p = 3$. In this section we shall give another proof of this fact by exploring the

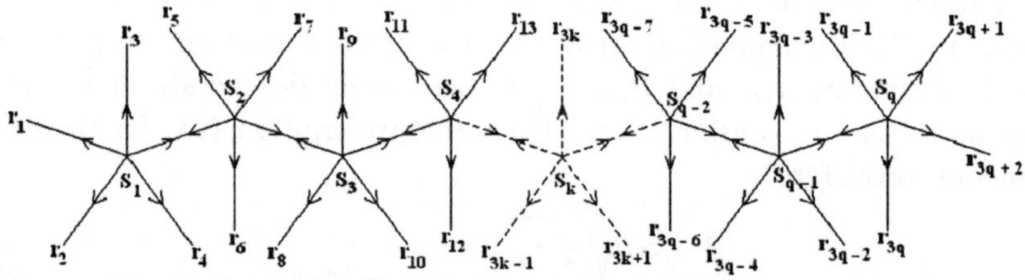

Figure 3. Geometrical scheme for a Steiner Problem with $p = 5$.

concept of a Steiner network with physical interaction among its vertices. The structure depicted at Figure 3 is a representative of the network which models the fundamental interactions inside a biomacromolecule. Let us consider the interaction of this structure with similar structures. Let the resulting interaction forces as applied to input and Steiner points be $\vec{f}_j$, $\vec{f}_{S_k}$, respectively, and let $l_{S_k r_j}$ be the length of an edge between a Steiner point and an input point on its neighbourhood. We have the following identities:

$$l_{S_k r_j} = \frac{1}{a_j}\vec{f}_j \cdot \overrightarrow{S_k r_j} = \frac{1}{a_j}\vec{f}_j \cdot (\vec{r}_j - \vec{f}_{S_k}) \tag{29}$$

$$l_{S_k S_{k+1}} = \frac{1}{a_{S_k}}\vec{f}_{S_k} \,.\, \overrightarrow{S_k S_{k+1}} = \frac{1}{a_{S_k}}\vec{f}_{S_k} \,.\, (\vec{S}_{k+1} - \vec{S}_k) \tag{30}$$

where a_j, a_{S_k}, $j = 1, 2, \ldots, n$, $k = 1, 2, \ldots, q$ stand for the modulus of the parallel components to the edges of the resulting forces $\vec{f}_j$, $\vec{f}_{S_k}$, respectively.

The total length of the tree above is,

$$l = \sum_{j=1}^{p-1} l_{S_1 r_j} + \sum_{j=p}^{2p-3} l_{S_2 r_j} + \sum_{j=2p-2}^{3p-5} l_{S_3 r_j} + \ldots + \sum_{j=n-p+2}^{n} l_{S_q r_j} + \sum_{k=1}^{q-1} l_{S_k S_{k+1}} \tag{31}$$

From Eqs. (29) and (30), we can write the total length in the form,

$$\begin{aligned} l = \sum_{j=1}^{n} \vec{r}_j \cdot \frac{\vec{f}_j}{a_j} - \vec{S}_1 \cdot \left(\frac{\vec{f}_{S_1}}{a_{S_1}} + \sum_{j=1}^{p-1} \frac{\vec{f}_j}{a_j} \right) \\ - \sum_{k=2}^{q-1} \vec{S}_k \cdot \left(\frac{\vec{f}_{S_k}}{a_{S_k}} + \frac{(-\vec{f}_{S_{k-1}})}{a_{S_{k-1}}} + \sum_{j=(k-1)p-2k+4}^{kp-2k+1} \frac{\vec{f}_j}{a_j} \right) \\ - \vec{S}_q \cdot \left(\frac{(-\vec{f}_{S_{q-1}})}{a_{S_{q-1}}} + \sum_{j=n-p+2}^{n} \frac{\vec{f}_j}{a_j} \right) \end{aligned} \tag{32}$$

We now specialize this set of applied forces at the vertices as being collinear with the edges joining them, or

$$\vec{f}_j = a_j \hat{f}_{jk||} \; ; \qquad \vec{f}_{S_k} = a_{S_k} \hat{f}_{S_{k+1,k}||} \tag{33}$$

where the double vertical stroke means "collinear with the edge" and the hat over a letter stands as unit vector.

We now assume that the forces along the edges are Hooke elastic forces

$$\vec{f}_{jk||} = -C(\vec{S}_k - \vec{r}_j) \quad \text{ or } \quad \hat{f}_{jk||} = \hat{r}_{jk} = \frac{\vec{r}_j - \vec{S}_k}{||\vec{r}_j - \vec{S}_k||} \tag{34}$$

$$\vec{f}_{S_{k+1,k}||} = -C(\vec{S}_k - \vec{S}_{k+1}) \quad \text{ or } \quad \hat{f}_{S_{k+1,k}||} = \hat{S}_{k+1,k} = \frac{\vec{S}_{k+1} - \vec{S}_k}{||\vec{S}_{k+1} - \vec{S}_k||} \tag{35}$$

where C is the elastic force constant.

The assumption of local equilibrium of these forces lead to the conditions:

$$\hat{S}_{2,1} + \sum_{j=1}^{p-1} \hat{r}_{j1} = 0 \tag{36}$$

$$\hat{S}_{k+1,k} - \hat{S}_{k,k-1} + \sum_{j=(k-1)p-2k+4}^{kp-2k+1} \hat{r}_{jk} = 0, \qquad k = 2, 3, \ldots, q-1 \tag{37}$$

$$-\hat{S}_{q,q-1} + \sum_{j=n-p+2}^{n} \hat{r}_{jq} = 0. \tag{38}$$

For this equilibrium configuration, Eq.(31) turns into

$$l = \sum_{j=1}^{n} \vec{r}_j \cdot \hat{f}_{j||}. \tag{39}$$

The stability of this equilibrium configuration under a variation of the applied forces can be tested by

$$\delta l = \sum_{j=1}^{n} \vec{r}_j \cdot \delta \hat{f}_{j||} = 0. \tag{40}$$

We take cartesian coordinates for the $\mathbb{R}^3$ vectors $\vec{r}_j = (x_i, y_i, z_i)$, $\vec{S}_k = (x_{S_k}, y_{S_k}, z_{S_k})$ and we consider the three independent variations $\delta_{x_{S_k}}$, $\delta_{y_{S_k}}$, $\delta_{z_{S_k}}$ in the coordinates of the Steiner points.

The corresponding variations in the length of the tree are of the form

$$\delta_{x_{S_k}} l = \delta x_{S_k} \sum_{j=1}^{n} \frac{-x_j(\vec{r}_j - \vec{S}_k)^2 + [(\vec{r}_j - \vec{S}_k) \cdot \vec{r_j}](x_j - x_{S_k})}{||\vec{r}_j - \vec{S}_k||^3} = 0 \quad (41)$$

and two other analogous expressions for the variations $\delta_{y_{S_k}}$, $\delta_{z_{S_k}}$.

From the arbitrariness of these variations we can write,

$$\sum_{j=1}^{n} \frac{(\vec{r}_j - \vec{S}_k) \times [(\vec{r}_j - \vec{S}_k) \times \vec{r}_j]}{||\vec{r}_j - \vec{S}_k||^3} = 0 \quad (42)$$

We can also write

$$\sum_{j=1}^{n} \frac{(\vec{r}_j \cdot \vec{S}_k)^2 - r_j^2 S_k^2}{||\vec{r}_j - \vec{S}_k||^3} = 0. \quad (43)$$

We now write the position vectors $\vec{r}_j$, $\vec{S}_k$ for the configuration depicted at Figure 3. The points can be taken as evenly spaced along right circular helices which radii are 1 and R_p, respectively. We have,

$$\vec{r}_j = (\cos(j-1)\omega, \sin(j-1)\omega, \alpha(j-1)\omega) \quad (44)$$

$$\vec{S}_k = (R_p \cos k\omega, R_p \sin k\omega, \alpha k\omega) \quad (45)$$

The function $R_p(\omega, \alpha)$ can be derived from the equilibrium conditions in Eqs. (36)–(38). For $p = 3$ there is only one solution given by

$$R_3(\omega, \alpha) = \frac{\alpha\omega}{\sqrt{A_1(A_1+1)}} \ ; \quad A_1 = 1 - 2\cos\omega. \quad (46)$$

This solution coincides with Eq. (7).

For the configuration given by Eqs. (44)–(45), Eq.(43) can be written as,

$$\sum_{j=1}^{n} T_{jkp}(\omega, \alpha) = 0$$

where the geometrical object T_{jkp} can be written in the coordinates of Eqs. (44) and (45) as,

$$T_{jkp} = \frac{[\cos^2(j-1-k)\omega - 1 - \alpha^2\omega^2(j-1)^2]R_p^2 + \alpha^2\omega^2[2R_p(j-1)k\cos(j-1-k)\omega - k^2]}{[1 + R_p^2 - 2R_p\cos(j-1-k)\omega + \alpha^2\omega^2(j-1-k)^2]^{3/2}} \quad (47)$$

If R_p is a smooth function in the allowed range for ω-values of this modelling, then to each k-value, there will be a term $j = k+1$ which dominates the sum above. However, we cannot have $j = k+1$ for $p > 3$. This can be seen from the fact for a vertex S_k $(k \neq 1, q)$ there are $(p-2)$ nearest external vertices r_j. The sequence of their consecutive position vectors is,

$$\vec{r}_{(k-1)p-2k+4}, \cdots\cdots, \vec{r}_{kp-2k+1} \tag{48}$$

and the requirement $j = k+1$ corresponds to an integer p-value only for $p = 3$.

This $p = 3$ case which is known to correspond to the most stable problem[5] has as a possible configuration the Figure 4.

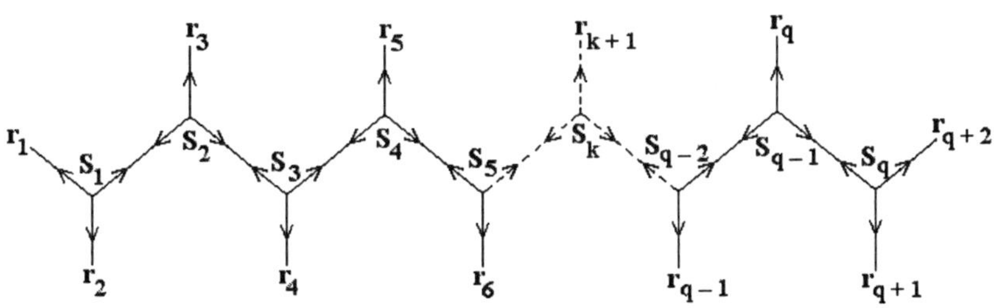

Figure 4. The stable structure of the $p = 3$ Steiner Problem.

3. The Measure of Chirality as the Constraint of a Constrained Optimization Problem

The study of biomacromolecular conformations and their geometrical structures, seems to be the key to the understanding of the processes which are essential to the emergence of life and its maintenance. One of the fundamental characterizations of macromolecular structures is the notion of chirality. After the conceptual definition by Kelvin – If we cannot make an object to coincide with its mirror image through rotations and translations, the object and its image will be chiral to each other – we just know how to characterize an object and its image as chiral or not. The definition says nothing about how chiral is an object to its mirror image. The works of Pasteur have contributed to shed light upon the chirality phenomenon. Chirality has a microscopic essence since the chirality of a substance is due to the existence of chiral molecules as well as chiral unit cells in crystals. This notion leads to think at monomeric units in the macromolecular organization. In Figure 5, the central configuration is formed from the points given by Eq.(5) with $\omega = \omega_R$, $\alpha = \alpha_R$ from Eqs. (26) and (27), after connecting them by edges in order to form regular tetrahedra glued together at common faces. The figure has 23 points (vertices) and 20 regular tetrahedra ("monomeric unit cells"). This structure is a piece of the 3-sausage structure[15] in which the Steiner Ratio Function achieves its lowest upper value given by Eq. (27). We can note that three right handed helices can be identified in this central structure. We can also note that the parts of the structure with 4 and 5 vertices or one and two tetrahedra are achiral. The chirality arises after 6 vertices (3 tetrahedra) since from this number onwards, there is no 3-fold axis of symmetry anymore.

We now look at the other structures depicted in Figure 5. The left hand side one has two right-handed helices (we are looking along the sequences of smallest edges). The structure of the right hand side has also two helices, but these are left-handed. These structures can be obtained from the central one by phase transition process still unknown. Actually Nature has chosen to drive a phase transition in the part of the Universe in which we live only between the central structure and the left-hand side one. The right-hand side structure is then disconnected of the other two in terms of evolution in our region of the Universe.

It is worth to note that the two lateral structures are chiral ones for any number of vertices. Their monomeric units are irregular tetrahedra cells which are themselves chiral. If we adopt the unit tetrahedra cells as a model for aminoacids according to the chemical literature[6], these will be levogirous for the central and left-hand side structures in Figure 5 (three and two right handed helices, respectively) and dextrogirous for the right-hand side structure (two left handed helices). The central and left hand side structures are then good models for studying protein chirality as well as for modelling the A and B forms of DNA.

There are many proposals for geometric chirality measure in the scientific literature[4]. Its importance for understanding intermolecular interaction and the formation of old and new drugs is universally accepted. As far as a serious and falsifiable scientific theory of geometric chirality is still missing, this is an open problem. In the present contribution, we emphasize the modelling approach of the study of Steiner Ratio Functions and their minimization as a representative of the minimum energy configuration of biomacromolecular structures.

Figure 5. The central structure and the left hand side have three and two right handed helices on their structures. The right hand side structure has two left handed helices.

In order to proceed along these guidelines, we now introduce our proposal for geometric chirality measure[12,9].

According Pasteur's ideas, we can concentrate our analysis in the unit tetrahedra cells of our model configurations. An elementary characterization of a chirality measure could be done by the volume measure of these unit cells. The variation of this volume would be due to the internal motions of the macromolecule on its way to a most stable configuration. The pseudoscalar nature of the volume measure is specially adapted to our requirement of a chirality measure.

We take the position vectors $\overrightarrow{P_j}$ of the points given into Eq. (5) and we form the vectors

$$\Delta\overrightarrow{P_l} = \overrightarrow{P_{j+l}} - \overrightarrow{P_j} \ , \quad l = 1, 2, 3 \tag{49}$$

The volume of the tetrahedral cell formed from the vectors above is,

$$V_1(\omega,\alpha) = \frac{1}{6}\Delta\overrightarrow{P_1} \times \Delta\overrightarrow{P_2} \cdot \Delta\overrightarrow{P_3} = \frac{1}{6}\alpha\omega\sin\omega(A_1+1)^2 \tag{50}$$

There is another chain of tetrahedral cells which vertices are the Steiner points given by Eq. (6). The volume of one of these unit cells is

$$V_r(\omega,\alpha) = \frac{1}{6}r^2(\omega,\alpha)\alpha\omega\sin\omega(A_1+1)^2 \tag{51}$$

where $r(\omega,\alpha)$ is given by Eq. (7).

Our chirality measure proposal is simply given by the difference of these volumes, or,

$$\begin{aligned}\chi(\omega,\alpha) &= V_r(\omega,\alpha) - V_1(\omega,\alpha)\\ &= \frac{1}{6}\alpha\omega\sin\omega\left(\frac{A_1+1}{A_1}\right)(\alpha^2\omega^2 - A_1(A_1+1))\end{aligned} \tag{52}$$

We shall test the effectiveness of this chirality measure as the constraint of a constrained optimization problem. This problem corresponds to search for local minima in the neighbourhood of the minimum (ω_R, α_R) of Eq. (25). A new cost function will be formed by taking instead Eq. (25), its convex envelope function $\rho_1(\omega,\alpha)$ (Eq. (26) for $m=1$) in the neighbourhood of (ω_R,α_R). We have[1]:

$$G(\omega,\alpha,T) = (1+T)\rho_1(\omega,\alpha) - T\chi(\omega,\alpha) \tag{53}$$

where T is a Lagrange multiplier.

In Figure 6, we show two solutions of the minimization problem of Eq. (53). These solutions are given by,

$$\begin{aligned}(\omega_1,\alpha_1,T_1) &= (2.465780770, 0.2435156215, 0.3651324727)\\ (\omega_2,\alpha_2,T_2) &= (3.817404537, 0.1572943425, 0.3651324727)\\ \rho(\omega_1,\alpha_1) &= \rho(\omega_2,\alpha_2) = 0.7621675689\end{aligned}$$

From Figure 6, we can see that the present proposal for chirality constraint is efficient at searching for minima around the global minimum given by Eqs. (27)–(28). These local minima correspond to chiral configurations in which the unit cells are already chiral. If the chirality function to be used in the optimization problem could be derived with a deep knowledge of molecular structure, the search for the local minima would lead to more realistic results. However, the best we can do now is to explore the molecular structure by good insights for the chirality functions as we have done in the example above.

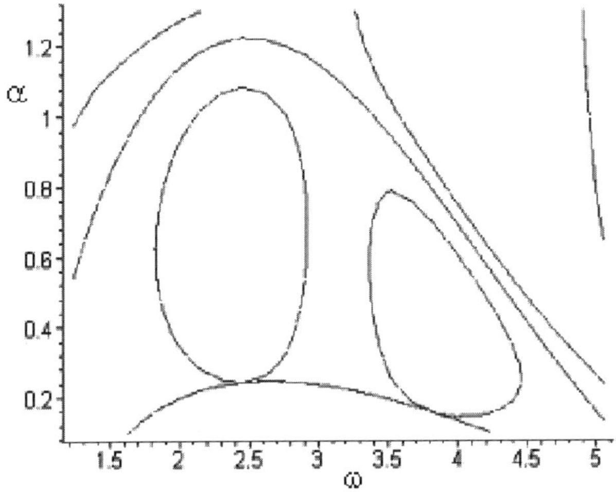

Figure 6. The two solutions of the optimization problem, Eq. (53).

4. Concluding Remarks

We have stressed on some past publications that there is a self-consistent treatment[9,11] of the intramolecular organization of biomacromolecules in terms of Steiner networks. This representation is able at deriving information concerning its stability and evolution. The supporting facts for stability are now well-established and the ideas related to the evolution of macromolecules are in their way to be developed and accepted as a preliminary theory of molecular evolution. The missing subject is a full description of geometric chirality and in order to unveil some of its properties, we have proposed to study the influence of some proposals for chirality measure on the dynamics of optimization problems. These are aimed at studying the structures which energy is around the assumed energy of the minimum solution and the variation process of the chiral properties in the neighbourhood of this minimum. We think that this research line is worth of serious scientific work and should take advantage of the best efforts of very good scientific researchers for some years.

References

1. V. Alexeev, E. Galeev and V. Tikhomirov, *Recueil de Problémes d'Optimization, Editions Mir* (1987).

2. B. R. Brooks, R. E. Bruccoleri, B. D. Olafson, D. J. States, S. Swaminathan and M. Karplus, *CHARMM: A Program for Macromolecular Energy, Minimization and Dynamics Calculations. J. Comp. Chem.* **4**, 187 (1983).
3. C. A. Floudas, J. L. Klepeis, J. D. Lambris, D. Morikis, *De Novo Protein Design: Interplay of Global Optimization, Mixed-Integer Optimization and Experiments. Proc. BIOMAT Symp. 4* **2**, 141–166 (2005).
4. G. Gilat, *On Quantifying Chirality – Obstacles and Problems towards Unification. J. Math. Chem.* **15**, 197–205 (1994).
5. E. N. Gilbert and H. O. Pollak, *Steiner Minimal Trees. SIAM J. Appl. Math.* **16**, 1 (1968).
6. J. P. Glusker, M. Lewis and M. Rossi, *Crystal Structure Analysis for Chemists and Biologists. VHC Publications Inc.* (1994).
7. J. MacGregor Smith, and B. Toppur, *Euclidean Steiner Minimal Trees, Minimal Energy Configurations, and the Embedding Problem of Weighted Graphs in E^3. Discret. Appl. Math.* **71**, 187–215 (1996).
8. R. P. Mondaini, *The Disproof of a Conjecture on the Steiner Ratio in E^3 and its consequences for a Full Geometric Description of Macromolecular Chirality. Proc. BIOMAT Symp.* **2**, 101–177 (2003).
9. R. P. Mondaini, *The Euclidean Steiner Ratio and the Measure of Chirality of Biomacromolecules. Gen. Mol. Biol.* **27**, 4, 658–664 (2004).
10. R. P. Mondaini and N. V. Oliveira, *The State of Art on the Steiner Ratio Value in $\mathbb{R}^3$. Tendências em Matemática Aplicada e Computacional - TEMA* **5**, 2, 249–258 (2004).
11. R. P. Mondaini, *The Steiner Ratio and the Homochirality of Biomacromolecular Structures. Nonconvex Optimization and its Applications Series – Kluwer Acad. Publ.* **74**, 373–390 (2004).
12. R. P. Mondaini, *Proposal for Chirality Measure as the Constraint of a Constrained Optimization Problem. Proc. BIOMAT Symp.* **3**, 2, 65–74 (2004).
13. R. P. Mondaini and N. V. Oliveira, *A New Approach to the Study of the Smith + Smith Conjecture.* http://www.arxiv.org/math-ph/0506050 (2005).
14. R. P. Mondaini, *Modelling the Biomacromolecular Structure with Selected Combinatorial Optimization Techniques.* http://www.arxiv.org/math-ph/0502051 (2005).
15. W. D. Smith and J. MacGregor Smith, *On the Steiner Ratio in 3-Space. J. Comb. Theor.* **A69**, 301–332 (1995).

EXPLORING CHEMICAL SPACE WITH COMPUTERS: INFORMATICS CHALLENGES FOR AI AND MACHINE LEARNING

PIERRE BALDI

Institute for Genomics and Bioinformatics
School of Information and Computer Sciences
University of California, Irvine
Irvine, CA 92697-3435
E-mail: pfbaldi@ics.uci.edu

The penetration of informatics and modern computer metods in chemistry lags far behind their penetration in physics and biology. Rather than resulting from intrinsic peculiarities of chemistry as a science, this unfortunate state of affairs is more likely to be the product of historical accidents. As was the case for bioinformatics, we argue that two key ingredients are essential for the large-scale development of chemoinformatics. First, the development of large-scale, publicly available, database and data sets of chemical information, including compounds, reactions, and annotations. Second the development of algorithms and resulting open source software for a variety of computational tasks, prominently including the mathematical quantification and efficient implementaton of methods for measuring chemical similarity.

1. Introduction

In spite of its central role between physics and biology, chemistry has remained in a backward state of informatics development compared to its two close relatives. Computers, public databses, and large collaborative projects have become the pervasive hallmark of research in physics and biology. The Human Genome Project, for instance, required collaboration among dozens if not hundreds of scientists across the world. And the resulting human DNA sequence, as well as a wealth of other biological information, are available for anyone to download from public repositories on the Web such as GenBank, Swissprot, the PDB, and PubMed. Virtually every biologist today uses publicly available tools, such as BLAST, to search sequence databases and analyze high-throughput data. Similar observations can be made in physics with large collaborative efforts in, for instance astronomy or high-energy physics. The Web itself was born at CERN, a European consortium with over half a century of history, and the world largest particle physics laboratory. In stark contrast, large collaborative efforts and public databses and software are comparatively absent from chemical research.

This is not to say that chemists do not use computers or databases at all. Of course they do, but these uses have remained limited and peripheral to the chemical sciences. By and large, chemistry has remained the business of individual investigators and their laboratories. Suffice it to say that to this date there is no publicly available repository of all known molecules and no large-scale collaborative effort to annotate any significant portion of chemical space.The equivalent of BLAST for chemistry remains to be created.

This unfortunate state of affairs and the overall conservatism of the chemistry community is unlikely to result from some intrinsic properties of chemistry as a science. Rather, it is likely to be the product of complex historical and sociologial factors, that can be traced back at least to the middle ages and the secretive research projects of the alchemists, in their quest for a recipe for converting vulgar metals to gold. In more modern times, the American Chemical Society has certainly played a role in the current state of affairs[11] by controlling and profiting from the dissemination of chemical information through journal and database ownership and commercilization.

It is therefore likely that this state of affairs can and will be changed. In fact there are clear signs that rapid change is on its way and it is only a matter of when, and not if, computers will become one of the main tools of modern chemistry. To further drive this point, we focus for concreteness on a particular area of chemistry, namely samll molecules in organic chemistry.

2. Small Molecules in Organic Chemistry

Small molecules with at most a few dozen atoms play a fundamental role in organic chemistry and biology. They can be used as combinatorial building blocks for chemical synthesis[14,1], as molecular probes for perturbing and analyzing biological systems in chemical genomics and systems biology[15,17,6], and for the screening, design, and discovery of useful compounds. These include of course new drugs[10,9], the majority of which are small molecules. Furthermore, huge arrays of new small molecules can be produced in a relatively short period of time[7,14].

Table 1. Comparison of astronomical and chemical spaces.

	Astronomical	Chemical
Visited Universe	0-1 star	10^7 compounds
Existing Universe	10^{22} stars	10^7 compounds
Virtual Universe	10^{22} stars	10^{60} compounds
Travel	very difficult	relatively easy

It is worth comparing small-molecule chemical space to our own astronomical space. Astronomical space contains on the order of 10^{22} stars, roughly 10^{11} galaxies, each containing 10^{11} stars. The number of known small molecules, encountered so far in nature, or synthesized by man, is on the order of 10^7 (the ACS databse, the largest chemical database, currently contains 26 million compounds). However, estimates in the literature of the size of the virtual space of small molecules that could be created vary between 10^{18} and 10^{200}, with 10^{60} being currently one of the more cited estimates[4]. Thus by any of these estimates, chemical space remains rather unexplored and uncharted. A second essential difference between chemical and astronomical space is that chemical space is comparatively easier to travel, both virtually and in reality. Small molecules can be recursively enumerated in silico and synthesized in vitro from known building blocks and known reactions. Of course we do not mean to imply that chemical synthesis is a trivial matter–it is not. But general guiding principles and tools are available. And would you rather have to synthesize a new small molecule or travel to a new galaxy thousands of light years away from Earth? In short, with 10^{60} enumerable and synthesizable compounds remaining to be explored, it is hard to see how the computer could avoid becoming the chemoscope–e.g. the central tool of future chemical astronomers.

3. Chemoinformatics Challenges

The key challenge for computational methods then is not traveling through chemical space per se, but rather to be able to focus traveling expeditions in a vast chemical space towards interesting regions, and to be able to recognize interesting stars and galaxies when they are encountered. The notion of what is interesting may vary of course with the task (e.g. drug discovery, reaction discovery, polymer discovery). But at the most fundamental level what is needed are tools to predict the physical, chemical, and biological properties of small molecules and reactions in order to focus searches and filter search results.

Computational methods in chemistry can be organized along a spectrum ranging from Schrodinger equation, to molecular dynamics, to statistical machine learning methods. Quantum mechanical methods, or even molecular dynamics methods, are computationally intensive and do not scale well to large datasets. These methods are best applied to specific questions on focused small datasets. Statistical and machine learning methods are more likely to yield successfull approaches for rapidly sifting through large datasets of chemical information.

Because in the absence of large public database and datasets, chemoinformatics is in a state reminiscent of bioinformatics two or three decades ago, it may be productive to adapt the lessons learnt from bionformatics to chemoinformatics, while maintaining also a perspective on the fundamental differences between these two relatively young interdisciplinary sciences. If this analogy is correct, two key ingredients were essential for unlocking the large-scale development of bioinformatics and the application of modern statistical machine learning methods to biological data[2], data and similarity measures. data and similarity measures. In bioinformatics,such as Genbank, Swissprot, and the PDB while alignment algorithms have provided robust similarity measures with their fast BLAST implementation becoming the workhorse of the field. Mutatis mutandis, the same is likely to be true in chemoinformatics.

4. Data: Database, Datasets, and Annotations

Limited catalogs of small molecules are available in digital format from many vendors across the world, as well as a number of public Web sites. As datasets of small molecules become increasingly available, it is important to develop computational methods to both organize these data in rapidly searchable databases and to extract or predict useful information for each molecule, including its physical, chemical, and biological properties. Conversely, large and well-annotated datasets are essential for developing statistical machine learning methods in chemoinformatics, whether supervised or unsupervised, including predictive classification, regression, and clustering of small molecules and their properties[12,13]. Aggregation and organization of datasets of chemical information allows for massive in silico processing that would be impractical or even impossible in a traditional experimental setting.

Several parallel efforts have emerged recently to start to address the data bottleneck, including PubChem (http://pubchem.ncbi.nlm.nih.gov), the Harvard Chembank[18], UCSF's ZINC[8], and the UCI ChemDB[5]. The UCI ChemDB is a public database containing over 4M compounds as well as a repository of annotated datasets that can be used to develop statistical machine learning methods. Together, these datasets already pose important challenges for both supervised and unsuperviwsed machine learning methods, from clustering to kenel methods[13,19]

In the longer run, a critical challenge is going to be the annotation of these databases. Can annotation be carried by chemistry laboratories in a concerted way across the world and deposited in a central public repository? Can data gathered over the years by large pharmaceutical companies become public? How much annotation can be derived more or less automatically from the literature using automated information retrieval methods? Can new annotation models be implemented (e.g. using organic chemistry classes to produce annotations)? This area is likely also to produce new challenges for database technology due to the sheer size of chemical space.

Another very important related set of challenges has to do with the creation of a public repository of chemical reactions, currently in progress at UCI. Such a repository is essential for a variety of tasks ranging from reaction discovery, to automatics determination and optimization of synthetic pathways to travel chemical space.

5. Similarity Measures and Kernels

Good similarity metrics between compounds, or between reactions, are essential to rapidly search large databases of compounds or reactions. Consider, for instance, a classical drug discovery problem where the starting point is a protein of known structure and perhaps a corresponding ligand (Figure 1). With a good database of small molecules, the discovery process can proceed from both ends. Starting from the protein, one can dock millions of small molecules to the protein in silico[16]. In fact, with sufficient computing power, one ought to be able to dock all known small molecules to all proteins with known structure contained in the PDB[3]. Producing such a matrix ought to be a significant goal for systems bioogy and pharmacology in the coming years. On the other hand, starting from the ligand, one can search the database of small molecules for compounds that are "similar" to the known ligand(s), where similarity can be defined in different ways. In both approaches, additional filters can be used to eliminate molecules that are, for instance, poorly soluble, too flexible, or toxic[19]. Furthermore, in silico chemical reactions applied to the molecules in the database can further expand the space of interesting molecules being screened or designed. Thus similarity to a single compounds, or a set of compounds, must be defined precisely in ways that can be computed efficiently together with a statistical theory for assessing the significance of the hits (c.f. "the e-values of BLAST").

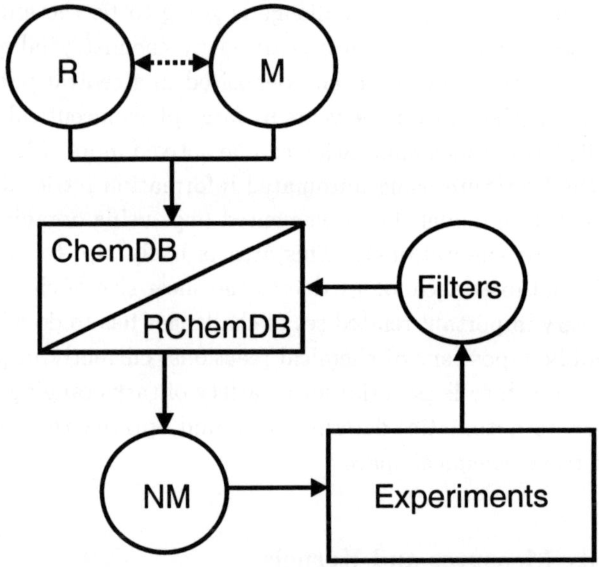

Figure 1. High-level view of a basic drug screening/design pipeline. R = receptor protein; M = molecular ligand(s); NM = new molecular ligands; RChemDB = set of compounds derived from ChemDB using a library of reactions. NM is obtained by molecular docking applied to R, or by constrained similarity searches applied to M. Computational filters can be used to predict and constrain molecular properties (e.g. flexibility, solubility, toxicity).

But similarity is also essential to develop predictive machine learning methods to predict the physical, chemical, and biological properties of compounds. This is not too surprising since, given an annotated training set of molecules (e.g. toxic/non-toxic), the properties of a new molecule ought to be inferred from its similarities to the molecules in the training set. This is precisely the basic idea behind kernels methods, one of the leading methods in machine learning. To our advantage, compounds (and reactions) can be represented in many ways, including 1D SMILES strings, 2D graphs of bonds, and 3D structures. Good kernels can be derived for each one of these representations. Spectral kernels in particular, counting the number of occurrences of each possible substructure, lead to efficient molecular "fingerprints" and similarity measures that are useful both in database searches and statistical machine learning applications[13,19].

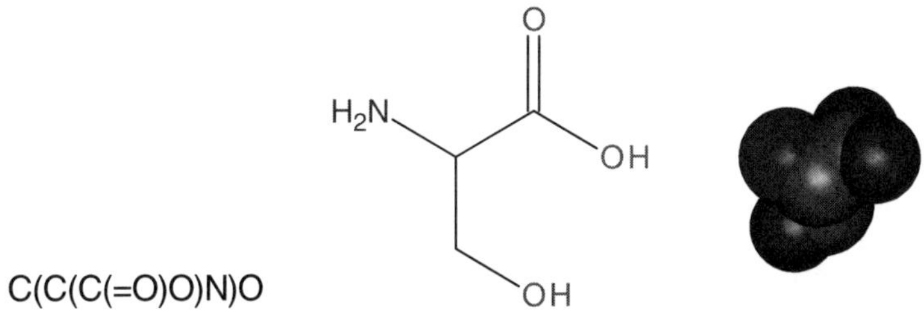

(a) (b) (c)

Figure 2. Three representations for the amino acid Serine: (a) 1D SMILES string; (b) 2D graph of atoms and bonds; and (c) 3D space-filling model.

Additional similarity measures and kernels can be derived baseed on molecular surfaces (2.5D), pharmacophores (3D), and even beyond (4D) using conformers, isomers, or dynamic evolution. Challenges remain in developing and testing these similarity measures, their complementary values and usages, as well as their statistical properties and extreme value distributions.

Acknowledgments

Work supported in part by grants from the NIH, NSF, and a Laurel Wilkening Faculty Innovation Award.

References

1. D. K. Agrafiotis, V. S. Lobanov, and F. R. Salemme. Combinatorial informatics in the post-genomics era. *Nature Reviews Drug Discovery*, 1:337–346, 2002.
2. P. Baldi and S. Brunak. *Bioinformatics: the machine learning approach.* MIT Press, Cambridge, MA, 2001. Second edition.
3. H. M. Berman, J. Westbrook, Z. Feng, G. Gilliland, T. N. Bhat, H. Weissig, I. N. Shindyalov, and P. E. Bourne. The Protein Data Bank. *Nucl. Acids Res.*, 28:235–242, 2000.

4. R. S. Bohacek, C. McMartin, and W. C. Guida. The art and practice of structure-based drug design: a molecular modelling perspective. *Medicinal Research Reviews*, 16(1):3–50, 1996.
5. J. Chen, S. J. Swamidass, Y. Dou, and J. Bruand P. Baldi. ChemDB: a public database of small molecules and related chemoinformatics resources. *Bioinformatics*, 2005. In Press.
6. C. M. Dobson. Chemical space and biology. *Nature*, 432:824–828, 2004.
7. R. A. Houghten. Parallel array and mixture-based synthetic combinatorial chemistry: tools for the next millenium. *Annual Review of Pharmacology and Toxicology*, 40:273–282, 2000.
8. J. J. Irwin and B. K. Shoichet. ZINC–a free database of commercially available compounds for virtual screening. *Journal of Chemical Information and Computer Sciences*, 45:177–182, 2005.
9. S. O. Jonsdottir, F. S. Jorgensen, and S. Brunak. Prediction methods and databases within chemoinformatics: Emphasis on drugs and drug candidates. *Bioinformatics*, 21:2145–2160, 2005.
10. C. Lipinski and A. Hopkins. Navigating chemical space for biology and medicine. *Nature*, 432:855–861, 2004.
11. E. Marris. Chemistry society goes head to head with NIH in fight over public database. *Nature*, 435(7043):718–719, 2005.
12. A. Micheli, A. Sperduti, A. Starita, and A. M. Biancucci. A novel approach to QSPR/QSAR based on neural networks for structures. In H. Cartwright and L. M. Sztandera, editors, *Soft Computing Approaches in Chemistry*, pages 265–296. Springer Verlag, Heidelberg, Germany, 2003.
13. L. Ralaivola, S. J. Swamidass, H. Saigo, and P. Baldi. Graph kernels for chemical informatics. *Neural Networks*, 2005. Special issue on Neural Networks and Kernel Methods for Structured Domains. In press.
14. S. L. Schreiber. Target-oriented and diversity-oriented organic synthesis in drug discovery. *Science*, 287:1964–1969, 2000.
15. S. L. Schreiber. The small-molecule approach to biology: chemical genetics and diversity-oriented organic synthesis make possible the systematic exploration of biology. *Chemical and Engineering News*, 81:51–61, 2003.
16. B. K. Shoichet. Virtual screening of chemical libraries. *Nature*, 432:862–865, 2004.
17. B. R. Stockwell. Exploring biology with small organic molecules. *Nature*, 432:846–854, 2004.
18. R. L. Strauseberg and S. L. Schreiber. From knowing to controlling: a path from genomics to drugs using small molecule probes. *Science*, 300(5617):294–295, 2003.
19. S. J. Swamidass, J. Chen, J. Bruand, P. Phung, L. Ralaivola, and P. Baldi. Kernels for small molecules and the prediction of mutagenicity, toxicity, and anti-cancer activity. *Bioinformatics*, 21(Supplement 1):i359–368, 2005. Proceedings of the 2005 ISMB Conference.

OPTIMIZATION OF BETWEEN GROUP ANALYSIS OF GENE EXPRESSION DISEASE CLASS PREDICTION

FLORENT BATY, MICHEL P. BIHL AND MARTIN BRUTSCHE

Pulmonary Gene Research,
University Hospital Basel,
Basel CH-4031, Switzerland
E-mail: florent.baty@unibas.ch

AEDÍN C. CULHANE

Bioinformatics Conway Institute,
University College Dublin,
Dublin 4, Ireland
E-mail: Aedin.Culhane@ucd.ie

GUY PERRIÈRE

Laboratoire de Biométrie et Biologie Évolutive, UMR CNRS 5558,
Université Caude Bernard - Lyon I,
43 bd. du 11 Novembre 1918,
F-69622 Villeurbanne Cedex, France
E-mail: perriere@biomserv.univ-lyon1.fr

Recent publications described a supervised classification method for microarray data: Between Group Analysis (BGA). This method associated with multivariate ordination Correspondence Analysis (COA) proved to be very efficient for both classification of samples into pre-defined groups and disease class prediction of new unknown samples. Classification and prediction with BGA are classically performed by using the whole set of genes and no variable selection is required. We hypothesize that an optimized selection of highly discriminating genes can improve the prediction power of this method.
We propose an optimization of BGA with a jackknife-based gene selection procedure. The objective of this procedure is to select a subset of highly discriminative genes that optimize disease class prediction. It is a backward optimization method. The least influential genes are removed one by one from the analysis by maximizing the percentage of between group inertia. We applied this optimization on two datasets and compared it to other classification methods. The results showed a considerable improvement in the predictive accuracy of BGA when tested on the classification of independent data sets and cross-validation.
The R code is available on request and supplementary information is accessible at: http://pulmogene.unibas.ch/articles/optimization

1. Introduction

Gene expression microarrays enable the simultaneous measurements of the expression level of thousands of genes. Supervised classification of gene expression data aim to identify which combinations of genes enable the best discrimination of groups of samples specified in advance. For such methods, which are classically used in disease class prediction, identification of a subset of discriminating genes can be critical[1,2]. Indeed, a large proportion of genes are generally non-informative in terms of disease class prediction. A gain in classification and prediction performance is expected when predictors are built upon a subset of highly discriminating genes[3,4].

Several algorithms capable of selecting a subset of predictive genes were recently proposed[5]. These methods include a genetic algorithm[6], support vector machine[7,8], shrunken centroids technique[2,9] and those which make use of discriminant functions[10]. However, two issues remain: firstly different subsets of genes may provide comparable optimal discriminations[1]; secondly it is generally difficult to decide which number of genes is optimal for the discrimination[11,12]. This number may vary according to different parameters such as the number of individuals in the training set, the number of groups to discriminate, the method used for classification and prediction.

Dolédec and Chessel[13] developed a supervised classification approach, the Between Group Analysis (BGA) which was recently applied to microarray data[14].The authors specified several key features of BGA that make it a method of choice for sample classification and class prediction. In BGA, all genes participate to the discrimination. Consequently, no gene selection step is required. On the other hand, BGA calculates group means and is therefore sensitive to outliers. Our objective was to improve the robustness of BGA by optimizing the number of discriminating genes supporting the analysis.

In this study, we propose a new jackknife-based algorithm that optimizes selection of the most robust discriminating genes in order to improve the accuracy of disease class prediction. This algorithm is applied to the BGA but it could also be associated with other supervised methods.

We tested the efficiency of the algorithm on two datasets using independent test sets and leave-one-out cross-validation (LOOCV). We compared our approach to different classification methods.

2. Methods

2.1. *Data sets*

Sarcoidosis data. The gene expression study was carried out on 12 healthy controls (H), 7 sarcoidosis stage I patients (SI) and 5 sarcoidosis stage II/III patients (SII). This dataset was previously published and details can be found in Rutherford et al.[16]. These 24 samples correspond to the sarcoidosis training set. In addition, 8 follow-up chips were done six months later in some of the sarcoidosis patients. Among these patients, three had still active sarcoidosis stage II/III and five were recovered from sarcoidosis stage I. These 8 supplementary samples correspond to the sarcoidosis test set. The expression level of 12,626 probe sets was measured on Affymetrix GeneChip® (HG-U95Av2). The complete dataset and the raw files have been deposited in NCBIs Gene Expression Omnibus (GEO, http://www.ncbi.nlm.nih.gov/geo/) and are accessible through GEO Series accession number GSE1907.

Tumour data. This dataset was published by Khan et al.[17]. The authors measured the expression of 6567 genes in four types of small round blue cell tumours (NB: neuroblastoma; RMS: rhabdomyosarcoma; BL: Burkitt's lymphoma; EWS: Ewing's sarcoma). A filtered dataset containing the expression level of 2308 genes is publicly accessible (http://research.nhgri.nih.gov/microarray/Supplement). The whole dataset contains 88 samples split into a training set (63 samples) and a test set (25 samples).

Software and statistical analysis. The gene selection algorithm was written in R (version 1.9.1), an open-source statistical software. The algorithm is freely accessible at http://pulmogene.unibas.ch/articles/optimization. Some specific R packages were used in this study: the Bioconductor packages for microarray analysis[18]; ADE-4[19] and MADE-4[20] for multivariate analysis. The sarcoidosis dataset was normalized using the *vsn* algorithm[21].

Overall method description. The optimization method follows four main steps described in Figure 1:

- Select n most discriminating genes of a BGA.
- Determine recursively the influence of each gene by a jackknife-based procedure.
- Remove successively the less influential genes.
- End the procedure when the minimum number of genes required

to perform a BGA is reached.

The percentage of between group inertia is maximized during this optimization procedure. Two parameters are monitored at each step of the procedure: the percentage of samples correctly classified by LOOCV and the variance of between group inertia.

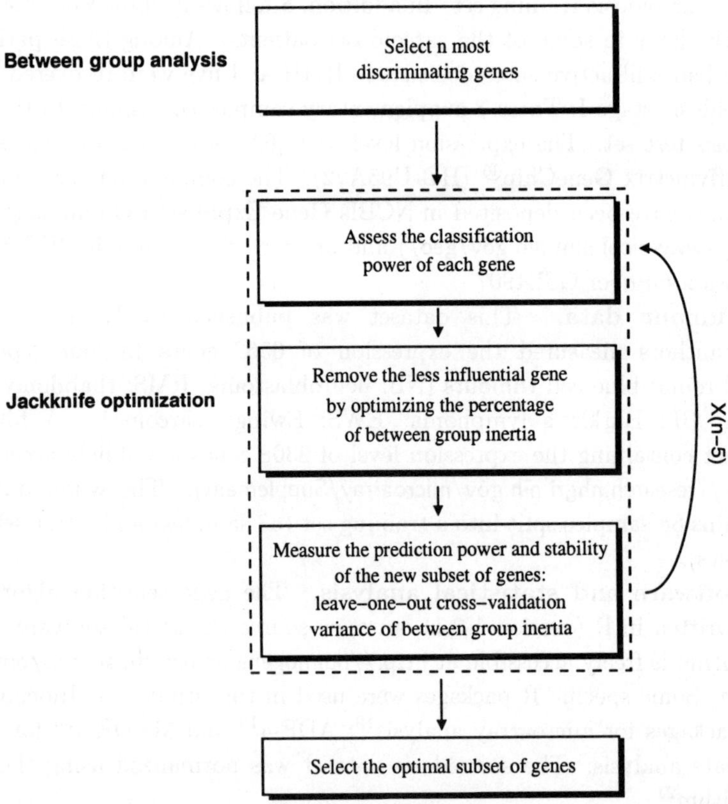

Figure 1. Description of the algorithm for the optimization of gene selection

Between group analysis. BGA is a particular extension of conventional ordinating methods such as principal component analysis (PCA) or correspondence analysis (COA) where sample grouping are specified in advance[13]. The association of COA with BGA is particularly powerful as COA has been shown to have several advantages over PCA in analysis of gene expression data[22,23]. In order to simplify the notations in the rest of the paper, the acronym BGA will refer to the between-group correspondence analysis.

Let us consider K the number of specified groups, a typical BGA yields $K - 1$ discriminating axes that ordinate the groups of sample by maximizing the between group variance (see[14] for mathematical details). Linear discriminant analysis is a related method which aims to maximize the percentage of variance explained by the grouping but which has different constraints and which cannot be applied to tables where the number of variables exceeds the number of samples[24]. Genes and samples ordinated by BGA can be projected on discriminating axes and visualized simultaneously on a biplot. The most discriminating genes are projected at the extremity of each axis whereas less informative genes are projected near the origin of each axis.

2.2. *Optimization of gene selection for disease class prediction*

Pre-selection of discriminating genes. In order to reduce the number of calculations, a subset of discriminating genes is chosen by selecting those genes projected at the extremity of the axes of a preliminary BGA (including all genes). This initial gene selection could be achieved using one of two approaches. If samples were classified into two groups, one could select an equal amount of genes from the extremity of each end of the single discriminating BGA axis. On the other hand, when samples are classified into more than two groups, genes projected at the periphery of each pair of discriminating axes may be selected by using a "peeling" function (successive 2D convex hulls). We applied this second procedure as samples in our two data sets were subdivided into three and four groups.

The next steps of the algorithm may be rather time consuming. For reasons of computational limitations, the number of genes selected in this initial subset should be in the order of two hundred (the optimization of 150 genes and 24 samples required 1h50min on a computer Pentium IV 2.66 GHz).

Optimization criterion. The objective of the algorithm is to improve the discrimination efficiency of BGA by increasing the between group inertia and decreasing the within group inertia of samples classified by genes with BGA.

Let N the number of samples (x_i is the ith sample and ω_i its weight),

$\mathrm{d}(x_i, x_j)$ the squared Euclidean distance between two samples x_i and x_j, K the number of groups (G_k is the kth group) and N_k the number of individuals in the kth group. By using a weighted pair-group average calculation, the total inertia can be decomposed as follow:

- Within group inertia:

$$\mathrm{inertia}_W = \frac{1}{\sum_{k=1}^{K} \mathrm{C}_{N_k}^2} \sum_{k=1}^{K} \sum_{i,j \in G_k} \omega_i \omega_j \mathrm{d}(x_i, x_j)$$

- Between group inertia:

$$\mathrm{inertia}_B = \frac{1}{\mathrm{C}_N^2 - \sum_{k=1}^{K} \mathrm{C}_{N_k}^2} \sum_{k=1}^{K} \sum_{\substack{i \in G_k \\ j \notin G_k}} \omega_i \omega_j \mathrm{d}(x_i, x_j)$$

The algorithm aims to maximize the percentage of between group inertia:

$$\%\mathrm{inertia}_B = \mathrm{inertia}_B / (\mathrm{inertia}_W + \mathrm{inertia}_B) \qquad (1)$$

Selection of the most discriminating genes. From the initial subset of discriminating genes, the influence of each gene is individually assessed using a jackknife leave-one-out procedure. Each gene is removed from the dataset and a BGA is carried out with the remaining genes. The influence of each gene is measured by the difference in the percentage of between group inertia before and after removing the gene from the analysis. At each step of the algorithm, the gene that contributes least to the between group inertia is removed from the dataset and another jackknife procedure is carried out. The algorithm runs until the minimum number of genes required for a BGA is reached.

Accuracy of disease class prediction. At each step of the algorithm, the performance of the predictive genes is assessed by LOOCV. In jackknife cross-validation a sample is removed from the dataset and a BGA is performed on the remaining samples. The excluded sample is projected on to the BGA and classified. This is iteratively performed until all samples have been subjected to cross-validation. The percentage of samples predicted correctly by cross-validation is calculated. This parameter measures the prediction accuracy of the subset of genes.

In addition, we projected the blind test samples on to the BGA and calculated the percentage of new test samples correctly classified. Importantly, this parameter, which gives the effective measure of the prediction efficiency of the subset of genes, was not taken into account in the optimization procedure.

Stability and robustness of the optimization. The stability of the optimization method was assessed in order to prevent over-fitting. During the backward optimization, when the number of genes included in the classifier is getting lower, the effect of removing a gene from the analysis tends to increase the variance of the percentage of between group inertia. This variance is measured at each step of the algorithm. At the same time, the statistical significance of the BGA is evaluated with a Monte-Carlo permutation test.

Identification of the optimized subset of genes for disease class prediction. A diagram that summarizes the different steps of the algorithm was used to determine the optimal number of genes. The optimal subset of genes should have simultaneously an optimal accuracy (i.e. the best rate of cross-validation) and an optimal stability (minimal variance).

3. Results

3.1. *Sarcoidosis data*

Between group analysis. BGA was applied to the sarcoidosis whole training data set. The biplot representation shows that BGA separates the three phenotypes with no overlap (Figure 2A). The first axis separates the healthy controls from the sarcoidosis patients. The second axis separates the two stages of sarcoidosis. The efficacy to classify new samples was measured by LOOCV. Seventy five percent of the 24 samples were classified correctly. However, we observed some large discrepancies between the three phenotypes. All healthy controls, 6 out of 7 stage I, but none of the stage II/III sarcoidosis patients were correctly re-classified by jackknife. When classifying the 8 follow-up patients by BGA using the whole set of genes, only 50% of these test samples were correctly classified. Four out of five patients, which recovered six months after they were diagnosed with a stage I sarcoidosis, were classified in the healthy group. On the other hand, none of the three still active sarcoidosis stage II/III were correctly classified.

Optimization of gene selection. From the initial BGA, the 105 most discriminating genes were selected using the above mentioned peeling procedure (Figure 2C). The optimization procedure was applied on this subset of genes. The least influential genes, when assessed using the percentage of between group inertia criterion were removed one by one.

Figure 2B shows the evolution of the classification parameters, the per-

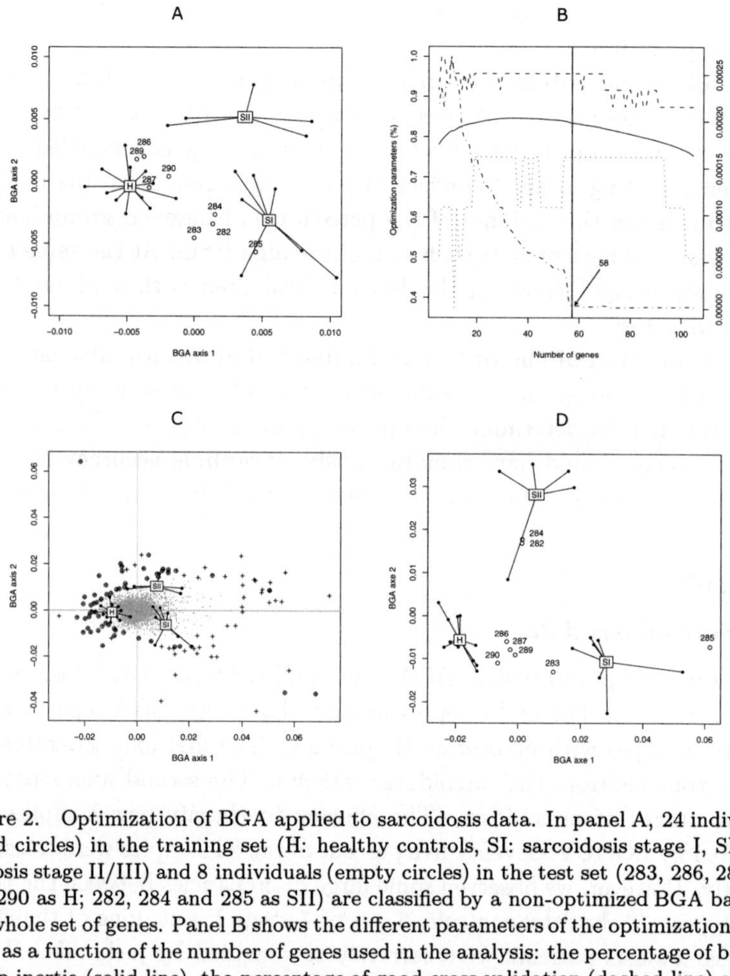

Figure 2. Optimization of BGA applied to sarcoidosis data. In panel A, 24 individuals (solid circles) in the training set (H: healthy controls, SI: sarcoidosis stage I, SII: sarcoidosis stage II/III) and 8 individuals (empty circles) in the test set (283, 286, 287, 289 and 290 as H; 282, 284 and 285 as SII) are classified by a non-optimized BGA based on the whole set of genes. Panel B shows the different parameters of the optimization procedure as a function of the number of genes used in the analysis: the percentage of between group inertia (solid line), the percentage of good cross-validation (dashed line) and the variance of between group inertia (dot-dashed line). For indication, the percentage of test samples correctly predicted is represented by a dotted line. This parameter was not used in optimization of the training model. The vertical line shows the optimal number of genes. In panel C, the 105 most discriminating genes (initial subset) are located at the periphery of the biplot (black crosses) and the 58 optimal genes are highlighted (circled crosses). In panel D, 8 test-samples are classified using a BGA based on the 58 optimal genes.

centage of between group inertia, together with the percentage of correct cross-validation and the variance of between group inertia. During the optimization process, the percentage of between group inertia increases when the number of genes decreases until it reaches an optimum, then it decreases when the number of genes fell below this optimum threshold. The percentage of correct cross-validation is stable in a range of 20-70 genes. When the number of genes further decreases, it starts to oscillate. The variance of between group inertia is very low for subsets of more than 58 genes. This parameter considerably increases for subsets of genes smaller than 57. Finally, the dotted line represents the evolution of the percentage of test sets correctly classified (this parameter was not considered during optimization).

The optimal subset of genes was determined as the subset with the best cross-validation efficiency and least variable percentage of between group inertia. Therefore, we decided to choose a subset of 58 genes (Figure 2C). Figure 2D shows the projection of the 8 follow-up samples predicted by this subset of classifiers. The LOOCV obtained with this optimized subset of genes was clearly improved since 96% of samples were correctly classified (100, 80 and 100% respectively in sarcoidosis stage I, stage II/III and healthy controls). Predictions obtained from the 8 follow-up patients were also improved since 2/3 of the new samples issued from sarcoidosis stage II/III were correctly associated to their group, whereas 4/5 of the patients in remission from a stage I sarcoidosis were classified as healthy. Patient 283, which was mis-classified, clinically recovered from a sarcoidosis stage I. Signal of gene activity that is specific to stage I sarcoidosis might still be detectable in this patient.

3.2. *Tumour data*

Between group analysis. BGA clearly separates the 4 different types of tumours with no overlap (Figure 3A). Based upon the complete set of 2308 genes, the LOOCV shows that 92.6% of the 63 samples were correctly cross-validated and 19/20 of the test sets were correctly predicted. The most discriminating genes associated with the different groups are identifiable at the periphery of the BGA biplot (Figure 3C).

Optimization of gene selection. From the initial BGA, the 245 most discriminating genes were selected and the optimization started from this initial subset. By optimizing the percentage of between group inertia, the optimal number of genes was approximately equal to 90 (Figure 3B). The LOOCV of the 63 training samples BGA based upon the 90 optimal genes increased to 100% and 20/20 test sets were correctly classified according to this subset of genes. These optimal genes are plotted in Figure 3C and one

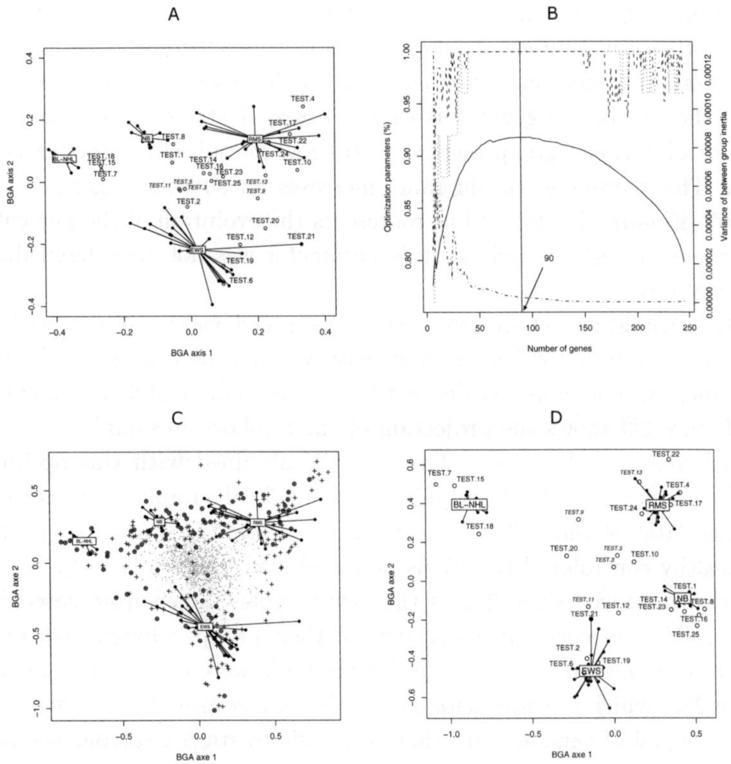

Figure 3. Optimization of BGA applied to tumour data. In panel A, 63 samples (solid circles) of the training set (BL: Burkitt's lymphoma, EWS: Ewing's sarcoma, NB: neuroblastoma, RMS: rhabdomyo sarcoma) and 25 samples (empty circles) of the test set (7, 15 and 18 as BL-NHL; 2, 6, 12, 19, 20 and 21 as EWS; 1, 8, 14, 16, 23 and 25 as NB; 4, 10, 17, 22 and 24 as RMS; 3, 5, 9, 11 and 13 as control samples that do not belong to one of the 4 groups) are classified by the non-optimized BGA based on the whole set of genes. Panel B shows the different parameters of the optimization procedure as a function of the number of genes used in the analysis: the percentage of between group inertia (solid line), the percentage of good cross-validation (dashed line) and the variance of between group inertia (dot-dashed line). For indication, the percentage of test samples correctly predicted is represented by a dotted line. This parameter was not used in optimization of the training model. The vertical line shows the optimal number of genes. In panel C, the 245 most discriminating genes are represented with small crosses and the 90 optimal genes are highlighted (circled crosses). In panel D, the 25 test-samples are classified using a BGA based on the 90 optimal genes.

can see in Figure 3D that the test sets were correctly classified.

Stability of the optimization. We tested the statistical significance of the BGA at each step of the algorithm by carrying out a Monte-Carlo permutation test which proved to be constantly significant for both datasets (estimated p-value = 0.001). This result documents the reliability of our method. On the other hand, the stability of our optimization procedure was checked in order to control the risk of over-fitting due to the algorithm. The stability of the optimal subset of genes was defined by measuring the variance of the percentage of between group inertia. This parameter is of great importance as it is preferable that the classification is not predominantly influenced by a few genes.

Comparison with other algorithms. First, we analyzed the gain in sensitivity and specificity of the optimized BGA compared with non-optimized BGA by carrying out a ROC analysis. For both the sarcoidosis (Figure 4A) and the tumour datasets (Figure 4B), the optimized approach represents an improvement of sensitivity and specificity according to both types of validation (LOOCV and independent dataset). We compared our method with three other gene selection methods recently described: the genetic algorithm with k-nearest neighbors (GA/KNN)[6], the maximal margin linear programming (MAMA)[15] and the nearest shrunken centroid[9]. Results (Table 1) document that our method outperforms these other approaches.

4. Discussion

The identification of genes that optimize the prediction power of a classifier is a significant and difficult challenge in microarray data analysis. Firstly due to the fact that most of the discriminative functions require more cases than variables. Secondly due to the considerable amount of noise in microarray data. A feature of BGA is that it could be applied to the complete datasets without prior gene selection. Pre-pruning of genes may potentially be based on arbitrary selection criteria. However, we showed that an optimized selection considerably improve the prediction power of BGA. In applying our jackknife based algorithm we tested the robustness of BGA discriminating genes and excluded weaker discriminators, thus optimizing both the performance of BGA and reducing the number of discriminating genes.

The algorithm presented here might be rather time consuming depending on the size of the initial subset of genes taken. We recommend an initial number of genes that is the range of 100-200 genes per analysis. Increasing the number of genes in the initial dataset ensures that more potentially dis-

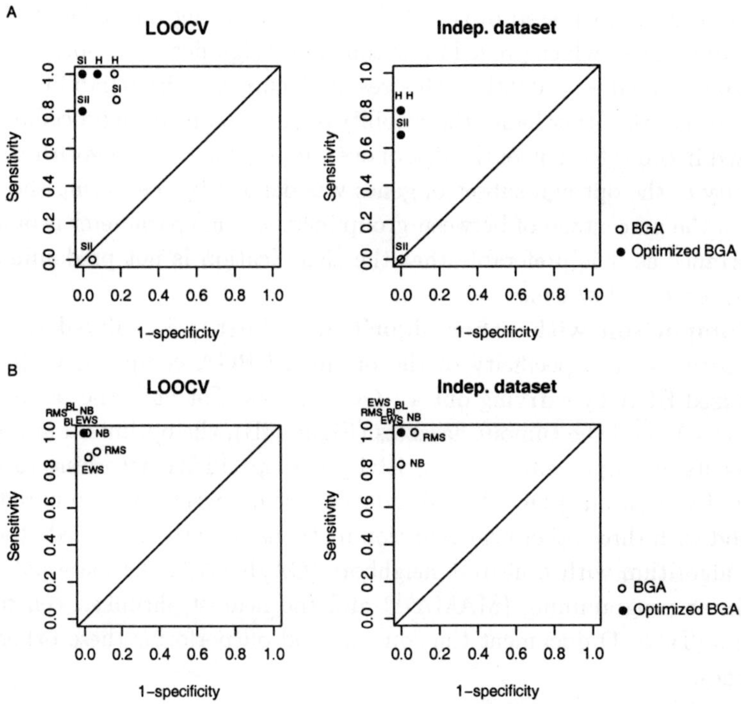

Figure 4. The sensitivity and the specificity of the optimized BGA (solid circles) were compared with those of the non-optimized BGA (empty circles). ROC analysis was performed in the case of sarcoidosis dataset (A) after LOOCV (left panel) and after classification of the independent dataset (right panel). ROC analysis was also applied in the case of the tumour dataset (B) after LOOCV (left panel) and after classification of the independent dataset (right panel).

criminative genes are present in the analysis. However, the time required for the optimization process increases significantly. We assessed the percentage gain in between group inertia obtained by increasing the number of genes in the initial subset. This number depends on the dataset and on the number of groups to discriminate. This number is around 100 for the sarcoidosis data and around 150-200 for the tumour data (Figure 5).

In our approach, the choice of the initial subset of genes from which the algorithm starts remains critical and alternative procedures might be used. For example, the genetic algorithm proposed by Li et al.[6] could be associated to our optimization and might provide some improvements of performance. However results might be less consistent than those exclu-

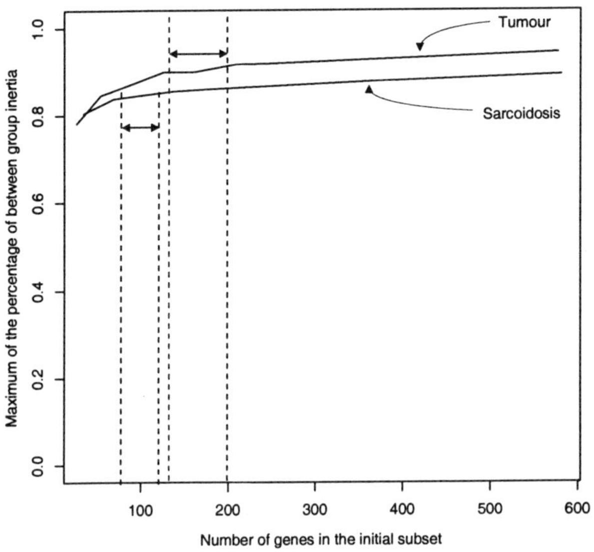

Figure 5. Number of genes included in the initial subset. This plot shows the maximum of the percentage of between group inertia reached by optimization procedure as a function of the number of genes present in the initial subset of genes (higher curve: tumour data; lower curve: sarcoidosis data).

sively obtained from the BGA. We also assessed the relevance of the peeling selection compared with a random selection of genes in the performance of our algorithm. The optimized parameters were found to be significantly improved when using the peeling procedure both for the sarcoidosis dataset (Figure 6A) and the tumour dataset (Figure 6B).

We decided to choose a backward optimization procedure as this seemed to be more adapted for taking possible gene interactions into account. The prediction power of a single gene might be negligible on itself while it might be preponderant when it is associated with one or a few other genes. Removing a gene that jointly participates with other genes to the group discrimination will have an impact which is measurable by a backward approach whereas no evidence might be found by using a forward optimization.

Our results show that an improvement in the discriminative and predictive power of BGA can be achieved by reducing the number of predictors in the analysis to a small subset of highly discriminative genes. These genes contribute to improve the percentage of between group inertia. In

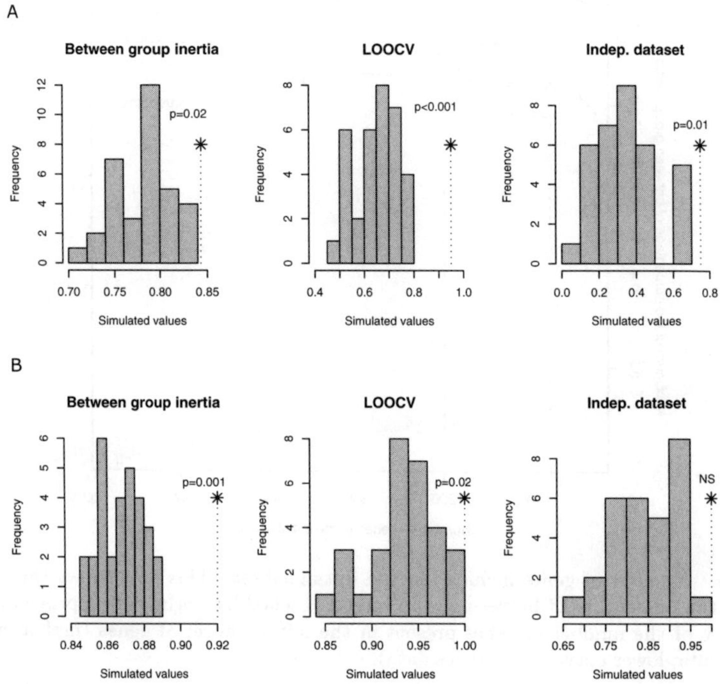

Figure 6. The optimized parameters obtained after a peeling selection of genes (stars) were compared with the optimized parameters obtained from a set of randomly selected genes in the sarcoidosis dataset (A) and in the tumour dataset (B) (left: percentage of between group inertia; center: percentage of correct LOOCV; right: percentage of independent dataset correctly predicted).

this study, two criteria were used to define the optimal subset of genes: a positive criterion, the percentage of correct classification by jackknife cross-validation and a negative criterion the variance of the between group inertia. Generally, a range of the near optimal number of genes was found. In the sarcoidosis dataset, it was around 60 genes, whereas in the tumour dataset around 90 genes were found to be optimal. By using a method that associates a genetic algorithm with the k-nearest neighbors technique (GA/KNN) on a lymphoma dataset[6], concluded that using only a few discriminating genes may not be reliable whereas using too many genes will add noise to the classification. These authors suggested to keep 50-200 genes in the analysis to get an optimal result. Our work tends to confirm this.

Table 1. Comparison of the classification efficiency of the optimized BGA with different classification methods.

Dataset	Method	% LOOCV	% Test samples
Sarcoidosis	Optimized BGA	96%	75%
	BGA	75%	50%
	GA/KNN	92%	62.5%
	MAMA	67%	62.5%
	Shrunken centroids	79[1]%	62.5[1]%
Tumour	Optimized BGA	100%	100%
	BGA	92.6%	95%
	GA/KNN	100%	95%
	MAMA	98%	76%
	Shrunken centroids	100[2]%	90[2]%

[1]Threshold of 2.801
[2]Threshold of 2.459

In conlusion, we described a new algorithm for the optimization of gene selection that improve both the classification and predictive power of BGA. This algorithm proved to outperform alternative gene selection and classification procedures. Our approach is flexible and further developments will be done in order to adapt it to other classification methods.

Acknowledgements

This study was sponsored by the Krebsliga beider Basel.

Bibliography

1. L. Li, L.G. Pedersen, T.A. Darden and C.R. Weinberg, In *Proceedings of The Atlantic Symposium on Computational Biology, Genome Information Systems and Technology*, Duke University (2001).
2. K.Y. Yeung and R.E. Bumgarner, *Genome Biol.* **4**, R83 (2003).
3. W. Li, F. Sun and I. Grosse, *J. Comput. Biol.* **11**, 215-226 (2004).
4. Y. Tan, L. Shi, W. Tong and C. Wang, *Nucleic Acids Res.* **33**, 56-65 (2005).
5. M. Xiong, L. Jin, W. Li and E. Boerwinkle, *Biotechniques* **29**, 1264-1270 (2000).
6. L. Li, C. Weinberg, T. Darden and L. Pedersen, *Bioinformatics* **17**, 1131-1142 (2001).
7. M. Brown, W. Grundy, D. Lin, N. Cristianini, C. Sugnet, T. Furey, M. Ares and D. Haussler, *Proc. Natl. Acad. Sci. USA* **97**, 262-267 (2000).
8. T. Furey, N. Cristianini, N. Duffy, D. Bednarski, M. Schummer and D. Haussler, *Bioinformatics* **16**, 906-914 (2000).

9. R. Tibshirani, T. Hastie, B. Narasimhan and G. Chu, *Proc. Natl. Acad. Sci. USA* **99**, 6567-6572 (2002).
10. S. Dudoit, J. Fridlyand and T.P. Speed, *J. Amer. Stat. Assoc.* **97**, 77-87 (2002).
11. W. Li and Y. Yang, In *Methods of Microarray Data Analysis*, S. Lin and K.F. Johnson (eds.), Kluwer Academic Publishers, 137-150 (2002).
12. L. Ein-Dor, I. Kela, G. Getz, D. Givol and E. Domany, *Bioinformatics* **21**, 171-178 (2005).
13. S. Dolédec and D. Chessel, *Acta Oecologica Oecologia Generalis* **8**, 403-426 (1987).
14. A.C. Culhane, G. Perrière, E.C. Considine, T.G. Cotter and D.G. Higgins, *Bioinformatics* **18**, 1600-1608 (2002).
15. A.V. Antonov, I.V. Tetko, M.T. Mader, J. Budczies and H.W. Mewes, *Bioinformatics* **20**, 644-652 (2004).
16. R.M. Rutherford, F. Staedtler, J. Kehren, S.D. Chibout, L. Joos, M. Tamm, J.J. Gilmartin and M.H. Brutsche, *Sarcoidosis Vasc. Diffuse Lung. Dis.* **21**, 10-18 (2004).
17. J. Khan, J.S. Wei, M. Ringner, L.H. Saal, M. Ladanyi, F. Westermann, F. Berthold, M. Schwab, C.R. Antonescu, C. Peterson and P.S. Meltzer, *Nature Med.* **7**, 673-679 (2001).
18. R.C. Gentleman, V.J. Carey, D.M. Bates, B. Bolstad, M. Dettling, S. Dudoit, B. Ellis, L. Gautier, Y. Ge, J. Gentry, K. Hornik, T. Hothorn, W. Huber, S. Iacus, R. Irizarry, F. Leisch, C. Li, M. Maechler, A.J. Rossini, G. Sawitzki, C. Smith, G. Smyth, L. Tierney, J.Y. Yang and J. Zhang, *Genome Biol.* **5**, R80 (2004).
19. D. Chessel, A.B. Dufour and J. Thioulouse, *R News* **4**, 5-10 (2004).
20. A.C. Culhane, J. Thioulouse, G. Perrière and D. G. Higgins, *Bioinformatics* **21**, 2789-2790 (2005).
21. W. Huber, A. von Heydebreck, H. Sultmann, A. Poustka and M. Vingron, *Bioinformatics* **18** suppl. 1, S96-104 (2002).
22. K. Fellenberg, N.C. Hauser, B. Brors, A. Neutzner, J.D. Hoheisel and M. Vingron, *Proc. Natl. Acad. Sci. USA* **98**, 10781-10786 (2001).
23. L. Wouters, H.W. Gohlmann, L. Bijnens, S.U. Kass, G. Molenberghs and P.J. Lewi, *Biometrics* **59**, 1131-1139 (2003).
24. G. Perrière and J. Thioulouse, *Comput. Methods Programs Biomed.* **70**, 99-105 (2003).

ON BICLUSTERING WITH FEATURE SELECTION FOR MICROARRAY DATA SETS

PANOS M. PARDALOS, STANISLAV BUSYGIN AND OLEG A. PROKOPYEV

Department of Industrial and Systems Engineering
University of Florida, Gainesville, FL 32611, USA
E-mail: pardalos@ufl.edu; busygin@ufl.edu; oap4ripe@ufl.edu

Let a set of training and test samples be given, and the samples from the training set be partitioned into a number of classes, while classification of the test samples is unknown. The classification problem consists in determining classes of the test samples utilizing the information provided by the training set. Supervised biclustering is a specific type of classification problems, where we simultaneously partition both the set of samples and the set of their features. Samples and features classified together are supposed to have a high relevance to each other which can be observed by intensity of their expressions. Moreover, not all features of the data set are informative, and we need to find a subset of features relevant to the classification of interest. This task is called the feature selection. In this paper we applied supervised biclustering with feature selection to several microarray data sets. Computational results indicate that the obtained solution provides a reliable feature selection and the test set classification based on it.

1. Introduction

Let a data set of n samples and m features be given as a rectangular matrix $A = (a_{ij})_{m\times n}$, where the value a_{ij} is the expression of i-th feature in j-th sample. We consider classification of the samples into classes

$$\mathcal{S}_1, \mathcal{S}_2, \ldots, \mathcal{S}_r,\ \mathcal{S}_k \subseteq \{1\ldots n\},\ k = 1\ldots r,$$

$$\mathcal{S}_1 \cup \mathcal{S}_2 \cup \ldots \cup \mathcal{S}_r = \{1\ldots n\},$$

$$\mathcal{S}_k \cap \mathcal{S}_\ell = \oslash,\ k,\ell = 1\ldots r,\ k \neq \ell.$$

The set of samples is divided into the *training* and *test* sets. For the samples from the training set the classification is known, while for the samples from the test set it has to be performed utilizing the information provided by the training set classification. Generally, the classification should be done so that samples from the same class share certain common properties characterizing the classes.

This is one of the major problems of data mining theory and applications, and in practice it is frequently complicated by the fact that not all features of the data are informative for discovering the classification, and a subset of features determining it should be found. This task is called the *feature selection*. The principle we use for feature selection is based on simultaneous clustering of samples and features of the data set. In other words, the classification should be done so that samples from the same class share certain common properties (features). Suppose there exists a partition of features into r classes

$$\mathcal{F}_1, \mathcal{F}_2, \ldots, \mathcal{F}_r,\ \mathcal{F}_k \subseteq \{1 \ldots m\},\ k = 1 \ldots r,$$

$$\mathcal{F}_1 \cup \mathcal{F}_2 \cup \ldots \cup \mathcal{F}_r = \{1 \ldots m\},$$

$$\mathcal{F}_k \cap \mathcal{F}_\ell = \oslash,\ k, \ell = 1 \ldots r,\ k \neq \ell$$

such that features of class $\mathcal{F}_k$ are highly expressed in the samples of class $\mathcal{S}_k$. We will call the set of class pairs

$$\mathcal{B} = ((\mathcal{S}_1, \mathcal{F}_1), (\mathcal{S}_2, \mathcal{F}_2), \ldots, (\mathcal{S}_r, \mathcal{F}_r)) \tag{1}$$

a *biclustering* of the data set.

Biclustering (or co-clustering) of samples and features was a topic of interest in several papers, including biclustering of expression data investigated by Y. Cheng and G.M. Church [14], a paper of I.S. Dhillon on textual biclustering [6], double conjugated clustering algorithm by S. Busygin, G. Jacobsen and E. Krämer[10], and spectral biclustering of microarray data by Y. Kluger, R. Basri, J.T. Chang, and M. Gerstein[15]. However, all these works are dealing with *unsupervised* biclustering and do not allow one to use the information provided by training data. In this chapter, we describe the application of the approach for *supervised* biclustering developed in Pardalos et al.[11] for mining several well-known microarray data sets. Computational results indicate that the obtained solution provides a reliable feature selection and the test set classification based on it.

The remainder of the chapter is organized as follows. In Section 2 we briefly describe necessary formulations and the optimization-based algorithm for supervised biclustering with feature selection. In Section 3 we present our computational experiment results on three microarray data sets.

2. Problem Statement and the Algorithm

First, let us describe the formal setup for performing the feature selection. Let each sample be already assigned somehow to one of the classes $\mathcal{S}_1, \mathcal{S}_2, \ldots, \mathcal{S}_r$. Introduce a 0–1 matrix $S = (s_{jk})_{n \times r}$ such that $s_{jk} = 1$ if $j \in \mathcal{S}_k$, and $s_{jk} = 0$ otherwise. The sample class *centroids* can be computed as the matrix $C = (c_{ik})_{m \times r}$:

$$C = AS(S^T S)^{-1}, \tag{2}$$

whose k-th column represents the centroid of the class $\mathcal{S}_k$. Each value c_{ik} in the matrix C gives us the average expression of the i-th feature in the sample class $\mathcal{S}_k$. The principle used for feature selection constraints is based on simultaneous clustering of samples and features of the data set. Similarly to the matrices S and C defined above, we introduce the 0–1 matrix $F = (f_{ik})_{m \times r}$ such that $f_{ik} = 1$ if $i \in \mathcal{F}_k$ and $f_{ik} = 0$ otherwise, and the matrix of feature class centroids $D = (d_{jk})_{n \times r}$:

$$D = A^T F(F^T F)^{-1}, \tag{3}$$

whose k-th column represents the centroid of the class $\mathcal{F}_k$. Now the value d_{jk} gives us the average feature expression in the sample j among features of the class $\mathcal{F}_k$. The condition of up-regulation of the features of a class $\mathcal{F}_k$ in the samples of the class $\mathcal{S}_k$ implies

$$i \in \mathcal{F}_{\hat{k}} \;\Rightarrow\; \forall k = 1 \ldots r,\; k \neq \hat{k} :\; c_{i\hat{k}} \geq c_{ik}, \tag{4}$$

and, symmetrically,

$$j \in \mathcal{S}_{\hat{k}} \;\Rightarrow\; \forall k = 1 \ldots r,\; k \neq \hat{k} :\; d_{j\hat{k}} \geq d_{jk}. \tag{5}$$

If the biclustering $\mathcal{B}$ satisfies both (4) and (5), we will call it *consistent* [11].

For the purpose of feature selection, when the classes of samples $\mathcal{S}_1, \mathcal{S}_2, \ldots, \mathcal{S}_r$ are already given, we construct the classes of features $\mathcal{F}_1, \mathcal{F}_2, \ldots, \mathcal{F}_r$ according to (4). Then, to obtain a consistent biclustering, we remove some features from the data set in order to satisfy (5). Considering now the variables x_i to be 0–1 (i.e., fractional feature weights are impossible), we arrive at the following feature selection constraints:

$$\frac{\sum_{i=1}^{m} a_{ij} f_{i\hat{k}} x_i}{\sum_{i=1}^{m} f_{i\hat{k}} x_i} \geq \frac{\sum_{i=1}^{m} a_{ij} f_{ik} x_i}{\sum_{i=1}^{m} f_{ik} x_i}, \tag{6}$$

for all $j \in \mathcal{S}_{\hat{k}}$, $\hat{k}, k = 1 \ldots r$, $\hat{k} \neq k$.

We formulate the feature selection problem as an optimization task and use the objective function to minimize the information loss. In other words the goal is to select as many features as possible and the objective function may be expressed as

$$\max \sum_{i=1}^{m} x_i \tag{7}$$

The objective function is linear but the constraints are fractional 0–1 functions, and in order to be tackled by industrial optimization solvers, they need to be linearized. Unfortunately, while the linearization works nicely for small-size problems, it often creates instances, where the gap between the integer programming and the linear programming relaxation optimum solutions is very big for larger problems. As a consequence, the instance of a linear mixed 0–1 programming problem obtained after linearization can not be solved exactly in a reasonable time even with the best techniques implemented in modern integer programming solvers. Therefore, in order to solve our fractional 0–1 programming problem we have applied a heuristic approach, which is presented in detail in Pardalos et al.[11].

Moreover, we can strengthen the class separation by introduction of a coefficient greater than 1 for the right-hand side of the inequality (6). In this case, we improve the quality of the solution modifying (6) as

$$\frac{\sum_{i=1}^{m} a_{ij} f_{i\hat{k}} x_i}{\sum_{i=1}^{m} f_{i\hat{k}} x_i} \geq (1+t) \frac{\sum_{i=1}^{m} a_{ij} f_{ik} x_i}{\sum_{i=1}^{m} f_{ik} x_i}, \tag{8}$$

for all $j \in \mathcal{S}_{\hat{k}}$, $\hat{k}, k = 1 \ldots r$, $\hat{k} \neq k$, and $t > 0$ is a constant that becomes a parameter of the method, which we call *the parameter of separation.*

After the feature selection is done, we perform classification of test samples according to (5). That is, if $b = (b_i)_{i=1\ldots m}$ is a test sample, we assign it to the class $\mathcal{S}_{\hat{k}}$ satisfying

$$\frac{\sum_{i=1}^{m} b_i f_{i\hat{k}} x_i}{\sum_{i=1}^{m} f_{i\hat{k}} x_i} \geq \frac{\sum_{i=1}^{m} b_i f_{ik} x_i}{\sum_{i=1}^{m} f_{ik} x_i} \tag{9}$$

for all $k = 1 \ldots r$, $\hat{k} \neq k$.

3. Computational Experiments

3.1. *ALL vs. AML data set*

We applied supervised biclustering to a well-researched microarray data set containing samples from patients diagnosed with *acute lymphoblastic leukemia* (ALL) and *acute myeloid leukemia* (AML) diseases[7]. It has been the subject of a variety of research papers, e.g. [3,4,12,13]. This data set was also used in the CAMDA data contest[1]. It is divided into two parts – the training set (27 ALL, 11 AML samples), and the test set (20 ALL, 14 AML samples). According to the described methodology, we performed feature selection for obtaining a consistent biclustering using the training set, and the samples of the test set were subsequently classified choosing for each of them the class with the highest average feature expression. The parameter of separation $t = 0.1$ was used. The algorithm selected 3439 features for class ALL and 3242 features for class AML. The obtained classification contains only one error: the AML-sample 66 was classified into the ALL class. To provide the justification of the quality of this result, we should mention that the support vector machines (SVM) approach delivers up to 5 classification errors on the ALL vs. AML data set depending on how the parameters of the method are tuned[12]. Furthermore, the perfect classification was obtained only with one specific set of values of the parameters.

The heatmap for the constructed biclustering is presented in Figure 1.

3.2. *HuGE Index data set*

Another computational experiment that we conducted was on feature selection for consistent biclustering of the Human Gene Expression (*HuGE*) Index data set[2]. The purpose of the HuGE project is to provide a comprehensive database of gene expressions in normal tissues of different parts of human body and to highlight similarities and differences among the organ systems. We refer the reader to Hsiao et al.[8] for the detailed description of these studies. The data set consists of 59 samples from 19 distinct tissue types. It was obtained using oligonucleotide microarrays capturing 7070 genes. The samples were obtained from 49 human individuals: 24 males with median age of 63 and 25 females with median age of 50. Each sample came from a different individual except for first 7 BRA samples that were from the different brain regions of the same individual and 5th LI sample, which came from that individual as well. We applied to the data set Algorithm 1 with the parameter of separation $t = 0.1$.

The obtained biclustering is summarized in Table I and its heatmap is presented in Figure 2. The distinct block-diagonal pattern of the heatmap evidences the high quality of the obtained feature classification. We also mention that the original studies of HuGE Index data set in Hsiao et al.[8] were performed without 6 of the available samples: 2 KI samples, 2 LU samples, and 2 PR samples were excluded because their quality was too poor for the statistical methods used. Nevertheless, we may observe that

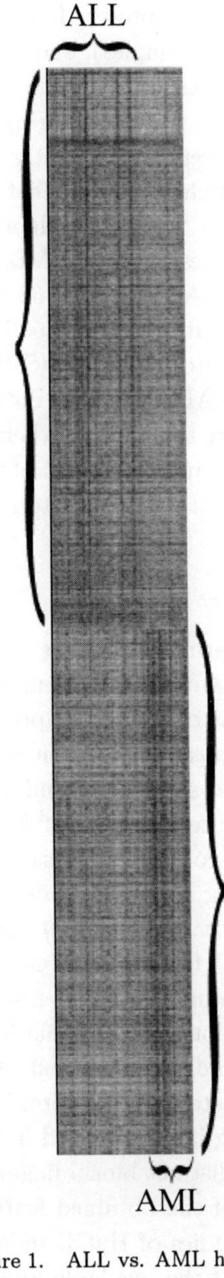

Figure 1. ALL vs. AML heatmap

none of them distorts the obtained biclustering pattern, which confirms the robustness of our method.

Tissue type	Abbreviation	#samples	#features selected
Blood	BD	1	472
Brain	BRA	11	614
Breast	BRE	2	902
Colon	CO	1	367
Cervix	CX	1	107
Endometrium	ENDO	2	225
Esophagus	ES	1	289
Kidney	KI	6	159
Liver	LI	6	440
Lung	LU	6	102
Muscle	MU	6	532
Myometrium	MYO	2	163
Ovary	OV	2	272
Placenta	PL	2	514
Prostate	PR	4	174
Spleen	SP	1	417
Stomach	ST	1	442
Testes	TE	1	512
Vulva	VU	3	186

3.3. *GBM vs. AO data set*

Finally, the last experiment we conducted was on a microarray data set containing samples from patients diagnosed with *glioblastoma* (GBM) and *anaplastic oligodendroglioma* (AO) diseases[5]. Malignant gliomas are one of the most common types of brain tumor and result in about 13,000 deaths in USA annually[9]. While glioblastomas are very resistant to many of the available therapies, anaplastic oligodendrogliomas are more compliant to treatment (for more details, see Betensky et al.[5]). Therefore, classification of GBM vs. AO is a task of crucial importance. The data set, which we used, was divided into two parts – the training set (21 classic tumors with 14 GBM and 7 AO samples), and the test set (29 non-classic tumors with 14 GBM and 15 AO samples). The total number of features was 12625.

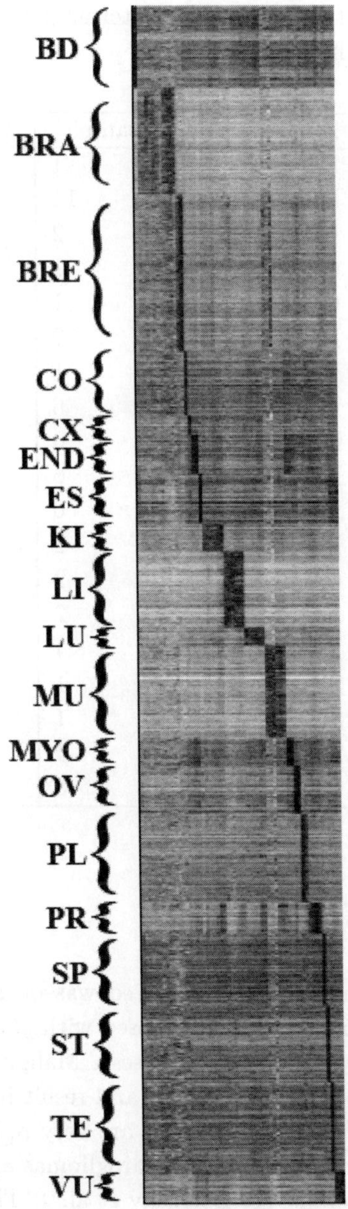

Figure 2. HuGE Index heatmap

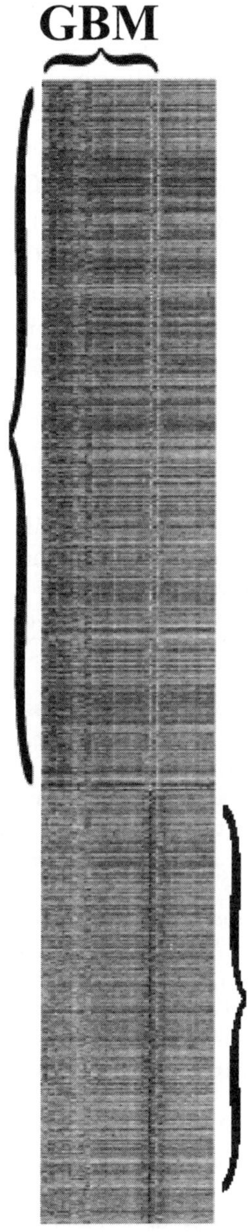

Figure 3. GBM vs. AO heatmap

According to the described methodology, we performed feature selection for obtaining a consistent biclustering using the training set, and the samples of the test set were subsequently classified choosing for each of them the class with the highest average feature expression. The parameter of separation $t = 15$ was used. The algorithm selected 3875 features for the class GBM and 2398 features for the class AO. The obtained classification contained only 4 errors: two GBM samples (Brain_NG_1 and Brain_NG_2) were classified into the AO class and two AO samples (Brain_NO_14 and Brain_NO_8) were classified into the GBM class.

The heatmap for the constructed biclustering is presented in Figure 3.

References

1. *CAMDA 2001 Conference.* http://bioinformatics.duke.edu/camda/camda01/.
2. *HuGE Index.org Website.* http://www.hugeindex.org.
3. A. Ben-Dor, L. Bruhn, I. Nachman, M. Schummer, and Z. Yakhini. Tissue classification with gene expression profiles. *J. Comput. Biol.*, 7:559–584, 2000.
4. A. Ben-Dor, N. Friedman, and Z. Yakhini. Class discovery in gene expression data. In *Proc. Fifth Annual Inter. Conf. on Computational Molecular Biology (RECOMB)*, 2001.
5. R.A. Betensky, P. Tamayo, J.G. Cairncross, C. Ladd, U. Poh,l C. Hartmann, M.E. McLaughlin, T.T. Batchelor, P.M. Black, A. von Deimling, S.L. Pomeroy, T.R. Golub, C.L. Nutt, D.R. Mani and D.N. Louis. Gene expression-based classification of malignant gliomas correlates better with survival than histological classification. *Technical Report, http://www.broad.mit.edu/cgi-bin/cancer/datasets.cgi*, 2004.
6. I.S. Dhillon. Co-clustering documents and words using bipartite spectral graph partitioning. In *Proceedings of the 7th ACM SIGKDD International Conference on Knowledge Discovery and Data Mining(KDD)*, 2001.
7. T.R. Golub, D.K. Slonim, P. Tamayo, C. Huard, M. Gaasenbeek, J.P. Mesirov, H. Coller, M.L. Loh, J.R. Downing, M.A. Caligiuri, C.D. Bloomfield, and E.S. Lander. Molecular classification of cancer: class discovery and class prediction by gene expression monitoring. *Science*, 286:531–537, 1999.
8. L-L Hsiao, F. Dangond, T. Yoshida, R. Hong, RV. Jensen, J. Misra, W. Dillon, KF. Lee, KE. Clark, P. Haverty, Z. Weng, G. Mutter, MP. Frosch, ME. MacDonald, EL. Milford, CP. Crum, R. Bueno, RE. Pratt, M. Mahadevappa, JA. Warrington, G. Stephanopoulos, G. Stephanopoulos, and SR. Gullans. A compendium of gene expression in normal human tissues. *Physiol. Genomics*, 7:97–104, 2001.
9. M.C. Zlatescu D.K. Lisle D.M. Finkelstein R.R. Hammond J.S. Silver P.C. Stark D.R. Macdonald Y. Ino D.A. Ramsay J.G. Cairncross, K. Ueki and D.N. Louis. Specific chromosomal losses predict chemotherapeutic response and survival in patients with anaplastic oligodendrogliomas. *J Natl Cancer*

Inst, 90:1473–1479, 1998.

10. E. Krämer S. Busygin, G. Jacobsen. Double conjugated clustering applied to leukemia microarray data. In *SDM 2002 Workshop on Clustering High Dimensional Data and its Applications*, 2002.
11. P.M. Pardalos S. Busygin, O.A. Prokopyev. Feature selection for consistent biclustering via fractional 0–1 programming. *Journal of Combinatorial Optimization*, 10/1:7–21, 2005.
12. J. Weston, S. Mukherjee, O. Chapelle, M. Pontil, T. Poggio, and V. Vapnik. Feature selection for svms. In *Proc. NIPS Conf.*, 2001.
13. E.P. Xing and R.M. Karp. Cliff: Clustering of high-dimensional microarray data via iterative feature filtering using normalized cuts. *Bioinformatics Discovery Note*, 1:1–9, 2001.
14. G.M. Church Y. Cheng. Biclustering of expression data. In *Proceedings of the 8th International Conference on Intelligent Systems for Molecular Biology*, pages 93–103, 2000.
15. J.T. Chang M. Gerstein Y. Kluger, R. Basri. Spectral biclustering of microarray data: coclustering genes and conditions. *Genome Res*, 13:703–716, 2003.

SIMPLE AND EFFECTIVE CLASSIFIERS TO MODEL BIOLOGICAL DATA

ROGERIO L. SALVINI,* INÊS C. DUTRA

Department of Systems Engineering and Computer Science
COPPE, Federal University of Rio de Janeiro, Brazil
E-mail: {rsalvini,ines}@cos.ufrj.br

VIVIANE A. MORELLI

ONS, Brazil
E-mail: viviane@ons.org.br

Inductive Logic Programming (ILP) systems have been quite successful in extracting comprehensible models of relational data. Most Inductive Logic Programming (ILP) systems use a greedy covering algorithm to find a set of clauses that best model positive examples. This set of clauses is called a theory and can be seen as an *ensemble* of clauses. It turns out that the search for a theory within the ILP system is very time consuming and often yields overly complex classifiers. One alternative approach to obtain a theory is to use the ILP system to non deterministically learn one clause at a time, several times, and to combine the obtained clauses using ensemble methods. This can be a much faster approach, as this work shows, and can produce better classifiers than the theories produced with greedy covering algorithms.

1. Introduction

Inductive Logic Programming (ILP) systems have been quite successful in extracting comprehensible models of relational data. Indeed, for over a decade, ILP systems have been used to construct predictive models for data drawn from diverse domains. These include the sciences[19], engineering[8], language processing[34], environment monitoring[11], software analysis[4], pattern learning and link discovery[24,25]. Most ILP systems use a greedy covering algorithm that repeatedly examines candidate clauses (the "search space") to find good rules (or theories). Ideally, the search will stop when

*This work is supported by Capes, Brazil

the rules cover nearly all positive examples with only a few negative examples being covered.

This algorithm poses some challenges, since the search space can grow very quickly, sometimes turning unfeasible the search for a good solution. Several techniques have been proposed to improve search efficiency of ILP algorithms. Such techniques include improving computation times at individual nodes[3,28], better representations of the search[2], sampling the search space[29,30,33], parallelism[6,16,23,32,33,9,13,12], and utilisation of ensemble methods[10].

Ensembles are classifiers that combine the predictions of multiple classifiers to produce a single prediction[7]. Several researchers have been interested in the use of ensemble-based techniques for ILP. To our knowledge, the original work in this area is Quinlan's work on the use of boosting in FOIL[27]. His results suggested that boosting could be beneficial for first-order induction. More recently, Hoche proposed confidence-rated boosting for ILP with good results[18]. Zemke proposed bagging as a method for combining ILP classifiers with other classifiers[35]. Dutra *et al*[10] studied bagging in the context of the ILP system Aleph[31] learning theories. Their results showed that the applications benefitted from ensembles to a limited extent.

In this work we argue that learning **a single clause at a time** rather than learning **whole theories** (or sets of clauses) at a time can be more cost-effective and produce simpler classifiers. Our alternative approach then is to use *ensembles* of **clauses**. To some extent, an induced theory is an ensemble of clauses. However, finding an induced theory is very time consuming and can produce very complex classifiers. In this work we learn single clauses, and use ensemble methods to combine them, which produce classifiers that are better than any single clause or theory, in much less time than finding a theory.

The paper is organised as follows. First, we present in more detail the ensemble method used in this work. Next, we discuss our experimental setup and the applications used in our study. We then discuss how ensembles of clauses compare with ensembles of theories. Last, we offer our conclusions and suggest future work.

2. Ensemble Methods

In general, ensemble methods work by combining the predictions of several (hopefully different) weak classifiers to produce one final strong classifier.

Ensemble generation assumes two distinct phases: (1) training, and (2) combining the classifiers. The ensemble methods vary according to the constraints imposed to the training phase, and to the kind of combination used.

Figure 1 shows the structure of an ensemble of logic programs. This structure can also be used for classifiers other than logic programs. In the figure, each program $P_1, P_2, \ldots, P_N$ is trained using a set of training instances. At classification time each program receives the same input and executes on it independently. The outputs of each program are then combined and an output classification reached. Figure 1 illustrates that in order to obtain good classifiers one must address three different problems:

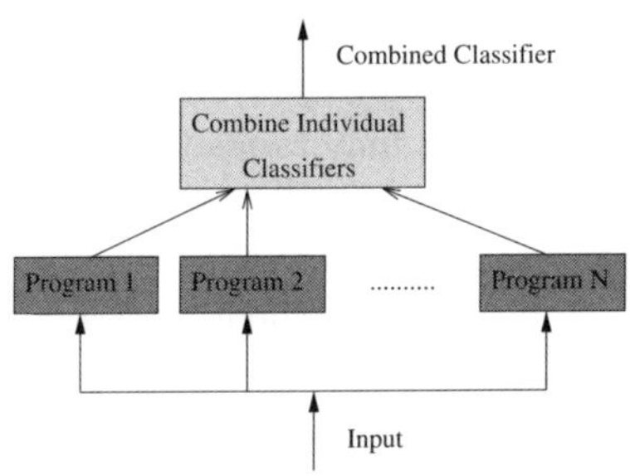

Figure 1. An Ensemble of Classifiers.

- how to generate the individual programs;
- how many individual programs to generate;
- how to combine their outputs.

Regarding the first problem, research has demonstrated that a good ensemble is one where the individual classifiers are accurate and make their errors in different parts of the instance space[21,26]. Obviously, the output of several classifiers is useful only if there is disagreement between them. Hansen and Salamon[17] proved that if the average error rate is below 50% and if the component classifiers are independent, the combined error rate can be reduced to 0 as the number of classifiers goes to infinity. In this work, we argue that ensembles of clauses produce rather independent classifiers than ensembles of theories as studied by Dutra *et al*[10].

Methods for creating the individual classifiers therefore focus on producing classifiers with some degree of diversity. In the present work, we follow two approaches to produce such classifiers. We produce clauses and theories (sets of clauses). We believe that clauses have a greater degree of diversity than theories.

The second issue we had to address was the choice of how many individual classifiers to combine. Previous research has shown that most of the reduction in error for ensemble methods occurs with the first few additional classifiers[17]. Larger ensemble sizes have been proposed for decision trees, where gains have been seen up to 25 classifiers.

The last problem concerns the combination algorithm. An effective combining scheme is often to simply average the predictions of the network[1,5,21,22]. An alternate approach relies on a pre-defined *voting threshold.* If the number of clauses or theories that cover an example is above or equal to the threshold, we say that the example is positive, otherwise the example is negative. Thresholds may range from 1 to the ensemble size. A voting threshold of 1 corresponds to a classifier that is the disjunction of all theories. A voting threshold equal to the ensemble size corresponds to a classifier that is the conjunction of all theories. We used this voting scheme in [10].

Individual classifiers that compose an ensemble can be obtained from different samples of the dataset or from the same dataset. They can also be obtained from one single ILP algorithm (homogeneous classifiers) or from different ILP algorithms (heterogeneous classifiers). In this work we use homogeneous classifiers and obtain classifiers from different samples of the datasets. The classifiers can be independent or dependent, depending on the ensemble method employed.

Several methods have been presented for ensemble generation. The two most popular are *bagging* and *boosting*[14]. Bagging works by training each data set on a random sample from the training set. Classifiers in this case are totally independent from each other. Boosting works by assigning penalties to misclassified examples, and refining the search for clauses that try to cover the misclassified examples. Therefore, in boosting, the classifiers are dependent. The current classifier is dependent on the previous in the training sequence.

In this work we evaluate bagging. We then contrast the results obtained with clause-based learning to the results obtained with theory-based learning.

2.1. *Bagging*

Bagging classifiers are obtained by training each classifier on a random sample of the training set. Each classifier's training set is generated by randomly, uniformly drawing K examples with *replacement*, where K is the size of the original training set. Thus, many of the original examples may be repeated in the classifier's training set.

Table 1. Example of bagging training sets.

Training Sets	Examples Included					
1	6	2	6	3	2	5
2	1	4	6	5	1	6
3	1	3	3	5	2	3
4	6	4	1	4	3	2
5	6	4	2	3	2	3
6	5	5	2	1	5	4

Table 1 shows six training sets randomly generated from a set with examples numbered from 1 to 6. We can notice that each bagging training set tends to focus on different examples. The first training set will have two instances of the second and sixth examples, while having no instances of the second and fourth example. The second example will have instead two occurrences of the first and sixth example, while missing the second and third example. In general, accuracy for each individual bagging classifier is likely to be lower than for a classifier trained on the original data. However, when combined, bagging classifiers can produce accuracies higher than that of a single classifier, because the diversity among these classifiers generally compensates for the increase in error rate of any individual classifier.

3. Methodology

We use the ILP system Aleph[31] in our study. Aleph assumes (a) background knowledge B in the form of a Prolog program; (b) some language specification L describing the hypotheses; (c) an optional set of constraints I on acceptable hypotheses; and (d) a finite set of examples E. E is the union of a nonempty set of "positive" examples E^+, such that none of the E^+ are derivable from B, and a set of "negative" examples E^-.

Aleph tries to find one hypothesis H in L, such that: (1) H respects the constraints I; (2) The E^+ are derivable from B, H, and (3) The E^- are not derivable from B, H. By default, Aleph uses a simple greedy set cover procedure that constructs such a hypothesis one clause at a time. The final classifier is a collection of clauses (a theory). In the search for any single clause, Aleph randomly selects an uncovered positive example as the seed example, saturates this example, and performs an admissible search over the space of clauses that subsume this saturation, subject to a user-specified clause length bound. As it was mentioned before, this is a very time-consuming process. We contrast this approach with the approach of generating one single clause and stopping, using a randomly chosen example as a seed.

These steps are done through a script that runs outside the Aleph code.

We have elected to perform a detailed study on three biological datasets, corresponding to non-trivial ILP applications. We created bagged training sets from the original set, and called Aleph once for each training set.

Aleph allows the user to set a number of parameters. We always set the following parameters as follows:

- search strategy: `search`. We set it to be breadth-first search, `bf`. This enumerates shorter clauses before longer ones. At a given clauselength, clauses are re-ordered based on their evaluation. This is the Aleph default strategy that favours shorter clauses to avoid the complexity of refining larger clauses.
- evaluation function: `evalfn`. We set this to be coverage. Clause utility is measured as $P - N$, with P and N being the number of positive and negative examples covered by the clause, respectively.
- chaining of variables: `i`. This Aleph parameter controls variable chaining during saturation: chaining depth of a variable that appears for the first time in a literal L_i, is 1 plus the maximum chaining depth of all variables that appear in previous literals $L_j, j < i$. We used a value of 5 instead of the default value of 2 in order to obtain more complex relations between literals in a clause.
- max number of nodes allowed: `maxnodes`. This corresponds to the number of clauses in the search space. We set this to be 100,000.
- maximum number of literals in a clause: `maxclauselength`. This was chosen based on some previous experiments and was set to 5.
- minimum clause accuracy: this is the minimum accuracy acceptable when generating a new clause. It was set to 0.9 (i.e., accuracy of 90%).

We used the same set of parameters used by Dutra *et al*[10] in order to compare some results.

Next, we organise our discussion of methodology into (a) experimentation and (b) evaluation.

Experimentation. Our experimental methodology employs five-fold cross-validation. For each fold, we consider ensembles with size varying from 1 to 25, when learning theories. Thus each application ran 25 times. This step generates 25 files with one theory per file (set of clauses). For the experiment that learns clauses, we varied the size of the ensemble from 1 to the average size of the theory, per application, per fold. Therefore we can compare the performance of obtaining one theory using a greedy covering algorithm with the performance of obtaining an ensemble of clauses (with the same theory size), without using a greedy covering algorithm.

Evaluation. For the evaluation phase, we used one popular metric to evaluate the quality of the ensembles, the accuracy. We studied how average accuracy varies with ensemble size. We present accuracy as $\frac{T_p+T_n}{Total\_of\_exs}$, where T_p and T_n, are respectively, the number of positive and negative examples, and $Total\_of\_exs$ is the dataset size.

We wish to test the effectiveness of different sizes of ensembles. Again, we do not repeat the ILP runs themselves to learn entirely new theories for each different ensemble size. Rather, we use the theories from the previous step. Because our results may be distorted by a particularly poor or good choice of these theories, we repeat this selection process 30 times and average the results.

Figure 2 shows the general algorithm to perform the evaluation step. The loop that goes from line 2 to 8 computes the points necessary to produce the accuracies, where `Tn` is the rate of true negatives, `Fn` is the rate of false negatives, `Fp` is the rate of false positives and `Tp` is the rate of true positives. This is done for 30 sets, where each set contains `ensSize` theories that are selected randomly from the 25. This builds a table per fold, per ensemble size, where each line represents the error sum for the 30 sets, for each voting threshold.

This is repeated twice, once for a tuning set, from where we extract the best threshold for each ensemble size, and again for the validation phase, where the voting threshold used is the one that was the best for the tuning phase. The final results are obtained by averaging the numbers across all folds.

The same procedure is repeated for the experiment that generates clauses and do not use the greedy covering algorithm, but at this time, the ensemble sizes can vary from 1 to the average size of the theories.

```
for fold = 1 to NumFolds do
   for ensSize = 1 to NumIterations do
      randomly select 30 sets of size ensSize;
      for threshold = 1 to ensSize do
         for s = 1 to 30 do
            errorSum[fold,ensSize,threshold] += (Tn, Fn, Fp, Tp
         endfor
      endfor
   endfor
endfor
```

Figure 2. General algorithm used in the evaluation step.

Bagging was implemented straightforwardly. We only needed to generate the bags and run the training/test experiments independently.

All experiments were performed using Yap Prolog 4.4.2, and Aleph 3.0. We used two machines to run all experiments: (1) AMD 2.8 GHz, with 512 MBytes of RAM, and (2) Intel Pentium IV 2.8 GHz, with 1 GBytes of RAM, both running Mandrake Linux 10.0. Although these machines have different characteristics, all experiments for the same application were performed in the same machine. Slower runs were launched in the more powerful machine.

3.1. *Benchmark Datasets*

Our benchmark set is composed of three datasets that correspond to non-trivial ILP applications that are also used in other works. We next describe the characteristics of each dataset with its associated ILP application, and present a dataset summary table.

Amine. Our first learning task is to predict amine re-uptake inhibition to discover new drugs for the Alzheimer disease[20]. We were given some positive examples of drugs that were effective against the disease, and some negative examples. The task is to build a model for an active drug that distinguishes between active and inactive drugs.

Choline. This application is also related to drug discovery for the Alzheimer disease. The learning task is to identify the inhibition of the aceto-choline-esterase enzyme.

Protein. Our last dataset consists of a database of genes and features of the genes or of the proteins for which they code, together with information about which proteins interact with one another and correlations among gene expression patterns. This dataset is taken from the function prediction task of KDD Cup 2001. While the KDD Cup task involved 14 different protein functions, our learning task focuses on the challenging function of "metabolism": predicting which genes code for proteins involved in metabolism. This is not a trivial task for an ILP system.

Table 2. Datasets characteristics.

Dataset	E+	E-	BK Size
Amine	343	343	232
Choline	663	663	232
Protein	172	690	6,913

Table 2 summarises the main characteristics of each application. The second and third columns correspond to the size of the full datasets. Bags are created by randomly picking elements, with replacement, from the full dataset. The last column indicates the size of the background knowledge (number of facts and rules). The sets for each 5-fold cross-validation experiment are obtained by a round-robin distribution of the full dataset.

4. Results

This section presents our results and analyses the performance of each application.

Accuracies presented in the graphs are averaged across all folds, and are related to the test sets. For each ensemble size, accuracy values for the test sets are obtained varying the voting threshold. The result plotted uses the threshold that produced the best accuracy in the tuning set where we trained on 3 of the 4 training data of each fold to obtain the best threshold. For clarity's sake, these parameter values are not shown in the curves.

Figure 3 shows the average accuracies for the three applications when learning theories.

According to Figure 3, only two applications (Choline and Protein) have a modest (not significant) improvement when we compare the accuracy of an ensemble of size 1 with larger ensemble sizes. Amine (Figure 3(a)) does not show any improvement when we compose theories in ensembles with larger sizes. In contrast, ensembles of clauses have a clear improvement as the ensemble sizes increase, as shown in Figure 4.

One important thing to note at this point is that, for each application,

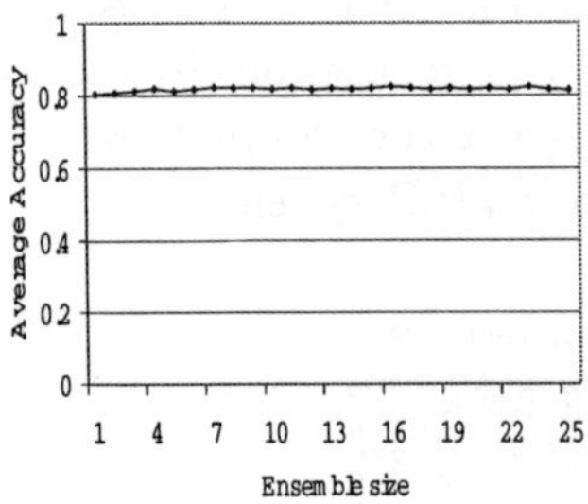

(a) Amine

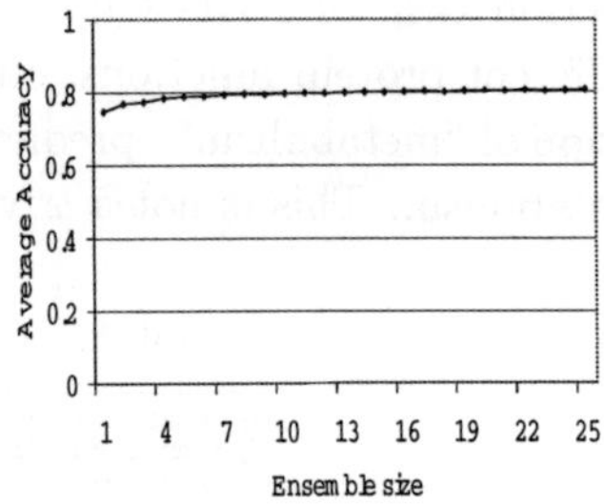

(b) Choline

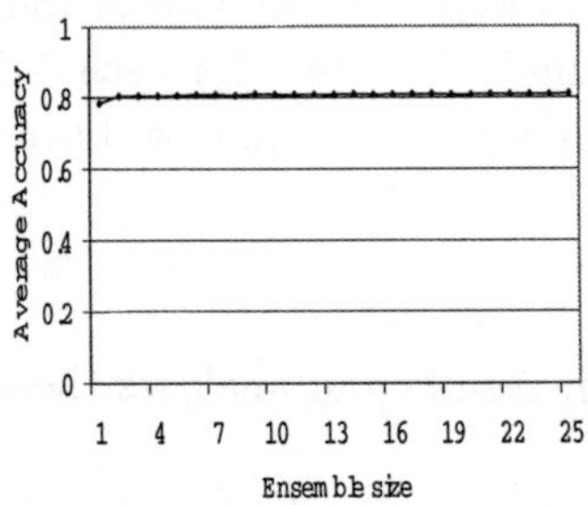

(c) Protein

Figure 3. Average accuracies for the three applications (theories).

each individual theory has an average size of 60 (Amine), 139 (Choline), and 63 (Protein) clauses. When we generate the ensemble of theories, the classifier can become very complex and difficult to interpret. In the case of clause-based learning, each individual classifier is composed by only one clause. To obtain an accuracy comparable to the best ensemble of theories we need far less clauses. For example, only two clauses are sufficient to produce the same accuracy of one theory of size 139, for application Protein

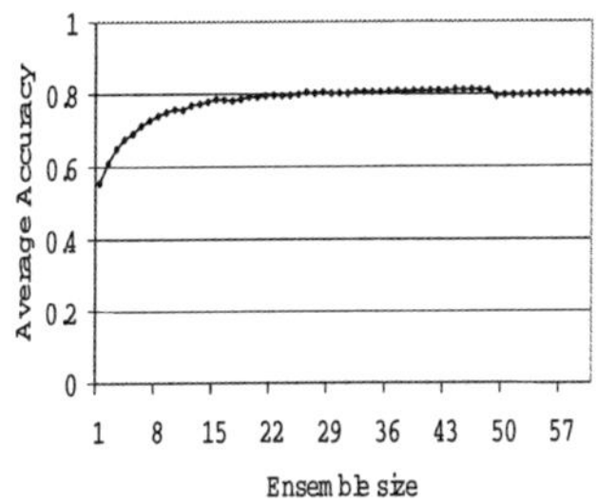

(a) Amine

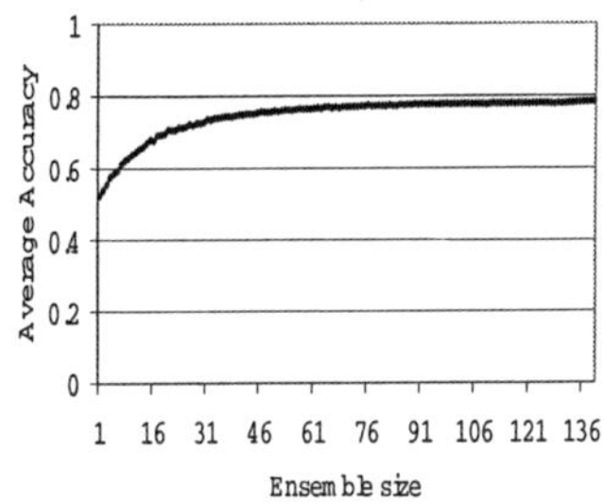

(b) Choline

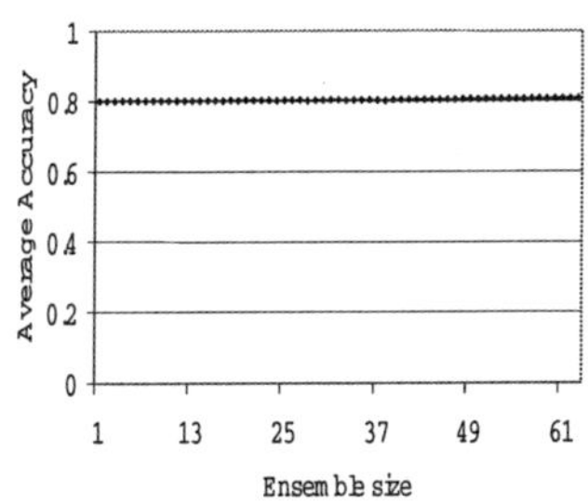

(c) Protein

Figure 4. Average accuracies for the three applications (clauses).

(Figures 4(c) and 3(c)). And the difference in execution time is extremely high: while it takes 3 hours and 42 minutes to generate the ensemble of theories, it takes only 0.75 minutes to generate the ensemble of two clauses (0.34% of the time spent to generate the ensemble of theories).

In order to obtain the same accuracy of one theory with average size 60, for application Amine, it is sufficient to combine only 46 clauses.

Ensembles of theories can produce slightly better results than ensembles of clauses. For example, Amine with ensemble of clauses never produces a classifier with better accuracy than ensemble of theories. However with only 46 clauses, and in much less time (0.4% of the time spent to find the same level of accuracy) it can obtain an accuracy that is only 1.5% worse than an ensemble of theories.

Table 3 shows that the price to pay to have this better value is too high when we look at the time spent by the ILP system to obtain the ensembles of theories.

In this table we show a summary of execution time and best accuracy achieved (at a given ensemble size), for the three applications. The execution times correspond to the total time to produce ensembles of all sizes. The column "One theory" shows the time needed to find only one theory in the process of cross-validation. The accuracy shown in this column corresponds to the average obtained across all folds.

We can observe that the accuracies obtained with ensembles of theories are slightly better than the ones obtained with ensembles of clauses, however the execution time to obtain the theories are much higher. On the other hand, we can obtain accuracies (with ensembles of clauses) very close to the best achieved by ensembles of theories in far less time.

Table 3 shows the execution times of the cross-validation for all applications. We are not counting the time to generate the final classifier, through the voting mechanism. This would add a proportional extra time to all slots of the table.

Table 3. Execution times.

	Ensemble theories		One theory		Ensemble of clauses	
App	time	best acc	time	avg acc	time	best acc
Amine	10h41min	0,824 (16)	25,68min	0.802	2,62min	0,812 (46)
Choline	5h03min	0,806 (25)	12,12min	0.761	34min	0,783 (139)
Protein	3h42min	0,810 (25)	8,88min	0.783	23,5min	0,807 (63)

The datasets Choline and Protein were executed on the AMD 2.8 GHz machine and Amine was executed on the Pentium IV, 2.8 GHz.

5. Discussion

The results shown demonstrate several important facts in ILP learning:

- Ensembles of clauses learned with ILP are very powerful methods to obtain good accuracies in a small amount of time;

- Bagging is effective on obtaining good quality classifiers, even when the individual classifier is as weak as a single clause;
- Ensemble sizes greater than 25 may produce better accuracies for methods that learn clauses;
- Learning theories can be unfeasible depending on the application, and learning clauses can produce equal or better results;
- Theories do not benefit much from ensembles. We believe that this is because a theory is already an ensemble;
- Very weak individual classifiers such as single clauses can benefit significantly from ensembles, and producing a final classifier takes time that is several orders of magnitude less than the time spent to learn theories;
- Clauses are classifiers simpler than theories. Consequently, an ensemble of clauses is simpler than an ensemble of theories, which may help an expert to interpret the results.

6. Conclusions and Future Work

This work presented an empirical study of ensemble methods, in the Inductive Logic Programming setting. We chose to apply ensembles to two different individual classifiers: (1) classifiers composed of one single clause, faster to obtain; and (2) classifiers composed of one or more clauses (theory), that take a huge amount of time to obtain. We applied ensemble methods to classifiers obtained with the Aleph system. We tested three ILP applications already used in the literature.

Our results show that ensembles built from single clauses are more cost-effective than ensembles built from theories. Clauses are much faster to be learned, and an ensemble of clauses produces more readable classifiers.

As future work we intend to extend this methodology to other applications and investigate other ensemble approaches. We also would like to investigate the theoretical implications of these results on the ensemble methods. In particular, we would like to study how is the behaviour of these ensemble methods on the light of bias plus variance decomposition of classification[15], in order to explain how an ensemble of clauses can be more powerful than a single theory and ensembles of theories.

References

1. E. Alpaydin. Multiple networks for function learning. In *IEEE International Conference on Neural Networks*, pages 9–14, 1993.
2. H. Blockeel, L. Dehaspe, B. Demoen, G. Janssens, J. Ramon, and H. Vandecasteele. Executing query packs in ILP. In J. Cussens and A. Frisch, editors, *Proceedings of the 10th International Conference on Inductive Logic Programming*, volume 1866 of *Lecture Notes in Artificial Intelligence*, pages 60–77. Springer-Verlag, 2000.
3. H. Blockeel, B. Demoen, G. Janssens, H. Vandecasteele, and W. Van Laer. Two advanced transformations for improving the efficiency of an ILP system. In J. Cussens and A. Frisch, editors, *Proceedings of the Work-in-Progress Track at the 10th International Conference on Inductive Logic Programming*, pages 43–59, 2000.
4. I. Bratko and M. Grobelnik. Inductive learning applied to program construction and verification. In S. Muggleton, editor, *Proceedings of the 3rd International Workshop on Inductive Logic Programming*, pages 279–292. J. Stefan Institute, 1993.
5. L. Breiman. Stacked Regressions. *Machine Learning*, 24(1):49–64, 1996.
6. L. Dehaspe and L. De Raedt. Parallel inductive logic programming. In *Proceedings of the MLnet Familiarization Workshop on Statistics, Machine Learning and Knowledge Discovery in Databases*, 1995.
7. T. Dietterich. Ensemble methods in machine learning. In J. Kittler and F. Roli, editors, *First International Workshop on Multiple Classifier Systems, Lecture Notes in Computer Science*, pages 1–15. Springer-Verlag, 2000.
8. B. Dolšak and S. Muggleton. The application of ILP to finite element mesh design. In S. Muggleton, editor, *Proceedings of the 1st International Workshop on Inductive Logic Programming*, pages 225–242, 1991.
9. I. C. Dutra, D. Page, V. Santos Costa, J. Shavlik, and M. Waddell. Toward automatic management of embarrassingly parallel applications. In *Euro-Par03, Klagenfurt, Austria, August 26 - 29*, pages 509–516, 2003.
10. I. C. Dutra, D. Page, and J. Shavlik V. Santos Costa. An empirical evaluation of bagging in inductive logic programming. In *Proceedings of the 12th International Conference on Inductive Logic Programming*, Lecture Notes in Artificial Intelligence. Springer-Verlag, September 2002.
11. S. Džeroski, L. Dehaspe, B. Ruck, and W. Walley. Classification of river water quality data using machine learning. In *Proceedings of the 5th International Conference on the Development and Application of Computer Techniques to Environmental Studies*, 1995.
12. Nuno A. Fonseca, Rui Camacho, and Fernando Silva. Strategies to Parallelize ILP Systems. In *To appear in Proceedings of the 15th International Conference on Inductive Logic Programming*, Lecture Notes in Artificial Intelligence. Springer-Verlag, 2005.
13. Nuno A. Fonseca, Fernando Silva, Vitor Santos Costa, and Rui Camacho. A pipelined data-parallel algorithm for ILP. In *To appear Proceedings of 2005 IEEE International Conference on Cluster Computing*. IEEE, Septem-

ber 2005.

14. Y. Freund and R. Shapire. Experiments with a new boosting algorithm. In *Proceedings of the 14th National Conference on Artificial Intelligence*, pages 148–156. Morgan Kaufman, 1996.
15. S. Geman, E. Bienenstock, , and R. Doursat. Neural networks and the bias/variance dilemma. *Neural Computation*, 4:1–58, 1992.
16. J. Graham, D. Page, and A. Wild. Parallel inductive logic programming. In *Proceedings of the Systems, Man, and Cybernetics Conference*, 2000.
17. L. Hansen and P. Salamon. Neural network ensembles. *IEEE Transactions on Pattern Analysis and Machine Intelligence*, 12(10):993–1001, October 1990.
18. S. Hoche and S. Wrobel. Relational learning using constrained confidence-rated boosting. In Céline Rouveirol and Michèle Sebag, editors, *Proceedings of the 11th International Conference on Inductive Logic Programming*, volume 2157 of *Lecture Notes in Artificial Intelligence*, pages 51–64. Springer-Verlag, September 2001.
19. R. King, S. Muggleton, and M. Sternberg. Predicting protein secondary structure using inductive logic programming. *Protein Engineering*, 5:647–657, 1992.
20. R. King, M. Sternberg, and A. Srinivasan. Relating chemical activity to structure: An examination of ilp successes. *New Generation Computing Journal*, 13(3&4):411–433, 1995.
21. A. Krogh and J. Vedelsby. Neural network ensembles, cross validation, and active learning. In G. Tesauro, D. Touretzky, and T. Leen, editors, *Advances in Neural Information Processing Systems*, volume 7, pages 231–238. The MIT Press, 1995.
22. N. Lincoln and J. Skrzypek. Synergy of clustering multiple backpropagation networks. In *Advances in Neural Information Processing Systems.* Morgan Kaufmann, 1989.
23. T. Matsui, N. Inuzuka, H. Seki, and H. Ito. Parallel induction algorithms for large samples. In S. Arikawa and H. Motoda, editors, *Proceedings of the First International Conference on Discovery Science*, volume 1532 of *Lecture Notes in Artificial Intelligence*, pages 397–398. Springer-Verlag, December 1998.
24. R. Mooney, P. Melville, L. P. Rupert Tang, J. Shavlik, I. Dutra, D. Page, and V. Santos Costa. Relational data mining with inductive logic programming for link discovery. In *Proceedings of the National Science Foundation Workshop on Next Generation Data Mining*, Baltimore, Maryland, USA, 2002.
25. R. J. Mooney, P. Melville, L. R. Tang, J. Shavlik, I. C. Dutra, D. Page, and V. S. Costa. Relational data mining with inductive logic programming for link discovery. In Hillol Kargupta, Anupam Joshi, K. Sivakumar, and Yelena Yesha, editors, *Data Mining: Next Generation Challenges and Future Directions.* AAAI/MIT Press, Berlin, 2003.
26. D. W. Opitz and J. W. Shavlik. Actively searching for an effective neural-network ensemble. *Connection Science*, 8(3/4):337–353, 1996.
27. J. R. Quinlan. Boosting first-order learning. *Algorithmic Learning Theory, 7th International Workshop, Lecture Notes in Computer Science*, 1160:143–

155, 1996.

28. V. Santos Costa, A. Srinivasan, and R. Camacho. A note on two simple transformations for improving the efficiency of an ILP system. In J. Cussens and A. Frisch, editors, *Proceedings of the 10th International Conference on Inductive Logic Programming*, volume 1866 of *Lecture Notes in Artificial Intelligence*, pages 225–242. Springer-Verlag, 2000.
29. M. Sebag and C. Rouveirol. Tractable induction and classification in first-order logic via stochastic matching. In *Proceedings of the 15th International Joint Conference on Artificial Intelligence*, pages 888–893. Morgan Kaufmann, 1997.
30. A. Srinivasan. A study of two sampling methods for analysing large datasets with ILP. *Data Mining and Knowledge Discovery*, 3(1):95–123, 1999.
31. A. Srinivasan. *The Aleph Manual*, 2001.
32. J. Struyf and H. Blockeel. Efficient cross-validation in ILP. In Céline Rouveirol and Michèle Sebag, editors, *Proceedings of the 11th International Conference on Inductive Logic Programming*, volume 2157 of *Lecture Notes in Artificial Intelligence*, pages 228–239. Springer-Verlag, September 2001.
33. F. Zelezny, A. Srinivasan, and D. Page. Lattice-search runtime distributions may be heavy-tailed. In *Proceedings of the 12th International Conference on Inductive Logic Programming*. Springer Verlag, July 2002.
34. J. Zelle and R. Mooney. Learning semantic grammars with constructive inductive logic programming. In *Proceedings of the 11th National Conference on Artificial Intelligence*, pages 817–822, Washington, D.C., July 1993. AAAI Press/MIT Press.
35. S. Zemke. Bagging imperfect predictors. In *Proceedings of the International Conference on Artificial Neural Networks in Engineering, St. Louis, MI, USA*. ASME Press, 1999.

INDEX